Lecture Notes in Computer Science   11613

More information about this series at http://www.springer.com/series/7412

Lucio Tommaso De Paolis ·
Patrick Bourdot (Eds.)

Augmented Reality, Virtual Reality, and Computer Graphics

6th International Conference, AVR 2019
Santa Maria al Bagno, Italy, June 24–27, 2019
Proceedings, Part I

 Springer

Editors
Lucio Tommaso De Paolis (iD)
University of Salento
Lecce, Italy

Patrick Bourdot (iD)
University of Paris-Sud
Orsay, France

ISSN 0302-9743   ISSN 1611-3349   (electronic)
Lecture Notes in Computer Science
ISBN 978-3-030-25964-8   ISBN 978-3-030-25965-5   (eBook)
https://doi.org/10.1007/978-3-030-25965-5

LNCS Sublibrary: SL6 – Image Processing, Computer Vision, Pattern Recognition, and Graphics

This Springer imprint is published by the registered company Springer Nature Switzerland AG
The registered company address is: Gewerbestrasse 11, 6330 Cham, Switzerland

Preface

Virtual Reality (VR) technology permits the creation of realistic-looking worlds where the user inputs are used to modify in real time the digital environment. Interactivity contributes to the feeling of immersion in the virtual world, of being part of the action that the user experiences. It is not only possible to see and manipulate a virtual object, but also to feel and touch them using specific devices.

Mixed Reality (MR) and Augmented Reality (AR) technologies permit the real-time fusion of computer-generated digital contents with the real world and allow for the creation of fascinating new types of user interfaces. AR enhances the users' perception and improves their interaction in the real environment. The virtual objects help users to perform real-world tasks better by displaying information that they cannot directly detect with their own senses. Unlike the VR technology that completely immerses users inside a synthetic environment where they cannot see the real world around them, AR technology allows them to see 3D virtual objects superimposed upon the real environment. AR and MR supplement reality rather than completely replacing it and the user is under the impression that the virtual and real objects coexist in the same space.

Human-Computer Interaction technology (HCI) is a research area concerned with the design, implementation, and evaluation of interactive systems that make more simple and intuitive the interaction between user and computer.

This book contains the contributions to the 6th International Conference on Augmented Reality, Virtual Reality and Computer Graphics (SALENTO AVR 2019) that has held in Santa Maria al Bagno (Lecce, Italy) during June 24–27, 2019. Organized by the Augmented and Virtual Reality Laboratory at the University of Salento, SALENTO AVR 2019 intended to bring together the community of researchers and scientists in order to discuss key issues, approaches, ideas, open problems, innovative applications, and trends in virtual and augmented reality, 3D visualization, and computer graphics in the areas of medicine, cultural heritage, arts, education, entertainment, military, and industrial applications. We cordially invite you to visit the SALENTO AVR website (www.salentoavr.it) where you can find all relevant information about this event.

We are very grateful to the Program Committee and Local Organizing Committee members for their support and for the time spent to review and discuss the submitted papers and for doing so in a timely and professional manner.

We would like to sincerely thank the keynote speakers who willingly accepted our invitation and shared their expertise through illuminating talks, helping us to fully meet the conference objectives. In this edition of SALENTO AVR we were honored to have the following invited speakers:

- Luigi Gallo – ICAR-CNR, Italy
- Danijel Skočaj – University of Ljubljana, Slovenia
- Pasquale Arpaia – University of Naples Federico II, Italy

We extend our thanks to the University of Salento and the Banca Popolare Pugliese for the enthusiastic acceptance to sponsor the conference and to provide support in the organization of the event.

SALENTO AVR attracted high-quality paper submissions from many countries. We would like to thank the authors of all accepted papers for submitting and presenting their works at the conference and all the conference attendees for making SALENTO AVR an excellent forum on virtual and augmented reality, facilitating the exchange of ideas, fostering new collaborations, and shaping the future of this exciting research field.

We hope the readers will find in these pages interesting material and fruitful ideas for their future work.

June 2019

Lucio Tommaso De Paolis
Patrick Bourdot

Organization

Conference Chair

Lucio Tommaso De Paolis   University of Salento, Italy

Conference Co-chairs

Patrick Bourdot   CNRS/LIMSI, University of Paris-Sud, France
Marco Sacco   ITIA-CNR, Italy
Paolo Proietti   MIMOS, Italy

Scientific Program Committee

Andrea Abate	University of Salerno, Italy
Giuseppe Anastasi	University of Pisa, Italy
Selim Balcisoy	Sabancı University, Turkey
Vitoantonio Bevilacqua	Polytechnic of Bari, Italy
Monica Bordegoni	Politecnico di Milano, Italy
Davide Borra	NoReal.it, Italy
Andrea Bottino	Politecnico di Torino, Italy
Pierre Boulanger	University of Alberta, Canada
Andres Bustillo	University of Burgos, Spain
Massimo Cafaro	University of Salento, Italy
Bruno Carpentieri	University of Salerno, Italy
Sergio Casciaro	IFC-CNR, Italy
Marcello Carrozzino	Scuola Superiore Sant'Anna, Italy
Mario Ciampi	ICAR/CNR, Italy
Pietro Cipresso	IRCCS Istituto Auxologico Italiano, Italy
Arnis Cirulis	Vidzeme University of Applied Sciences, Latvia
Mario Covarrubias	Politecnico di Milano, Italy
Rita Cucchiara	University of Modena, Italy
Yuri Dekhtyar	Riga Technical University, Latvia
Matteo Dellepiane	National Research Council (CNR), Italy
Giorgio De Nunzio	University of Salento, Italy
Francisco José Domínguez Mayo	University of Seville, Spain
Aldo Franco Dragoni	Università Politecnica delle Marche, Italy
Italo Epicoco	University of Salento, Italy
Ben Falchuk	Perspecta Labs Inc., USA
Vincenzo Ferrari	EndoCAS Center, Italy
Francesco Ferrise	Politecnico di Milano, Italy
Dimitrios Fotiadis	University of Ioannina, Greece

Antonio Emmanuele Uva   Polytechnic of Bari, Italy
Volker Paelke   Bremen University of Applied Sciences, Germany
Aleksei Tepljakov   Tallinn University of Technology, Estonia
Kristina Vassiljeva   Tallinn University of Technology, Estonia
Krzysztof Walczak   Poznań University of Economics and Business, Poland
Anthony Whitehead   Carleton University, Canada

Local Organizing Committee

Giovanna Ilenia Paladini   University of Salento, Italy
Silke Miss   Virtech, Italy
Valerio De Luca   University of Salento, Italy
Cristina Barba   University of Salento, Italy
Giovanni D'Errico   University of Salento, Italy

Keynote Speakers

Interactive Virtual Environments:
From the Laboratory to the Field

Luigi Gallo

ICAR-CNR, Italy

Virtual reality technology has the potential to change the way information is retrieved, processed and shared, making it possible to merge several layers of knowledge in a coherent space that can be comprehensively queried and explored. In realizing this potential, the human–computer interface plays a crucial role. Humans learn and perceive by following an interactive process, but the interaction occurs in different places and contexts, between people, and with people. Accordingly, the interactive system components have to be tailored to the application and its users, targeting robustness, multimodality, and familiarity. In this presentation, I aim to explore the promise and challenges of interactive virtual technologies by discussing generally the design and the evaluation results of selected real-world applications, developed by multi-disciplinary research groups, in very different domains, ranging from medicine and finance to cultural heritage.

Luigi Gallo is a research scientist at the National Research Council of Italy (CNR) – Institute for High-Performance Computing and Networking (ICAR). He graduated in 2006 in Computer Engineering, and received a PhD degree in Information Technology Engineering in 2010. Between 2012 and 2018, he worked as Adjunct Professor of Informatics at the University of Naples Federico II. He holds leadership roles within nationally funded research projects dealing with the development of ICT solutions for medical, financial, and cultural heritage applications. His fields of interest include computer vision, natural user interfaces, and the human interface aspects of virtual and augmented reality. He has authored and co-authored more than 90 publications in international journals, conference proceedings, and book chapters, and serves on the Organizing Committee of several international conferences and workshops.

Computer Vision as an Enabling Technology for Interactive Systems

Danijel Skočaj

University of Ljubljana, Slovenia

Computer vision research has made tremendous progress in recent years. Solutions for object detection and classification, semantic segmentation, visual tracking, and other computer vision tasks have become readily available. Most importantly, these solutions have become more general and robust, allowing for a significantly broader use of computer vision. Applications are less limited to constrained environments and are getting ready to be used in the wild. Efficient, robust, and accurate interpretation of scenes is also a key requirement of advanced interactive systems as well as augmented and mixed reality applications. Efficient understanding of scenes in 2D as well as in 3D and detection of key semantic elements in images allow for a wider use of such systems and more intelligent interaction with a user. In this talk, I will present some recent achievements in developing core computer vision algorithms as well as the application of these algorithms for interaction between a human, real environment and a virtual world.

Danijel Skočaj is an associate professor at the University of Ljubljana, Faculty of Computer and Information Science. He is the head of the Visual Cognitive Systems Laboratory. He obtained a PhD degree in computer and information science from the University of Ljubljana in 2003. His main research interests lie in the fields of computer vision, machine learning, and cognitive robotics; he is involved in basic and applied research in visually enabled cognitive systems, with emphasis on visual learning, recognition, and segmentation. He is also interested in the ethical aspect of artificial intelligence. He has led or collaborated in a number of research projects, such as EU projects (CogX, CoSy, CogVis), national projects (DIVID, GOSTOP, VILLarD), and several industry-funded projects. He has served as the president of the IEEE Slovenia Computer Society, and the president of the Slovenian Pattern Recognition Society.

Wearable Brain–Computer Interface
for Augmented Reality-Based Inspection
in Industry 4.0

Pasquale Arpaia

University of Naples Federico II, Italy

In the past two decades, augmented reality (AR) has gained great interest in the technical–scientific community and much effort has been made to overcome its limitations in daily use. The main industrial operations in which AR is applied are training, inspections, diagnostics, assembly–disassembly, and repair. These operations usually require the user's hands to be free from the AR device controller. Despite hand-held devices, such as tablets, smart glasses can guarantee hand-free operations with their high wearability. The combination of AR with a brain–computer interface (BCI) can provide the solution: BCI is capable of interpreting human intentions by measuring user neuronal activity. In this talk, the most interesting results of this technological research effort, as well as its further most recent developments, are reviewed. In particular, after a short survey on research at the University of Naples Federico II in cooperation with CERN, the presentation focuses mainly on state-of-the-art research on a wearable monitoring system. AR glasses are integrated with a trainingless non-invasive single-channel BCI, for inspection in the framework of Industry 4.0. A case study at CERN for robotic inspection in hazardous sites is also reported.

Pasquale Arpaia obtained a master's degree and PhD degree in Electrical Engineering at the University of Naples Federico II (Italy), where he is full professor of Instrumentation and Measurements. He is also Team Manager at the European Organization for Nuclear Research (CERN). He is Associate Editor of the Institute of Physics *Journal of Instrumentation*, Elsevier journal *Computer Standards & Interfaces*, and MDPI journal Instruments. He is Editor at Momentum Press of the book series *Emerging Technologies in Measurements, Instrumentation, and Sensors*. In the past few years, he was scientifically responsible for more than 30 awarded research projects in cooperation with industry, with related patents and international licences, and funded four academic spin-off companies. He acted as scientific evaluator in several international research call panels. He has served as organizing and scientific committee member in several IEEE and IMEKO conferences. He has been plenary speaker in several scientific conferences. His main research interests include digital instrumentation and measurement techniques for particle accelerators, evolutionary diagnostics, distributed measurement systems, ADC modelling and testing.

Contents – Part I

Medicine

Contents – Part II

Cultural Heritage

Education

Industry

Virtual Reality

Design of a SCORM Courseware Player Based on Web AR and Web VR

YanXiang Zhang[✉], WeiWei Zhang, and YiRun Shen

Department of Communication of Science and Technology,
University of Science and Technology of China, Hefei, Anhui, China
petrel@ustc.edu.cn,
{anna511,run1577}@mail.ustc.edu.cn

Abstract. In this paper, the authors developed a multimedia SCORM courseware player that supports Web AR and Web VR. It was designed based on the Sco units and integrated Web AR and Web VR. This courseware player made up for the flexibility of AR and VR applications in SCORM, and can also expand the possibility of application in teaching practice.

Keywords: SCORM courseware player · Web AR · Web VR · Learning management system (LMS)

1 Introduction

SCORM was widely used in online teaching as a good way to promote E-learning teaching. Neves et al. [1] gave four characteristics of SCORM standard: sustainability, reusability, interoperability and availability. Based on these principles, various forms of courseware are supported by SCORM [2, 3], such as video, image, audio etc. Hanisch and, Straβer have conducted a lot of research on the interaction of SCORM courseware [4]. In order to display multimedia content, especially interactive media, the authors designed a SCORM courseware player based on the Flash action script 3 language [5]. An immersive authoring tool using the augmented reality was presented by Jee, Lim, Youn, and Lee [6], which also identified the existence of libraries that allow applications with AR authorship to be constructed, such as ARToolKit. As the trend of future development in the field of new media education, the strong interaction between AR and VR content can bring better teaching results. [7]. However, the advantages of AR and VR are not fully applied into SCORM courseware.

Web AR and Web VR are an emerging technologies which can achieve AR and VR in the web environment. Compare with traditional AR and VR, Web AR and Web VR are open source, easy to use, low achievement barriers [8]. Based on these features, Web AR and Web VR can be well confirmed towards the SCORM standard and are consistent with the learning management system (LMS). Flexible forms of Web AR and Web VR provide educators with more possibilities for other courseware based on SCORM standard. Web AR together with Experience API allow teachers to create educational applications of AR which benefit the learning process [8]. Cubillo and Martín et al. developed an Augmented Reality Learning Environment (ARLE) system to integrate AR content into the theoretical content and learning practices of the classroom [9]. However, the majority

© Springer Nature Switzerland AG 2019
L. T. De Paolis and P. Bourdot (Eds.): AVR 2019, LNCS 11613, pp. 3–9, 2019.
https://doi.org/10.1007/978-3-030-25965-5_1

of teachers only need to display the AR/VR contents while teaching, for whom it is a quite difficult to understand the development processes. And there is no attempt to combine with Web AR and Web VR with SCORM courseware. In this paper, the authors bring some breakthroughs and references in the creation of E-learning courseware contained AR/VR contents, hoping to improve teaching effectiveness.

2   Design and Implementation

2.1   Implementation of Web AR & Web VR

The implementation of Web AR and Web VR are based on the rendering of virtual images through the browser. In terms of 3D rendering, there is some overlap between Web AR and Web VR. In the production of Web VR content, it only need to build the scene and set components on Three.js, and then realize the final integration on the webpage. A sample of Web VR is shown as Table 1. After creating the container, we established the perspective camera and added the corresponding controller. Next is the creation and setting of scenarios, such as the fog, hemisphere light, directional light and so on. Model loading and rendering are also required.

Table 1. A sample of Web VR

```
container = document.createElement( 'div' );
document.body.appendChild( container );

camera = new THREE.PerspectiveCamera( 10, window.innerWidth / win-
dow.innerHeight, 1, 2000 );
camera.position.set( 100, 200, 300 );

controls = new THREE.OrbitControls( camera );
controls.target.set( 0, 000, 0 );
controls.update();

scene = new THREE.Scene();
scene.background = new THREE.Color( 0xa0a0a0 );
scene.fog = new THREE.Fog( 0xa0a0a0, 200, 1000 );
```

As for Web AR, more steps are needed to implement AR content. In this paper, we used A-frame XR to realize. As the development tools for the 3D frame of Web VR, A-frame XR added the AR Module in the recently updated version. A sample of Web AR is shown as Table 2. A-frame containing AR.js is first introduced. After the definition of the dimensions of the body, an A-frame scene is created and the augmented reality content can also be added. Then add the camera and put the resources in <a-assets> placed in the scene. It ensures that resources are not lost visually and also avoids performance problems caused by scenarios where trying to capture assets while rendering.

Table 2. A sample of Web AR codes

```
<body style='margin : 0px; overflow: hidden; font-family: Monospace;'><div
style='position: fixed; top: 10px; width:100%; text-align: center; z-index: 1;'>
<br/>
</div>
<!-- <a-scene embedded arjs='sourceType: image;
sourceUrl:../../data/images/armchair.jpg;'> -->
<a-scene embedded arjs='sourceType: webcam;'>
<a-marker-camera type='pattern' url='pattern-ar.patt'></a-marker-camera>
<!-- Add your augmented reality here -->
<a-assets>
<a-asset-item id="tree-obj" src="grab.obj"></a-asset-item>
<a-asset-item id="tree-mtl" src="grab.mtl"></a-asset-item>
</a-assets>
<a-entity obj-model="obj: #tree-obj; mtl: #tree-mtl" scale="0.05 0.05
0.05"></a-entity>
</a-scene>
</body>
```

In order to adapt it to the SCORM, the implementation of Web AR and Web VR need to be defined as Sco. Based on the internet-based features, Web AR and Web VR are easy to integrate with Sco, which only need to call the HTML file generated whenever needed. In order to implement the effect of AR and VR on the website, we need to render them on the browser. Web AR and Web VR have some same 3D frameworks. The authors used Web GL framework Three.js to implement Web GL rendering (Fig. 1).

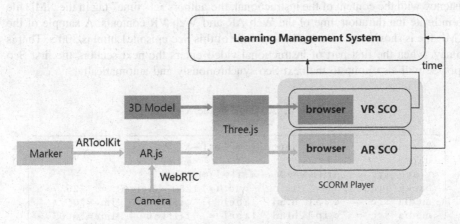

Fig. 1. Web VR and Web AR realized in the SCORM player

The difference between Web AR and Web VR is that Web VR can be implemented through the rendering of the model, while Web AR has more steps. The authors used

AR.js to identify and track marks. (1) ARToolKit can be used to train images as markers, moreover make sure the marker can be tracked and recognized. (2) Then, we need to get the corresponding video stream data through the computer camera. AR.js can call the built-in API, getUserMedia(), of WebRTC (Web real-time communication) in order to get camera permission of browser and collect the image information. (3) Next, we need to establish an identification and tracking system for the AR marks. On the network side, we use jsARToolKit to track the marker generated previously. At the same time, it can generate the corresponding coordinates, and the markers are locked. (4) At last, render the 3D content on the location of markers.

After the Sco content of Web VR and Web AR has been completed, it needs to be consistent with the learning management system by generate some information interactions, such as learning time, etc.

2.2 Design of Player

The player is composed of HTML nesting framework, which can be divided into three main frameworks: instructional video, menu and presentation. The Web AR and Web VR content are played in the demo area. By calling the relevant Web AR and Web VR html in the framework area, Web AR and Web VR content can be displayed. We set the instructional video as the primary object in this player. The instructional videos of each lesson has several Scos correspondingly, which are named according to its subject classification.

So far, the contents of AR and VR has developed in the form of computer-assisted instruction. In our design, instructional videos are also regarded as the main teaching content, taking into account the application of actual teaching situation. Therefore, in spite of the important role of Web AR and Web VR as Sco forms, users are not allowed to stay more time as they like at any Web AR and Web VR Sco. The Sco can be switched according to the progress of instructional video. In order to ensure the consistency with the content of the instructional, the authors set "time" tag in the XML file to indicate the duration time of the Web AR and Web VR contents. A sample of the XML file is shown as Table 3. The total time of this Sco episode1.html is 300 s. That is to say, when the first part of instructional video enters the next section, the first Sco episode will also jump to the next Sco synchronously and automatically.

Table 3. A sample of the XML file

```
<? xml version="1.0" encoding="utf-8" ?>
<data>
<video src = "ArtOfNewMedia.flv"/>
<background src = "UI.jpg" width="1280" height = "720"/>
<media src = "clip1.html" label = "tiltle01" time="0"/>
<media src = "clip2.html" label = "tiltle01" time="300"/>
<media src = "clip3.html" label = "tiltle01" time="564"/>
<media src = "clip4.html" label = "tiltle01" time="781"/>
</data>
```

One of the most important advantages of SCORM is that the SCORM standard is suitable with the LMS, which works well with each other. The SCORM-based courseware allows relevant information to be synchronized to the LMS, of which the learning progress becomes the main parameter. Recording the learning progress through the LMS not only facilitates the learner to return to the courseware, but also enables tracking of the learner's learning state. There is no doubt that it will improve the efficiency of learning. However, in traditional SCORM-based courseware, the learning status is assessed by the Sco units. It means that the learner can open the current Sco again, based on where he left the login earlier. But it is only suitable for the short-time Scos. If the duration of Scos is a bit long, the learner have no choice but to watch the same video once more from the starting point, which becomes a big restriction on the interaction between the traditional SCORM and LMS. In our SCORM-based courseware, we can intercept the current learning time point from the timeline of instructional video with JavaScript, and then write it into parameter "cmi.core.lesson_location" on the LMS server. When the learner open the courseware next time, it will get the parameter by using the function doLMSGetValue ("cmi.-core.lesson_location", time.point). As a result, the SCORM-based courseware will find the exact timepoint location of the teaching video.

3   Result

The authors designed a demo courseware about the structure of molecular. The menu of the knowledge points and the Scos' name was displayed on the left of the player, which can be chosen by the learner. The instructional video is located in the upper left corner of the player. The right area of the player was used to display a variety of media content, including captions, pictures and Web AR and Web VR materials for teaching. The Scos of this player was set to display in VR mode initially. There is a toggle button next to the progress bar, which could switch between the AR mode and VR mode. In addition, the permission of the camera is required in the AR mode. Therefore, the learner should allow the player to use the computer camera when the dialog box appears. Screenshots of the SCORM courseware running interface are shown as Figs. 2 and 3.

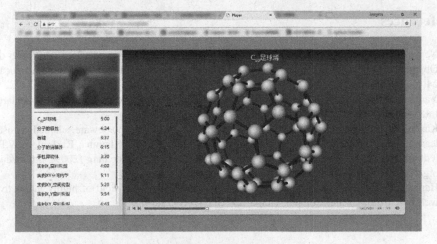

Fig. 2. SCORM courseware player when playing Web VR content

Fig. 3. SCORM courseware player when playing Web AR content

4 Conclusion

In this paper, the authors proposed a Web AR and Web VR supported browser player, which is helpful to improve teaching methods and promote teaching effect. It can not only support the mate-material for AR and VR with web form, but also can support other media. This courseware based on Web AR and Web VR is fitter with SCORM standards. It can also provide a more convenient possibility for new media courseware display. This player is browser-based, which makes it easier to realize cross-platform. On this basis, the application environment of SCORM courseware will become wider.

Acknowledgements. The research is supported by the Ministry of Education (China) Humanities and Social Sciences Research Foundation, number: 19A10358002.

It is also supported by Research Foundation of Department of Communication of Science and Technology University of Science and Technology of China.

References

1. Neves, D.E., Brandão, W.C., Ishitani, L.: Metodologia para recomendação e agregação de Objetos de Aprendizagem no padrão SCORM. Revista Brasileira de Informática na Educação **24**(1) (2016)
2. Yanhong, S.: Design of digital network shared learning platform based on SCORM standard. Int. J. Emerg. Technol. Learn. (iJET) **13**(07), 214–227 (2018)
3. Kazanidis, I., Satratzemi, M.: Efficient authoring of SCORM courseware adapted to user learning style: the case of ProPer SAT. In: Spaniol, M., Li, Q., Klamma, R., Lau, R.W.H. (eds.) ICWL 2009. LNCS, vol. 5686, pp. 196–205. Springer, Heidelberg (2009). https://doi.org/10.1007/978-3-642-03426-8_25
4. Hanisch, F., Straβer, W.: Adaptability and interoperability in the field of highly interactive web-based courseware. Comput. Graph. **27**(4), 647–655 (2003)

5. Zhang, Y., Zhang, W.: Design of an AR and VR supported SCORM courseware player. In: Proceedings of the 23rd International ACM Conference on 3D Web Technology. ACM (2018)
6. Jee, H.K., et al.: An immersive authoring tool for augmented reality-based e-learning applications. In: 2011 International Conference on Information Science and Applications. IEEE (2011)
7. Cheng, K.-H., Tsai, C.-C.: Affordances of augmented reality in science learning: suggestions for future research. J. Sci. Educ. Technol. 22(4), 449–462 (2013)
8. Barone Rodrigues, A., Dias, D.R.C., Martins, V.F., Bressan, P.A., de Paiva Guimarães, M.: WebAR: a web-augmented reality-based authoring tool with experience API support for educational applications. In: Antona, M., Stephanidis, C. (eds.) UAHCI 2017. LNCS, vol. 10278, pp. 118–128. Springer, Cham (2017). https://doi.org/10.1007/978-3-319-58703-5_9
9. Cubillo, J., et al.: A learning environment for augmented reality mobile learning. In: 2014 IEEE Frontiers in Education Conference (FIE) Proceedings. IEEE (2014)

Animated Agents' Facial Emotions: Does the Agent Design Make a Difference?

Nicoletta Adamo[1(✉)], Hazar N. Dib[1], and Nicholas J. Villani[2]

[1] Computer Graphics Technology, Purdue University, West Lafayette, IN, USA
{nadamovi; hdib}@purdue.edu
[2] Psychological Sciences, Purdue University, West Lafayette, IN, USA
nvillani@purdue.edu

Abstract. The paper reports ongoing research toward the design of multimodal affective pedagogical agents that are effective for different types of learners and applications. In particular, the work reported in the paper investigated the extent to which the type of character design (realistic versus stylized) affects students' perception of an animated agent's facial emotions, and whether the effects are moderated by learner characteristics (e.g. gender). Eighty-two participants viewed 10 animation clips featuring a stylized character exhibiting 5 different emotions, e.g. happiness, sadness, fear, surprise and anger (2 clips per emotion), and 10 clips featuring a realistic character portraying the same emotional states. The participants were asked to name the emotions and rate their sincerity, intensity, and typicality. The results indicated that for recognition, participants were slightly more likely to recognize the emotions displayed by the stylized agent, although the difference was not statistically significant. The stylized agent was on average rated significantly higher for facial emotion intensity, whereas the differences in ratings for typicality and sincerity across all emotions were not statistically significant. A significant difference in ratings was shown in regard to sadness (within typicality), happiness (within sincerity), fear, anger, sadness and happiness (within intensity) with the stylized agent rated higher. Gender was not a significant correlate across all emotions or for individual emotions.

Keywords: Affective animated agents · Character animation ·
Character design · Facial emotions · Affective multimodal interfaces

1 Introduction

Research has shown that animated pedagogical agents (APA) can be effective in promoting learning, but many questions still remain unanswered, particularly concerning the design of APAs. For instance, it is unclear which specific visual features of an agent, types of emotional expression, degree of embodiment and personalization, modes of communication, types of instructional roles and personas benefit a particular leaner population and why. To advance knowledge in this field and maximize the agent's positive impact on learning, there is a need to further investigate the effects of certain agent's features, and whether they are moderated by learner characteristics, learning topics, and contexts.

L. T. De Paolis and P. Bourdot (Eds.): AVR 2019, LNCS 11613, pp. 10–25, 2019.
https://doi.org/10.1007/978-3-030-25965-5_2

Although the preponderance of research on pedagogical agents tends to focus on the cognitive aspects of online learning and instruction, our research work explores the less-studied role of affective aspects. In particular, one of the research goals is to determine how to design agents that exhibit emotions that are believable and clearly recognizable, and that best foster student learning. With the growing understanding of the complex interplay between emotions and cognition, there is a need to develop life-like, convincing agents that not only provide effective expert guidance, but also con-vincing emotional interactions with the learners. The goal of our research is to improve the visual quality of the agents by identifying the design and animation features that can improve perception and believability of the emotions conveyed by the agents. The work described in the paper is a step in this direction. Its objective was to investigate the extent to which the agent's visual style, and specifically its degree of stylization, affects the perception of facial emotions and whether the effects are moderated by subjects' gender.

2 Related Work

2.1 Affective Pedagogical Agents

Pedagogical agents are animated characters embedded within a computer-based learning environment to facilitate student learning. Early examples of Animated Ped-agogical Agents (APA) are Cosmo [29], Herman [30, 31], STEVE [24], PETA [47], and the "Thinking Head" [14]. Animated signing agents have also been used to teach mathematics and science to young deaf children [2].

Affective agents are animated characters that display a specific emotional style and personality, and emotional intelligence, e.g. they can respond to the user emotional state. A few affective agents systems have been developed so far. The system by Lisetti and Nasoz [34] includes a multimodal anthropomorphic agent then adapts its interface by responding to the user's emotional states, and provides affective multi-modal feedback to the user. The IA3 system by Huang et al. [21] was an early attempt at developing Intelligent Affective Agents that recognize human emotion, and based on their understanding of human speech and emotional state, provide an emotive response through facial expressions and body motions. Autotutor [12, 13] and Simsei [40] use a multimodal sensing system that captures a variety of signals that are used to assess the user's affective state, as well as to inform the agent so she/he can provide appropriate affective feedback.

Many studies confirm the positive learning effects of systems using these agents [20, 25, 32, 33, 42, 52]. Studies also indicate that the manipulation of the APAs' affective states can significantly influence learner beliefs and learning efficacy [61]. A study by Kim et al. [28] showed that an agent's empathetic responses to the student's emotional states had a positive influence on learner self-efficacy for the task, whereas an agent's happy smiles per se did not have such an effect. A meta-analytic review that examined findings from studies on the efficacy of affective APAs in computer-based learning environments shows that the use of affect in APAs has a significant and moderate impact on students' motivation, knowledge retention and knowledge transfer [19].

Some researchers have investigated the effect of different APA's features on student's learning, engagement, and perception of self-efficacy. Mayer and DaPra [37] examined whether the degree of embodiment of an APA had an effect on students learning of science concepts. Findings showed that students learned better from a fully embodied human-voiced agent that exhibited human-like behaviors than from an agent who did not communicate using these human-like actions. A study by Adamo-Villani et al. [3] revealed that the visual style of an animated signing avatar had an effect on student engagement. The stylized avatar was perceived more engaging than the realistic one, but the degree of stylization did not affect the students' ability to recognize and learn American Sign Language signs. Other studies suggest that agent's features such as voice and appearance [15, 36], visual presence [49], non-verbal communication [7], and communication style [57] could impact learning and motivation.

A few researchers have investigated whether APAs are more effective for certain learner populations as compared to others. Kim and Lin's study [26] revealed that middle grade females and ethnic minorities improved their self-efficacy in learning algebraic concepts after working with the APA, and improved learning significantly compared to white males. High school students preferred to work with an agent with the same ethnicity more than with a different one [27, 41] College students of color felt more comfortable interacting with a similar agent than with a dissimilar one [41].

2.2 Stylized Versus Realistic Agent Design

In the book "The Illusion of Life" [54], Thomas and Johnston discuss how designers should construct the characters carefully, considering all features a character has, from its costume, body proportions, facial features, to surrounding environment. Some studies suggest that characters should be designed to look realistic [43, 48], while others suggest the opposite [3, 38, 50, 60]. In character design, the level of stylization refers to the degree to which a design is simplified and reduced. Several levels of stylization (or iconicity) exist, such as iconic, simple, stylized, realistic [6]. A realistic character is one that closely mimics reality and often photorealistic techniques are used. For instance, the body proportions of a realistic character closely resemble the proportions of a real human, the level of geometric detail is high and the materials and textures are photorealistic. A stylized character often presents exaggerated proportions, such as a large head and large eyes, and simplified painted textures. Figure 1 shows the realistic and stylized agents used in the study.

Both realistic and stylized agents have been used in interactive environments. A few researchers have conducted studies on realistic versus stylized agents with respect to interest and engagement effects in users. Welch et al. [58] report a study that shows that pictorial realism increases involvement and the sense of immersion in a virtual environment. Nass et al. [43] suggest that embodied conversational agents should accurately mirror humans and should resemble the targeted user group as closely as possible. McCloud [38] argues that audience interest and involvement is often increased by stylization. This is due to the fact that when people interact, they sustain a constant awareness of their own face, and this mental image is stylized. Thus, it is easier to identify with a stylized character. Mc Donnell et al. [39] investigated the effect of rendering style on perception of virtual humans. The researchers considered 11 types of rendering styles that ranged from realistic to stylized and used a variety of implicit and explicit measures to analyze subjects' perception. Results showed that cartoon characters were considered highly appealing, and were rated as more pleasant than characters with human appearance, when large motion artifacts were present. In addition, in general they were rated as more friendly than realistic characters, however not all stylized renderings were given high ratings. One of the stylized renderings used in the study evoked negative reactions from the participants, probably due to the lack of subjects' familiarity with the style. An interesting result of the study was that the speech and motions contributed to the interpretation of the characters' intention more than the rendering style. This finding suggests that rendering style does not play a major role in the interaction with virtual characters and therefore a realistic rendering style could be as effective as a cartoon one.

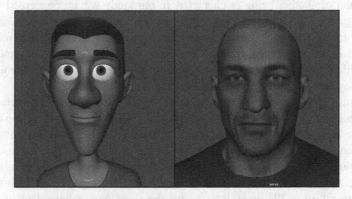

Fig. 1. Stylized agent (left); realistic agent (right)

2.3 Facial Emotion in Animated Agents: Expression and Perception

Several approaches for representing facial expressions in animated agents exist. Some computational frameworks are based on discrete representation of emotion; others on dimensional models; and others on appraisal theories [46]. Approaches that are based on the expression of standard emotions [17] compute new expressions as a mathematical combination of the parameters of predefined facial expressions [8, 45]. Approaches based on dimensional models use a 2 dimensional—valence and arousal [18] or 3 dimensional—valence, arousal, and power [5] representation of facial emotions. A few approaches use fuzzy logic to compute the combination of expressions of the six standard emotions, or the combination of facial regions of several emotions [46]. Some approaches are based on Scherer's appraisal theory [51] and model a facial expression as a sequence of the facial articulations that are displayed consecutively as a result of cognitive estimates [44].

Ongoing research suggests that the human vision system has dedicated mechanisms to perceive facial expressions [9] and categorizes facial perception into three types: holistic, componential and configural perception. Holistic perception models the face as a single unit whose parts cannot be isolated. Componential perception assumes that the human vision system processes different facial features individually. Configural perception models the spatial relations among different facial components (e.g. left eye-right eye, mouth-nose). It is possible that we use all these models when we perceive facial expressions [4].

Ekman and Friesen [16] suggest that there are three types of signals produced by the face: Static, Slow and Rapid. The static signals are the permanent or semi-permanent aspects of the face such as skin pigmentation, shape, bone structure. The slow signals include facial changes that occur gradually over time, such as permanent wrinkles, changes in muscle tone, skin texture, and even skin coloration. The rapid signals are the temporary changes in facial appearance caused by the movement of facial muscles [16]. The rapid signals are what the majority of people consider when thinking of emotion, for instance, the physical movement of the face to a smile or a frown. All three of these signals play an important role in how a viewer perceives the facial emotion of another being or character. In our study we are concerned with how the static signals of the face (e.g. face appearance and in particular the size, shape, and location of facial features such as brows, eyes, nose, mouth) affect perception of emotions, as in our experiment the slow and rapid signals were kept the same for both characters (the age of the agents is assumed to be approximately the same and the animation, e.g. the rapid signals, is identical for both characters).

A few studies that examined perception of emotion in animated characters can be found in the literature. A study by Mc Donnel et al. [39] examined perception of 6 basic emotions (sadness, happiness, surprise, fear, anger and disgust) from the movements of a real actor and from the same movements applied to 5 virtual characters (e.g. a low and high resolution virtual human resembling the actor, a cartoon-like character, a wooden mannequin, and a zombie-like character). Results of the experiment showed that subjects' perception of the emotions was for the most part independent of the character's body style. Although this study focused on perception of emotion from body movements (not from facial articulations), its findings suggest that character visual design might not affect perception of emotions in general, including facial emotions.

A study by Cissell [10] investigated the effect of character body style (cartoon and realistic) on perception of facial expressions. The study used a selection of animated clips featuring realistic and cartoon characters exhibiting 5 standard emotions. The clips were extracted from commercial animated movies. Results of the study showed that character body style did not have a significant effect on recognition of facial emotions; the emotions displayed by the cartoon characters were perceived on average more intense and sincere, while the ones displayed by the realistic character were perceived as more typical. The study is interesting, however, in our opinion, it has a flaw, as the pairs of animated clips used as stimuli did not show the same animation data for both character types. Hence, the differences in perception could be due to differences in static as well as rapid facial signals, and it is not possible to claim with confidence that the differences are due only to character design. Our study uses a similar evaluation framework as Cissell's experiment but improves on the design by comparing only static signals.

A study by Courgeon et al. [11] examined the effects of different rendering styles of facial wrinkles on viewers' perception of facial emotions. Findings showed that realistic rendering was perceived more expressive and was preferred by the subjects, however the rendering style did not have an impact on recognition of the facial emotions. A study by Hyde et al. [22] investigated the perceptual effects of damped and exaggerated facial motion in realistic and cartoon animated characters. In particular the researchers examined the impact of incrementally dampening or exaggerating the facial movements on perceptions of character likeability, intelligence, and extraversion. The results of the study are surprising, as they seem to contradict the principle of exaggeration. Participants liked the realistic characters more than the cartoon characters. Likeability ratings were higher when the realistic characters showed exaggerated movements and when the cartoon characters showed damped movements. The realistic characters with exaggerated motions were perceived as more intelligent, while the stylized characters appeared more intelligent when their motions were damped. Exaggerated motions improved perception of the characters as extraverted for both character styles. While Hyde's study focused on exaggerated versus damped facial motions, our study focuses on facial design and explores the effect of the exaggeration afforded by the degree of stylization on perception of emotion.

3   Description of the Study

The study examined the extent to which the degree of stylization of an animated affective agent (low versus high) affects the perception of facial emotions. The study used a within subjects design and a quantitative research approach. Data was collected in the form of answers to rating questions, which asked subjects to rate the typicality, sincerity, and intensity of the facial emotions exhibited by the agents. The study also collected data in the form of correct/incorrect answers to questions that asked the subjects to name the various facial emotions. In addition, the study investigated whether there was a significant difference in ratings by participants' gender.

The independent variable in the study was the degree of character stylization (low versus high), the dependent variables were recognition, typicality, intensity, and sincerity. Typicality refers to, "how often different variants of a facial expression are encountered in the real world" [56]. In other words, is the facial expression something you would see every day or is it in some way unusual? Typicality is also defined as, "having the distinctive qualities of a particular type of person or thing" [55]. So, to what extent does this expression of emotion have the distinctive qualities of a human's expression of this emotion? Intensity refers to, "Having or showing strong feelings…" [23]. In other words, how well does the character facial emotion strength match that of a human facial emotion strength? Sincere means, "free from pretense or deceit; proceeding from genuine feelings" [53]. Do the subjects feel the emotion being displayed is genuine or do they perceive it as not genuine, or deceitful?

3.1 Null Hypotheses

H1 The level of stylization does not have an effect on the subject's perceived typicality of the agent's emotion

H2 The level of stylization does not have an effect on the subject's perceived sincerity of the agent's facial emotion

H3 The level of stylization does not have an effect on the subject's perceived intensity of the agent's facial emotion

H4 The level of stylization does not have an effect on the subjects' ability to recognize the agent's facial emotion

H5 Subjects' gender is not a significant correlate

3.2 Subjects

Eighty-two subjects age 19–25 years, 42 males and 40 females participated in the study. All subjects were students at Purdue University in the departments of Computer Graphics Technology and Computer Science. None of the subjects had color blindness, blindness, or other visual impairments.

3.3 Stimuli

The characters used in the study were rigged using identical facial skeletal deformation systems. The facial skeleton, comprised of 30 floating joints with 55 DOF, is based on best practices in character animation, on the Facial Action Coding System (FACS) [17], on the AR Face Database [35] and on research on keyboard encoding of facial expressions [1].

The layout of the skeletal joints (represented in Fig. 2, frame 1) is derived from 4 face regions (Head, Upper Face, Nose, Lower Face) and 15 articulators: Head; Eyebrows, Upper Eyelids, Lower Eyelids, Eye gaze, Nose, Cheeks, Upper Lips, Lower Lips, Both Lips, Lip Corners, Tongue, Teeth, Chin, Ears. We control each articulator with 1 or more joints whose rotations and/or translations induce facial deformations or movements. The facial model allows for representing 36 Action Units (AU) of the FACS +tongue/teeth/chin/ears movements with naturalness and believability. Table 1 shows the list of face articulators, the number of joints controlling them, the joint DOFs, and the induced facial movements/deformations.

The animations were created by an expert animator with more than 20 years of experience in character animation. The expert animator animated the facial emotions on the realistic agent; the animation data was then retargeted to the stylized agent. Although the animation data was identical for both agents, some differences in facial deformations (especially in the eyebrows) can be noted. They are due to differences in facial design and facial geometry (e.g. facial proportions and mesh topology). Each animation clip was 2 s long, was rendered with a resolution of 720 × 480 pixels and was output to Quick Time format with a frame rate of 24 fps. Ten animation clips featured the stylized character, (2 for each of the 5 emotions) while the other 10 featured the realistic character (2 for each of the 5 emotions). The animations did not include any sound (speech was muted) and the characters were framed using the same camera angle and same lighting scheme. Figure 2 (frames 2–6) shows 5 pairs of frames extracted from the stimuli animations (one pair per emotion).

Table 1. List of face articulators, number of joints controlling them, joint DOFs, and induced facial movements/deformations

Articulator	Number of joints	Horizontal: R/L	Vertical: Up/Down	In/Out: z axis	AU (Induced facial movement/deformation)
Head	2	X	X	X	51, 52, 53, 54, 55, 56 (Head turn left, right, up, down, tilt left, right)
Eyebrows	6	X	X		1, 2, 4 (Inner and Outer Brow Raiser, Brow Lowerer)
Upper Eyelids	2		X		41, 42, 43, 44, 45, 46 (Lid droop, Slit, Eyes Closed, Squint, Blink, Wink)
Lower Eyelids	2		X		7, 44 (Lid tightener, Squint)
Eye gaze	2	X	X		61, 62, 63, 64
Nose	3		X		9 (Nose wrinkler)
Cheeks	2		X	X	6, 13 (Cheek raiser, Cheek Puffer)
Lips: Upper	1		X		10 (Upper Lip Raiser)
Lips: Lower	1	X	X	X (pout)	16, 20 (Lower Lip Depressor, Lip Stretcher)
Lips: Both	1	X	X	X	22, 23, 24 25, 26, 27 (Lip Funneler, Tightener, Pressor, and Part, Jaw Drop, and MouthStretch)
Lip corners	2	X	X		20, 15, 12 (Lip Stretcher, Lip Corner Depressor and Puller)
Tongue	2	X	X	X (forward)	No corresponding AU
BottomTeeth (jaw)	1	X	X	X (forward)	No corresponding AU
Chin	1			X (forward)	No corresponding AU
Ears	2		X		No corresponding AU

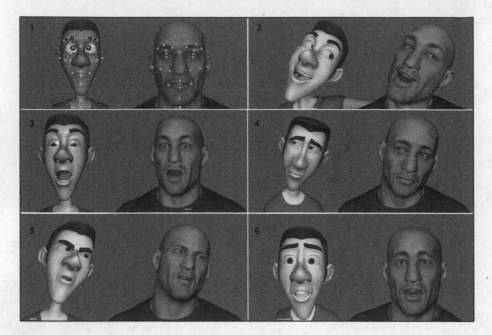

Fig. 2. Facial joints placement (frame 1); screenshots extracted from the stylized character and realistic character animations of the 5 emotions considered in the study: happiness (frame 2); surprise (frame 3); sadness (frame 4); anger (frame 5); fear (frame 6)

3.4 Procedure and Evaluation Instrument

Volunteers were recruited on the Purdue campus via email. Those who agreed to participate in the study were sent a link to an online survey they could access from any computer, and were asked to read about the research and complete a pre-survey, determining if they were eligible for participation. If eligible for participation, they proceeded to the full survey which included 21 screens: 1 screen with instructions and 2 demographics questions (age and gender) and 20 screens, each one showing one of the 20 animation clips followed by a series of questions. The order of presentation of the 20 screens was randomized. One question asked the participants to identify the emotion exhibited by the agent, and three questions asked the participants to rate the typicality, sincerity and intensity of the emotion using a 7-point scale (1 = low and 7 = high). Participants completed the on-line survey using their own computers and the survey remained active for 2 weeks. Two screenshots of the survey (and two videos) used in the study can be accessed at http://hpcg.purdue.edu/idealab/faceexpression.html.

3.5 Findings

For the analysis of the subjects' typicality, sincerity and intensity ratings we conducted a series of paired sample T-tests. With 10 pairs of animations for each subject, there were a total of 820 rating pairs.

Typicality. The mean of the ratings across all 5 emotions for animations featuring the realistic agent was 5.49, (SD = 1.76) and the mean of the ratings for animations featuring the stylized character was 5.41 (SD = 1.56). Using the statistical software SPSS, a probability value of .068 was calculated. At an alpha level of .05, H1 (e.g. stylization does not have an effect on the user's perceived typicality of agent's emotion) could not be rejected.

T-tests conducted for each individual emotion revealed a significant difference in typicality ratings for sadness, with the stylized agent rated significantly higher M (stylized) = 6.13; SD(stylized) = 1.4; M(realistic) = 5.6; SD(realistic) = 1.5; p-value = 0.04; the percentage of increase in rating scores for the stylized character was 7.5%. Gender was not a significant correlate, e.g. it did not have significant effect on typicality ratings across all emotions (p-value = 0.74) or for individual emotions.

Sincerity. The mean of the ratings across all 5 emotions for animations featuring the realistic agent was 5.69, (SD = 1.44) and the mean of the ratings for animations featuring the stylized character was 5.81 (SD = 1.65). Using the statistical software SPSS, a probability value of .057 was calculated. Since p-value > .05, H2 (e.g. stylization does not have an effect on the user's perceived sincerity of agent's emotion) could not be rejected.

T-tests conducted for each individual emotion revealed a significant difference in sincerity ratings for happiness, with the stylized agent rated significantly higher M (stylized) = 6.32; SD(stylized) = 1.5; M(realistic) = 5.8; SD(realistic) = 1.5; p-value = 0.041; the percentage of increase in rating scores for the stylized character was 7.8%. Gender did not have a significant effect on intensity ratings across all emotions (p-value = 0.67) or for individual emotions.

Intensity. The mean of the ratings across all 5 emotions for animations featuring the realistic agent was 5.78, (SD = 1.51) and the mean of the ratings for animations featuring the stylized character was 6.34 (SD = 1.73). Using the statistical software SPSS, a probability value of .045 was calculated. Since p-value < .05, H3 (e.g. stylization does not have an effect on the user's perceived intensity of agent's emotion) was rejected. Subjects perceived the emotions conveyed by the stylized agent significantly more intense than those exhibited by the realistic agent. The percentage of increase in rating scores for the stylized character was 7.6%.

T-tests conducted for each individual emotion revealed a significant difference in intensity ratings for fear, anger, and happiness. For fear, the stylized agent was rated significantly higher M(stylized) = 6.22; SD(stylized) = 1.5; M(realistic) = 5.62; SD (realistic) = 1.5; p-value = 0.04; the percentage of increase in rating scores for the stylized character was 9%. For anger, the stylized agent was rated significantly higher, M(stylized) = 6.27; SD(stylized) = 1.6; M(realistic) = 5.88; SD(realistic) = 1.6; p-value = .041; the percentage of increase in rating scores for the stylized character was 5.5%. For happiness the stylized agent was rated significantly higher, M(stylized) = 6.19; SD(stylized) = 1.34; M(realistic) = 5.67; SD(realistic) = 1.53; p-value = 0.042; the percentage of increase in rating scores for the stylized character was 7.5%. Gender did not have a significant effect on intensity ratings across all emotions (p-value = .09) or for individual emotions.

Emotion Recognition. The subjects were asked to enter the name of the emotion displayed by the agent in a text box. Based on the Feeling wheel [59], we considered the following terms correct: joy, excitement, glee, intrigue, and awe for happiness, frustration, hurt, disappointment, rage for anger, worried, depression, shame, boredom for sadness, and helplessness, insecurity, anxiety, confusion, for fear. For the emotion surprise, which is not included in the Feeling wheel, we considered the following terms acceptable, as they are commonly used synonyms of surprise: astonishment, bewilderment, amazement, consternation.

The emotion recognition rate for the stylized character was 96% across all emotions (happiness = 97%; surprise = 92%; sadness = 98%; anger = 96%; fear = 97%). The emotion recognition rate for the realistic agent was 94% across all emotions (happiness = 95%; surprise = 93%; sadness = 94%; anger = 93%; fear = 95%). The McNemar test, a variation of the chi-square analysis, which tests consistency in responses across two variables, was used to determine if the difference in emotion recognition between the two agents was statistically significant. Using SPSS software a p-value of .062 was calculated. At an alpha level of .05, a relationship between realistic and stylized agents and the subjects' ability to identify the emotions could not be determined. Our null hypothesis H5(0) (e.g. the presence of stylization does not affect the subjects' ability to recognize the facial emotion) could not be rejected. Gender was not a significant correlate for emotion recognition (p-value = .75) (nor for emotion ratings).

4   Discussion and Future Work

In this paper we have reported a study that explored the extent to which an animated agent's visual design affects students' perception of facial emotions. Findings show that subjects found the facial emotions exhibited by the stylized agent significantly more intense than those exhibited by the realistic one. In addition, subjects were more likely to recognize the emotions displayed by the stylized agent, even if the difference in recognition was not statistically significant. Analyses of ratings for individual emotions show that the stylized character was rated significantly higher in typicality in regard to sadness, in sincerity in regard to happiness, and in intensity in regard to happiness, anger, and fear. No significant differences in perception of typicality and sincerity across all emotions and no gender effects across all emotions or for individual emotions were found.

The overall higher ratings received by the stylized agent might be due to the fact that facial deformations appear more exaggerated on the stylized character because of its design, even if both agents use the same skeletal deformation system and same animation data (retargeted from the realistic character to the stylized one). Exaggeration, one of the 12 principles of animation, was used by Disney animators to present characters' motions and expressions in a wilder, more extreme form, while remaining true to reality. Exaggeration is often used with stylized characters for comedic effects, but also as a means to achieve the principle of staging, e.g. the presentation of an idea so that it is completely and unmistakably clear [54]. Stylized characters present a lower level of visual detail compared to realistic ones, especially in the face. Facial deformations that appear exaggerated are a way to compensate for the lack of facial details by making the expression clearly perceivable. The findings from our study suggest that the exaggeration effect afforded by the stylized character design is more effective at conveying facial emotions and their intensity than the higher level of visual detail of the realistic agent (e.g. realistic facial geometry and textures).

Overall, the recognition rate and the participants' ratings were high for both characters. These results suggest that both stylized and realistic character designs could be effective at conveying facial emotions and could be used for developing effective affective agents. However, the higher ratings of the stylized character suggest that a more simplified, caricatured design could benefit students' perception of the agent facial emotion. The results of our experiment are consistent with prior research [3, 10, 39] and we are inclined to believe that they would hold true for different types of stylized and realistic agents (e.g. agents showing different ages, facial features, ethnicity and gender), and for subjects from different age groups and educational backgrounds. However, in order to state with confidence that the benefit of a stylized design will generalize to most animated agents and for most participants, additional research is needed to address the limitations of the current study.

The study included a relatively small sample size, and a fairly homogenous group of participants in regard to age and educational background (college students enrolled in Computer Science and Technology programs). In the future, it would be interesting to conduct additional experiments with larger pools of subjects to farther investigate how the agent visual design effect is moderated by subjects' characteristics such as age, educational interest (interest in humanities and social sciences versus interest in STEM disciplines), and cultural backgrounds. Another intriguing future direction of research would be to investigate at what level of stylization the advantages of a stylized design disappear and a realistic design becomes more effective at conveying facial emotions. For instance would the same results hold true for an "iconic" character, e.g. a character that show a very high degree of stylization?

The cartoon character used in the study is not an exact stylized version of the realistic one, as the two characters look different. Hence, it is possible that the differences in expressivity between the two characters might be due to the intrinsic characteristic of the stylized design, rather than to the stylization of the realistic one. In other words, one could argue that the same differences in emotion expressivity could be found across two realistic characters, or across two stylized ones. To better control this variability, in future experiments, we will conduct a pre-test on the characters' design to assess the similarities between the two in order to verify to what extent the second character is recognized as a stylized version of the realistic one, rather than another different character.

The findings of the study have direct practical implications for character artists and instructional designers, as they can help them make more informed agent design decisions. The overall goal of our research is to develop an empirically grounded research base that will guide the design of affective pedagogical agents that are effective for different types of learners. Toward this goal, we will continue to conduct research studies to identify key design, modeling, and animation features that make up ideal affective animated pedagogical agents, and examine the extent to which the effects of these features are moderated by the learners' characteristics.

Acknowledgments. This work is supported by NSF-Cyberlearning award 1821894, and by Purdue University Instructional Innovation Grant 2017–2019. The authors would like to thank Purdue Statistical Consulting for their help with the statistical analysis of the data.

References

1. Adamo-Villani, N., Beni, G.: Keyboard encoding of facial expressions. In: 8th International Conference on Information Visualization-HCI Symposium (IV04) Proceedings, pp. 324–328. IEEE, London (2004)
2. Adamo-Villani, N., Wilbur, R.: Two novel technologies for accessible Math and Science Education. IEEE Multimedia Spec. Issue Access. **15**, 38–46 (2008)
3. Adamo-Villani, N., Lestina, J., Anasingaraju, S.: Does character's visual style affect viewer's perception of signing avatars? In: Vincenti, G., Bucciero, A., Vaz de Carvalho, C. (eds.) eLEOT 2015. LNICST, vol. 160, pp. 1–8. Springer, Cham (2016). https://doi.org/10.1007/978-3-319-28883-3_1
4. Adolphs, R.: Perception and emotion: How we recognize facial expressions. Current Dir. Psychol. Sci. **15**(5), 222–226 (2006)
5. Albrecht, I., Schroeder, M., Haber, J., Seidel, H.-S.: Mixed feelings: expression of non-basic emotions in a muscle-based talking head. Virtual Real. (special issue) **8**, 201–212 (2005)
6. Bancroft, T.: Creating Characters with Personality. Watson-Guptil, New York (2016)
7. Baylor, A., Kim, S.: Designing nonverbal communication for pedagogical agents: when less is more. Comput. Human Behav. **25**(2), 450–457 (2009)
8. Becker, C., Wachsmuth, I.: Modeling primary and secondary emotions for a believable communication agent. In: Proceedings of International Workshop on Emotion and Computing, in Conjunction with the 29th annual German Conference on Artificial Intelligence (KI2006), Bremen, Germany, pp. 31–34 (2006)
9. Calder, A., Rhodes, G., Johnson, M., Haxby, J.: Oxford Handbook of Face Perception. Oxford University Press, Oxford (2011)
10. Cissel, K.: A study of the effects of computer animated character body style on perception of facial expression. Purdue University e-Pub, MS thesis (2015). https://docs.lib.purdue.edu/cgttheses/28/
11. Courgeon, M., Buisine, S., Martin, J.C.: Impact of expressive wrinkles on perception of a virtual character's facial expressions of emotions. In: Proceedings of 9th International Conference, IVA 2009, Netherlands, pp. 201–214 (2009)
12. D'Mello, S., Graesser, A.: Emotions during learning with AutoTutor. In: Durlach, P.J., Lesgold, A.M. (eds.) Adaptive Technologies for Training and Education, pp. 117–139. Cambridge University Press, New York (2012)
13. D'Mello, S., Graesser, A.: AutoTutor and affective autotutor: learning by talking with cognitively and emotionally intelligent computers that talk back. ACM Transact. Interact. Intell. Syst. **2**(4), 23 (2012)
14. Davis, C., Kim, J., Kuratate, T., Burnham, D.: Making a thinking-talking head. In: Proceedings of the International Conference on Auditory-Visual Speech Processing (AVSP 2007), Hilvarenbeek. The Netherlands (2007)
15. Domagk, S.: Do pedagogical agents facilitate learner motivation and learning outcomes? The role of the appeal of agent's appearance and voice. J. Media Psychol. **22**(2), 84–97 (2010)
16. Ekman, P., Friesen, V.W.: Unmasking the Face. A guide to Recognizing Emotions from Facial Expressions. Malor Books, Cambridge (1975)

17. Ekman, P., Friesen, V.W.: Manual for the Facial Action Coding System. Consulting Psychologists Press (1977)
18. Garcia-Rojas, A., et al.: Emotional face expression profiles supported by virtual human ontology. Comp. Anim. Virtual Worlds **17**, 259–269 (2006). https://doi.org/10.1002/cav.130
19. Guo, Y.R., Goh, H.L.: Affect in embodied pedagogical agents: meta-analytic review. J. Educat. Comput. Res. **53**(1), 124–149 (2015)
20. Holmes, J.: Designing agents to support learning by explaining. Comput. Educ. **48**, 523–547 (2007)
21. Huang, T.S., Cohen, I., Hong, P., Li, Y.: IA3: Intelligent Affective Animated Agents. In: Proceedings of ICIG 2000 (2000)
22. Hyde, J., Carter, E., Kiesler, S., Hodgins, J.K.: Perceptual effects of damped and exaggerated facial motion in animated characters. In: Proceedings of the 10th IEEE International Conference on Automatic Face and Gesture Recognition (FG 2013) (2013)
23. Intense. (n.d.). In Merriam-Webster. https://www.merriam-webster.com/dictionary/intense
24. Johnson, W.L., Rickel, J.: Steve: an animated pedagogical agent for procedural training in virtual environments. SIGART Bull. **8**, 16–21 (1998)
25. Kim, S., Baylor, A.: Pedagogical agents as social models to influence learner attitudes. Educat. Technol. **47**(01), 23–28 (2007)
26. Kim, Y., Lim, J.: Gendered socialization with an embodied agent: creating a social and affable mathematics learning environment for middle-grade females. J. Educat. Psychol. **105**(4), 1164–1174 (2013)
27. Kim, Y., Wei, Q.: The impact of user attributes and user choice in an agent-based environment. Comput. Educat. **56**, 505–514 (2011)
28. Kim, Y., Baylor, A., Shen, E.: Pedagogical agents as learning companions: the impact of agent emotion and gender. J. Comput. Assist. Learn. **23**, 220–234 (2007)
29. Lester, J., Voerman, J., Towns, S.J., Callaway, C.: Cosmo: a life-like animated pedagogical agent with deictic believability. In: Notes of the IJCAI 1997 Workshop on Animated Interface Agents: Making Them Intelligent, Nagoya, Japan, pp. 61–70 (1997)
30. Lester, J., Stone, B., Stelling, G.: Lifelike pedagogical agents for mixed-initiative problem solving in constructivist learning environments. User Model. User-Adap. Inter. **9**(1–2), 1–44 (1999)
31. Lester, J., Towns, S.G., Fitzgerald, P.: Achieving Affective Impact: Visual Emotive Communication in Lifelike Pedagogical Agents. Int. J. Artif. Intell. Educ. **10**(3–4), 278–291 (1999)
32. Lester, J., Converse, S., Kahler, S., Barlow, T., Stone, B., Bhogal, R.: The persona effect: affective impact of animated pedagogical agents. In: Proceedings of CHI 1997, pp. 359–366 (1997)
33. Lester, J., Converse, S., Stone, B., Kahler, S., Barlow, T.: Animated pedagogical agents and problem-solving effectiveness: a large-scale empirical evaluation. In: Proceedings of the 8th World Conference on Artificial Intelligence in Education, pp. 23–30 (1997)
34. Lisetti, C., Nasoz, F.: MAUI: a multimodal affective user interface. In: Proceedings 10th ACM Int'l Conf. Multimedia (Multimedia 2002), pp. 161–170 (2002)
35. Martinez, A., Benavente. R.: The AR Face Database. CVC Technical Report #24 (1998)
36. Mayer, R.E.: Principles based on social cues in multimedia learning: Personalization, voice, image, and embodiment principles. In Mayer, R.E. (ed.) The Cambridge handbook of multimedia learning, 2nd ed, pp. 345–368. Cambridge University Press, New York (2014)
37. Mayer, R.E., DaPra, S.: An embodiment effect in computer-based learning with animated pedagogical agent. J. Exp. Psychol. Appl. **18**, 239–252 (2012)
38. McCloud, S.: Understanding Comics: The Invisible Art. Northampton, Mass (1993)

39. McDonnell, R., Jorg, S., JMcHugh, J., Newell, F., O'Sullivan, C.: Evaluating the emotional content of human motions on real and virtual characters. In: Proceedings of the 5th Symposium on Applied Perception in Graphics and Visualization, pp. 67–73 (2008)

40. Morency, L.P., et al.: Demonstration: a perceptive virtual human interviewer for healthcare applications. In: Proceedings of the Twenty-Ninth AAAI Conference on Artificial Intelligence (2015)

41. Moreno, R., Flowerday, T.: Students' choice of animated pedagogical agents in science learning: a test of the similarity attraction hypothesis on gender and ethnicity. Contemp. Educ. Psychol. **31**, 186–207 (2006)

42. Moreno, R., Mayer, R.E.: Interactive Multimodal Learning Environments. Educ. Psychol. Rev. **19**, 309–326 (2007)

43. Nass, C., Isbister, K., Lee, E.J.: Truth is beauty: researching embodied conversational agents. Embodied Conversational Agents, pp. 374–402 (2000)

44. Paleari, M., Lisetti, C.: Psychologically grounded avatars expressions. In: Proceedings of First Workshop on Emotion and Computing at KI 2006, 29th Annual Conf. on Artificial Intelligence, Bremen, Germany (2006)

45. Pandzic, I.S., Forcheimer, R.: MPEG4 facial animation – The standard, implementations and applications. John Wiley, Sons, Chichester (2002)

46. Pelachaud, C.: Modelling multimodal expression of emotion in a virtual agent. Phil. Trans. R. Soc. B **364**, 3539–3548 (2009)

47. Powers, D., Leibbrandt, R., Luerssen, M., Lewis, T., Lawson, M.: PETA-a pedagogical embodied teaching agent. In: Proceedings of PETRA08: 1st international conference on PErvasive Technologies Related to Assistive Environments, Athens (2008)

48. Riek, D.L., Rabinowitch, T.C., Chakrabarti, B., Robinson, P.: How anthropomorphism affects empathy toward robots. In: Proceedings of the 4th ACM/IEEE international conference on Human robot interaction, pp. 245–246 (2009)

49. Rosenberg-Kima, R.B., Baylor, A., Plant, A., Doerr, C.: Interface agents as social models for female students: the effects of agent visual presence and appearance on female students' attitudes and beliefs. Comput. Hum. Behav. **24**, 2741–2756 (2008)

50. Ruttkay, Z., Dormann, C., Noot, H.: Embodied conversational agents on a common ground. In: Ruttkay, Z., Pelachaud, C. (eds.) From Brows to Trust. HIS, vol. 7, pp. 27–66. Springer, Dordrecht (2004). https://doi.org/10.1007/1-4020-2730-3_2

51. Scherer, K.R.: Appraisal considered as a process of multilevel sequential checking. In: Scherer, K., Schorr, A., Johnstone, T. (eds.) Appraisal processes in emotion: theory, methods, research, pp. 92–119. Oxford University Press, New York (2001)

52. Schroeder, N.L., Adesope, O., Barouch Gilbert, R.: How effective are pedagogical agents for learning? A Meta-Analytic Review. J. Educ.Comput. Res. **49**(1), 1–39 (2013)

53. Sincere. (n.d). In Oxford Dictionaries Online. https://en.oxforddictionaries.com/definition/sincere

54. Thomas, F., Johnston, O.: The Illusion of Life. Disney Animation. Disney Editions, Los Angeles (1995)

55. Typical. (n.d.). In Oxford Dictionaries online. https://en.oxforddictionaries.com/definition/typical

56. Wallraven, C., Breidt, M., Cunningham, D.W., Bulthoff, H.H.: Evaluating the perceptual realism of animated facial expressions. ACM Transact. Appl. Percept. **23**, 1–20 (2007). http://dl.acm.org/citation.cfm?id=127876456

57. Wang, N., Johnson, L., Mayer, R.E., Rizzo, P., Shaw, E., Collins, H.: The politeness effect: pedagogical agents and learning outcomes. Int. J. Hum. Comput. Stud. **66**, 98–112 (2008)

58. Welch, R., Blackmon, T., Liu, A., Stark, L.: The effects of pictorial realism, delay of visual feedback, and observer interactivity on the subjective sense of presence. Presence Teleoperators Virtual Environ. **5**, 263–273 (1996)
59. Willcox, G.: The feeling wheel: a tool for expanding awareness of emotions and increasing spontaneity and intimacy. Transact. Anal. J. **12**(4), 274–276 (1982)
60. Woo, H.L.: Designing multimedia learning environments using animated pedagogical agents: factors and issues. J. Comput. Assist. Learn. **25**(3), 203–218 (2009)
61. Zhou, L., Mohammed, A., Zhang, D.: Mobile personal information management agent: Supporting natural language interface and application integration. Inf. Process. Manage. **48** (1), 23–31 (2012)

Identifying Emotions Provoked by Unboxing in Virtual Reality

Stefan H. Tanderup[1]([✉]), Markku Reunanen[2], and Martin Kraus[1] [ID]

[1] Aalborg University, Rendsburggade 14, 9000 Aalborg, Denmark
`stande14@student.aau.dk, martin@create.aau.dk`
[2] Aalto University, Otaniementie 14, 02150 Espoo, Finland
`markku.reunanen@aalto.fi`

Abstract. Unboxing a product, i.e. removing a newly acquired product from its packaging, often is a momentary strong emotional experience. Therefore, it is increasingly regarded as a critical aspect of packaging design. This study examines the user interaction when unboxing a product in Virtual Reality (VR) by comparing three plain textured boxes, each with a unique packaging design, and by inquiring the participants' emotions provoked by each design. The boxes were created based on the following interaction design guidelines: freedom of interaction, interaction pattern, and richness of motor actions. The findings indicate that there is a difference between the strength of certain positive emotions when unboxing the three boxes in VR. This study discusses the relationships of certain interactions in VR, and how they can shape people's expectations. These findings can serve as guidelines for how designers can improve the virtual experience through different interaction methods, and they provide insights into the differences between the unboxing experience in the real and the virtual world.

Keywords: Unboxing · Affective interaction · Virtual Reality · Emotions

1 Introduction

The utilitarian packaging for consumer products has changed, and what has taken its place is a strong drive towards viewing the packaging as an integral part of the overall experience. What this entails is a shift in a product's packaging design so that it takes into account the first step after retrieving a product, namely, the unboxing.

The appeal and expectations for a newly acquired product strongly rely on its packaging, and by altering its characteristics, the appeal and expectations will change as well. Beneath the visuals is the often overlooked aspect of interaction with the packaging itself. In the real world, the interaction with the package during unboxing may provide emotions of satisfaction, fascination and admiration [11], which strongly depend on the interaction to unbox the package.

© Springer Nature Switzerland AG 2019
L. T. De Paolis and P. Bourdot (Eds.): AVR 2019, LNCS 11613, pp. 26–35, 2019.
https://doi.org/10.1007/978-3-030-25965-5_3

Virtual Reality is another dimension that is ready to be explored, and the unboxing process poses the unanswered question of how this interaction provokes emotions. Unboxing is a core example of a situation that elicits strong emotions and relies on little exterior stimulation. Ideally, it can provide a basis for exploration of intense emotional interactions in VR.

This paper presents a virtual environment and interactions for unboxing a product to heighten presence. Based on this environment, an unboxing experiment is constructed and tested with the aim to contribute to the scientific understanding of interaction in VR.

2   Related Work

Unwrapping a newly acquired product often is a process of joy. As you unravel the casing to reveal what is underneath, you might experience a momentary intense emotion [11]. A survey from 2016 on packaging design [12] shows that good packaging can increase repeat products purchases from the same merchant by 52%. Furthermore, good packaging makes the brand seem more upscale and excites consumers about the product.

Unboxing has seen several descriptions, most focusing on the strong emotional value. Lazarrera [13] claimed that unboxing adds value through provision of memorable and shareable experiences. Furthermore, it has been described as a narrative experience as it is a process in which a product can speak for itself, and which gives the product a body and soul [15]. Similar views include ritual-like nature with intense emotional arousal [10] and a trigger for personal emotional response [8]. Several of these descriptions share the sentiment that unboxing is an impactful experience. As with most of such experiences, they are often exciting to watch on digital medias as well, which arguably has contributed to it gaining a large following on online video sharing platforms.

Most studies concern themselves with the visual appeal of the package, yet Kim et al. [11] and Wang et al. [16] claimed that multi-sensory aspects dominate the emotional influence during unboxing. The multi-sensory aspect is defined as a phase that offers the user short-term, hands-on, non-instrumental, realized and explorative interaction [16]. This study adopts a similar line of thought to further explore and understand the emotional response during unboxing, based on affective interaction in VR.

For this study, "affective interaction" describes an interaction that results in the experience of a feeling or emotion. Kim et al. [11] defined a multi-sensory appraisal experience as an aesthetic interaction, reasoning aesthetic to be a a provocation of higher-level pleasures of the mind, stimulating sensory modalities. Hashim [7] argued that aesthetic is more often used in a philosophical setting. Therefore, affective interaction is used to describe the multi-sensory appraisal experience. Höök et al. [9] proposed that well designed affective interaction should not infringe on privacy or autonomy, but instead empower users. Guribye et al. [6] drew inspiration from this definition, and used it for constructing a tangible affective interaction device. One of their findings is that such a device should invite squeezing and it should feel good to squeeze. These findings argued that visuals are not only for aesthetics, but for raising desire for interaction.

Combining the thoughts from the presented sources, we define unboxing as a strong positive affective, one-time experience, relying on visuals to raise desire and interaction to empower the user. If this is achieved, the experience should be memorable and spark a personal emotional response. Kim et al. [11] suggested three different box designs to achieve affective (aesthetic) interaction based on a series of principles on interaction by Djajadiningrat [4]. Djajadiningrat proposed three factors for affective interaction: *freedom of interaction*, *interaction pattern* and *richness of motor actions* [4].

To evaluate affective interaction, the Product Emotion Measurement Tool (PrEmo) by Desmont [3] has seen use in similar experiments [11], and is argued to capture enough emotions to cover a large array of feelings [2]. Therefore, PrEmo is the chosen method for evaluating affective interaction in this study. PrEmo is a non-verbal self-report instrument to measure emotional responses, consisting of 14 animations representing positive (i.e. desire, pleasant surprise, inspiration, amusement, admiration, satisfaction, fascination) and negative (indignation, contempt, disgust, unpleasant surprise, dissatisfaction, disappointment, and boredom) emotions. Each animation is coupled with a Likert scale from 0 (do not feel) to 4 (feel strongly).

In order to achieve an actual unboxing experience in VR, this study focuses on having participants feel strong *presence* [5] in the virtual world. To evaluate presence, this study uses the cross-media presence questionnaire ITC-SOPI [14], which consists of 61 questions that deal with four factors: *sense of physical space*, *engagement*, *ecological validity*, and *negative effects*. The questionnaire is not tuned for interaction, but still holds validity as one of the most researched presence measurement methods and provides a control group to compare the data to, in order to analyze presence. A study by Baños et al. [1] is used as control group, in which participants used a VR headset in an emotion-provoking experience with no interaction. The emotion that the study was aiming to provoke was sadness. Given the differences between the two experiments, the findings from this comparison may have limited validity.

3   Experiment Implementation

3.1   Overview of Artifact

The artifact for the experiment, i.e. the software, was developed with the game engine Unity using the assets SteamVR and VRTK to build the interactive VR experience. The hardware used is an HTC Vive head-mounted display with its standard wand-like controllers coupled with headphones for audio. The virtual environment aims to simulate a real environment, with a virtual agent representing a realistic male model with animated idle movement guiding the participant

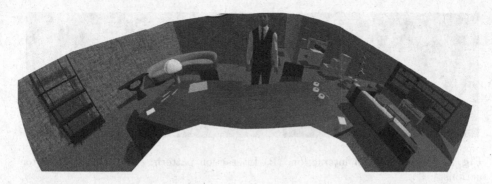

Fig. 1. Panorama picture of the environment created for the experiment, with the virtual agent.

along (see Fig. 1). To explain the concepts and to simulate a talking character, floating speech/thought bubbles are used. Similarly, to fill out the PrEmo questionnaire, a floating interactable UI was used.

When the software starts up, it presents users with the virtual world and allows them to explore it while providing a short introduction. During this process, users are asked to use the triggers on each controller and are shown how to interact with objects and UI elements. After the tutorial, they are presented with one of a total of three boxes and are asked to unbox it. Once unboxed, they are given a new box, until all three have been opened. After each unboxing, they fill out the PrEmo questionnaire (in VR), first showing the six positive emotions followed by the six negative emotions.

3.2 Box Designs

Each box was created with a different design guideline in mind: *Freedom of interaction*, *Interaction pattern* and *Richness of motor actions*. All boxes share the same dimensions and a plain paper texture. Furthermore, all outer boxes are fixed to the table, so as to not have the participants accidentally shove them onto the floor. Each box contains a computer mouse, which the participants are asked to unbox and place on an adjacent platform. If they were to lose or throw the mouse away, it would reappear in its original position in the box. The boxes can be seen in Fig. 2.

The boxes incorporate the design guidelines in the following way: Type A (freedom of interaction) is filled with balls similar to styrofoam. The material is widely used in packaging to absorb shock and protect the product. This design is meant to encourage participants to extract the product in any way they wish without a fixed order or sequence. Each HTC Vive controller was equipped with a transparent rectangular box, which would push away styrofoam if they moved the controller close to it. It was necessary to limit the amount of grains, and they were noticeably larger than typical packaging materials, as otherwise the simulation would have suffered from performance issues.

Fig. 2. (A) Freedom of interaction. (B) Interaction pattern. (C) Richness of motor actions.

Type B (interaction pattern) consists of a white box with a black pull-cord. Type B was designed so that if the cord was pulled, an animated sequence would rotate the inner box, revealing an open surface with the product, and if the user let go, the box would rewind back into its original position. This interaction was designed to represent an interaction pattern, as the users first needed to lift the cover, pull the string and then pick up the mouse. The timing, flow, and rhythm of the procedure were all linked to the action of unboxing.

Type C (richness of motor actions) was composed of a series of sequential procedures accompanied by corresponding tasks required to retrieve the packaged product. The transformative axis of each element differed to encourage the use of motor skills, providing participants an unboxing experience enriched by motor action. The implementation in VR poses some difficulties to this design, as it requires many physical surfaces and use of hinge joints, which are complicated to replicate properly in VR. To ease understanding of the unboxing process, instructions were added onto some of the walls of the box.

3.3   PrEmo

The PrEmo layout was created following a design proposed by Caicedo [2] with slight differences. Originally the questionnaire is to be answered on a computer or paper. In this study, it was recreate in VR to remove the hassle and confusion potentially caused by taking off the head-mounted display after each unboxing. The original layout of the questionnaire included all 14 emotions on the same screen. The design was changed to have the positive and negative emotion separated in different UIs to lessen the amount of information given at once.

3.4   Controller Interaction

Interaction with the virtual objects using the controller poses some challenges to the design of the setup. Participants should be able to open all boxes using only one button on each controller. The interaction should provide the feeling that the user is in control of the process. Furthermore, their behavior should avoid glitches that could affect the realism of the world. For the most part, the trigger button served as 'grab', meaning that if the controller is inside an object, and the button is pressed, the object will follow the translation and rotation of the controller.

Some boxes were of more complicated design and, thus, required a specific approach to interaction. These decisions were made based on the observations made in the internal pilot testing. Boxes B and C (Fig. 2) featured animations for pulling the chord (box B) and dragging the inner box out of the outer one (box C) using the controller. In addition, box C had an outer cover which the participants needed to lift and rotate over in order to access the inner box. This cover was set to not follow the rotation of the controller, making it easier to lift over the box.

4   Experiment Procedure

31 participants were recruited for this study. One of them did not complete the ITC-SOPI questionnaire, and was thus left out of the analysis. The final analyzed sample consisted of 30 participants, 8 female and 22 male, with age ranging from 21 to 45 years ($M = 25$). 10 different nationalities were part of the experiment, majority being Finnish ($n = 11$). All participants were either students or lecturers casually recruited from the School of Art and Design at Aalto University. None reported severe motor, hearing or visual impairments. All participants with visual impairments wore corrective glasses/lenses during the experiment.

Each participant was briefly told that the system was made to test certain means of interaction in VR. An initial tutorial, given by a virtual agent, explained how to interact in this experiment, along with the tasks that the participants had to complete. After this, they were presented with one of the three boxes, in random order, repeating until all three boxes had been completed. After having completed the virtual experiment, they were immediately asked to fill out the ITC-SOPI questionnaire on a tablet. During the tutorial stage, a few participants asked questions regarding how to 'grab' an object. No questions were raised concerning the UI. Only two participants asked questions during the unboxing stage: one regarding how to open box C, and another about where to place the mouse once it was unboxed.

The place of the simulation itself was inside a larger study room. The VR set was in a corner, free from any obstruction. There were other students present at all times scattered around the room, but it was generally quiet with no perceived distraction. The facilitation of the experiment itself was given by the virtual agent.

5   Results

5.1   PrEmO

Following the hypothesis that the boxes provoke different strength of certain emotions, the aim was to disprove the null hypothesis that they provoke an equal

bland, as to not interfere with the emotions elicited from the interaction, and differed instead based on the interaction required to unbox them. We found out that satisfaction is strongly affected by different designs, and gave hints on how to evoke positive emotions in a virtual environment. The null hypothesis ("Different interaction methods provoke no difference of emotions in VR") could, therefore, be rejected. The study further evaluated presence based on other similar experiments, finding that differences in spatial presence are significant, although it remains unclear if the cause is the introduction of interaction, or the kind of emotional engagement the experiment provoked. The findings contribute to the current knowledge on affective interaction in VR.

References

1. Baños, R.M., Botella, C., Alcañiz, M., Liaño, V., Guerrero, B., Rey, B.: Immersion and emotion: their impact on the sense of presence. Cyberpsychology Behav. **7**, 734–741 (2004). https://doi.org/10.1089/cpb.2004.7.734
2. Caicedo, D.G.: Designing the new premo (2009)
3. Desmet, P.: Measuring emotion: development and application of an instrument to measure emotional responses to products. Hum. Comput. Interact. Ser. **3**, 111–123 (2004)
4. Djajadiningrat, T., Wensveen, S., Frens, J., Overbeeke, K.: Tangible products: redressing the balance between appearance and action. Pers. Ubiquit. Comput. **8**(5), 294–309 (2004)
5. Gibson, J.: The Ecological Approach to Visual Perception. Houghton Mifflin, Boston (1979)
6. Guribye, F., Gjøsæter, T., Bjartli, C.: Designing for tangible affective interaction, pp. 30:1–30:10 (2016). https://doi.org/10.1145/2971485.2971547
7. Hashim, W.N.W., Noor, N.L.M., Adnan, W.A.W.: The design of aesthetic interaction: towards a graceful interaction framework, pp. 69–75 (2009). https://doi.org/10.1145/1655925.1655938
8. Henry, P.: The beauty of packaging (2014). http://whattheythink.com/articles/71215-beautiful-packaging/
9. Höök, K., Ståhl, A., Sundström, P., Laaksolaahti, J.: Interactional empowerment. In: Proceedings of the SIGCHI Conference on Human Factors in Computing Systems, CHI 2008, pp. 647–656 (2008)
10. Jacob-Dazarola, R., Torán, M.M.: Interactions for design - the temporality of the act of use and the attributes of products. In: Conference: 9th Norddesign Conference, At Aalborg, Denmark (2012)
11. Kim, C., Self, J.A., Bae, J.: Exploring the first momentary unboxing experience with aesthetic interaction. Des. J. **21**, 417–438 (2018). https://doi.org/10.1080/14606925.2018.1444538
12. Lazazzera, R.: 2013 ecommerce packaging survey (2016). https://dotcomdist.com/2016-eCommerce-Packaging-Survey/
13. Lazazzera, R.: How to create a memorable unboxing experience for your brand (2018). https://www.shopify.com/blog/16991592-how-to-create-a-memorable-and-shareable-unboxing-experience-for-your-brand
14. Lessiter, J., Freeman, J., Keogh, E., Davidoff, J.: A cross-media presence questionnaire: The ITC-sense of presence inventory. Presence: Teleoperators Virtual Environ. **10**, 282–297 (2001)

15. Pantin-Sohier, G.: The influence of the product package on functional and symbolic associations of brand image. Recherche et Appl. en Mark. (Engl. Ed.) **24**(2), 53–71 (2009). https://doi.org/10.1177/205157070902400203
16. Wang, R.W., Mu-Chien, C.: The comprehension modes of visual elements: how people know about the contents by product packaging. Int. J. Bus. Res. Manage. **1**, 1 (2010)

Collaborative Web-Based Merged Volumetric and Mesh Rendering Framework

Ciril Bohak$^{(\boxtimes)}$ ⬥, Jan Aleksandrov, and Matija Marolt ⬥

Faculty of Computer and Information Science, University of Ljubljana,
Ljubljana, Slovenia
{ciril.bohak,matija.marolt}@fri.uni-lj.si

Abstract. In this paper, we present a novel web-based collaborative data visualization approach combining the output of multiple rendering methods into a final seamless merged visualization. We have extended an existing web-based visualization framework Med3D with support for additional visualization plugins and an option for merging the outputs into a seamless final visualization. Such an approach allows users to place the anchors of 3D pinned annotations into volumetric data by ray casting mouse position to selected mesh geometry. The presented web-based collaborative volumetric data visualization framework is to the best of our knowledge the first of it's kind using state-of-the-art volumetric rendering system.

Keywords: Collaborative rendering · Volumetric rendering ·
WebGL 2.0 · Web-based rendering

1 Introduction

With the development of new technology, visualizations are becoming crucial in real-life decision-making activities. Most of us are making decisions based on different data visualizations displayed on our hand-held devices (e.g. checking the precipitation forecast, using navigation etc.). Moreover, we are also often discussing our decisions based on such visualizations with our friends, coworkers, and others through chat, video call or other means of communications. While many joint decisions are important for us, in some cases they are particularly important (e.g. the second opinion of a doctors decision). To make such decisions possible the collaborating party has to have a good insight into the discussed data and also the ability to share each others' view of the data as well.

Such scenarios are common in many areas; from the medical field to constructions, architecture, smart city infrastructure, military etc. An example from medical field are Multidisciplinary team meetings, which are a standard of care in modern hospital-based clinical practice for safe, effective patient management [2]. An information sharing framework for medical team meetings is presented in [4].

© Springer Nature Switzerland AG 2019
L. T. De Paolis and P. Bourdot (Eds.): AVR 2019, LNCS 11613, pp. 36–42, 2019.
https://doi.org/10.1007/978-3-030-25965-5_4

While most of such solutions offer to share of files containing documents, images or radiological data. Participants can also communicate via audio and/or video connection. Even though the discussion of the content of documents and images works well with the presented solution, it might be hard to show to other participants what exactly to look for in the imaging data. This is even more true in case of 3D radiological data (such as computed tomography- CT [1,3], magnetic resonance imaging - MRI [13], ultrasound - US [5] or positron emission tomography - PET [12]).

Some solutions already offer the use of web technologies for visualizing 3D medical data in the browser during the sessions, which can be achieved using A* Medical Imaging (AMI) Toolkit [11]. One of the downsides is that the visualization of the data does not use state-of-the-art volumetric rendering capabilities as are available in offline tools such as Exposure Render [6] or in online tools such as Volumetric Path Tracer [9]. Another downside is that individual users do not share the same view of the data. Also, in some cases, such a solution requires high network throughput due to the use of video streaming functionalities to display the individual's view to other participants. This was addressed in [7] with sharing the view with other participants in the session without streaming, and additionally with support for 3D annotations pinned to specific places in 3D models and in [8] with view-aligned annotations overlaid over the visualization of 3D models, the presented solutions are limited to the visualization of mesh geometry 3D models only.

In following sections of this paper, we present an extension of web-based real-time visualization framework with state-of-the-art volumetric rendering capabilities presented in [9] and is intended for sharing 3D visualizations with other collaborators via web-browser as presented in [7,8] allowing the users to add 3D pinned and view-aligned annotations to the direct volume rendering of radiological data. The rest of the paper consists of the presentation of used and developed methods in Sect. 2, the presentation of implementation results in Sect. 3, and the final conclusions and pointers for future work are presented in Sect. 4.

2   Methods

To achieve the presented goals, we make use of the already existing frameworks Med3D [7] and VPT [9], shortly presented in the following subsection. Next, we present the integration and extension of both frameworks: (1) system design outline, (2) data sharing, (3) view and rendering parameters sharing, and (4) merging the rendering outputs.

2.1   Med3D and VPT

Med3D is a web-based visualization framework with customizable rendering pipeline, where individual steps in the pipeline can be individually defined (e.g. mesh rendering, annotation rendering, 2D overlay etc.). As such it is a great base for developing a visualization system where we want to combine different visualization techniques for final image output. The framework supports loading mesh models (in OBJ format) and volumetric data (in MHD/RAW format) as well as conversion from volumetric to mesh data using Marching cubes [10]. The framework also allows the users to add two kinds of annotations to the scene: (a) 3D pinned annotations anchored to a specific point on 3D mesh model and (b) view-aligned annotations drawn by the user over the displayed data visualization displayed in Fig. 1.

Fig. 1. Examples of 3D pinned (left) and view-aligned (right) annotations in Med3D.

VPT is a web-based framework for direct volumetric rendering using state-of-the-art volumetric path-tracing methods. The framework allows users to define their own transfer functions for visualizing data from different domains and acquired using different scanning techniques (e.g. CT, MRI, US, PET). The framework supports the use of different visualization methods displayed in Fig. 2: (a) maximum intensity projection (MIP), (b) ISO surface rendering (ISO), (3) Emission-absorption model (EAM), and (4) Monte-Carlo single scattering path-tracing model (MCS).

For our purposes we used the functionalities of both frameworks and implemented our own extensions for achieving the desired goals.

2.2   System Design

The overall system is designed around Med3D, which was extended to support arbitrary rendering plugins. The output of rendering plugins can be used in one of the steps in the Med3D rendering pipeline. For plugins to create images for final merged visualization the plugins such as VPT need appropriate inputs. In case of VPT, the plugin needs camera parameters so that the view of the data is aligned with the view in Med3D, and the access to the desired data in different modality. The resulting output image from VPT plugin is then merged with internal Med3D rendering. Integration of VPT into Med3D framework is presented in Fig. 3.

2.3   Data Sharing

For collaborative visualization of the data, the data has to be shared with all the collaborating users. Med3D already covers the sharing of mesh-based data,

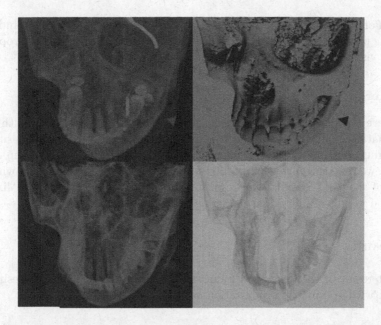

Fig. 2. Examples of different volumetric rendering techniques: (top-left) maximum intensity projection, (top-right) ISO surface, (bottom-left) Emission-absorption model, and (bottom-right) Monte-Carlo single scattering path-tracing model.

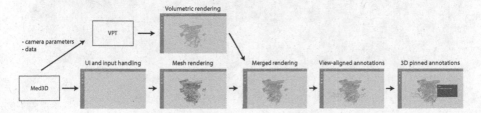

Fig. 3. The figure shows how data is shared between the frameworks and how the outputs are merged and displayed.

it's position and material properties. For collaborative visualization of volumetric data, we had to extend the sharing capabilities with support for binary volumetric data and it's descriptor meta-data. The data was encoded into an appropriate string (depending on the type of data) to include it into JSON description of the shared scene.

2.4 View and Rendering Parameters Sharing

While shared session already enabled view sharing (position and orientation of the camera) we extended the support for sharing visualization parameters as well. Med3D only offered mesh visualization option with predefined visualization properties (lighting, materials, etc.). The additional parameter set includes the selection of volumetric rendering methods (MIP, ISO, EAM and MCS) and their rendering properties. This allows users to share not only data and view properties but also visualization parameters.

2.5   Merging the Rendering Outputs

To achieve the desired goal of merging mesh geometry visualization with volumetric data visualization, we had to extend the existing pipeline with support for merging multiple types of rendering outputs. In this additional step of the pipeline, we take the output of internal mesh visualization and merge it with the output of the VPT plugin. The resulting image is alpha blended visualization where the weight of individual rendering is defined by the user.

3   Results

The first result of the developed framework is an option of merging mesh visualization with volumetric visualization as presented in Fig. 4.

Fig. 4. The result of merging mesh and volumetric rendering with use of different alpha blending option: 100% volumetric rendering - MIP (left), 67% MIP 33% mesh rendering (center) and 100% mesh rendering (right).

The second result is an option of overlaying volumetric rendering output over different mesh geometry. This is presented in Fig. 5 for mesh geometry extracted from volumetric data using Marching Cubes with a different thresholds. This allows users to add 3D pinned annotations on the geometry surface and thus effectively labeling features in the volumetric data.

The third result is an extension of the existing Med3D framework with support for direct volumetric rendering using different techniques and merging it with mesh geometry visualization. Example of such merged visualizations are presented in Fig. 6 where EAM output is merged with mesh geometry visualization.

Fig. 5. The result of overlaying volumetric rendering over different geometry: geometry created with lover threshold (right) and geometry created with higher threshold (left).

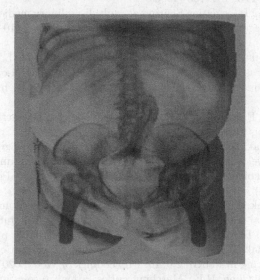

Fig. 6. Emission absorption model rendering overlaid on top of mesh geometry rendering, expressing the bones in the body.

4   Conclusion

In this work, we presented a novel web-based collaborative visualization framework with support for state-of-the-art volumetric rendering methods. The framework also allows users to easily insert 3D pinned annotations directly into volumetric data and share them with other users. An individual user in a collaboration session can individually change the view on the data and/or change parameters of visualization.

In the future, we are planning to further improve collaboration options by integrating voice and video chat into the platform. We also plan to integrate the options for volumetric data segmentation and support for remote rendering capabilities with the use of the state-of-the-art volumetric rendering techniques.

References

1. Crawford, C.R., King, K.F.: Computed tomography scanning with simultaneous patient translation. Med. Phys. **17**, 967–982 (1990)
2. Hong, N.J.L., Wright, F.C., Gagliardi, A.R., Paszat, L.F.: Examining the potential relationship between multidisciplinary cancer care and patient survival: An international literature review. J. Surg. Oncol. **102**(2), 125–134 (2010). https://doi.org/10.1002/jso.21589
3. Kalender, W.A., Seissler, W., Klotz, E., Vock, P.: Spiral volumetric CT with single-breath-hold technique, continuous transport and continuous scanner rotation. Radiology **176**, 181–183 (1990)
4. Kane, B., Luz, S.: Information Sharing at multidisciplinary medical team meetings. Group Decis. Negot. **20**(4), 437–464 (2011). https://doi.org/10.1007/s10726-009-9175-9
5. Krakow, D., Williams, J., Poehl, M., Rimoin, D.L., Platt, L.D.: Use of three-dimensional ultrasound imaging in the diagnosis of prenatal-onset skeletal dysplasias. Ultrasound Obstet. Gynecol. **21**(5), 467–472 (2003). https://doi.org/10.1002/uog.111
6. Kroes, T., Post, F.H., Botha, C.P.: Exposure render: an interactive photo-realistic volume rendering framework. PLoS ONE **7**(7), 1–10 (2012)
7. Lavrič, P., Bohak, C., Marolt, M.: Spletno vizualizacijsko ogrodje z možnostjo oddaljenega sodelovanja. In: Zbornik petindvajsete mednarodne Elektrotehniške in računalniške konference ERK 2016, 19–21 September 2016, Portorož, Slovenija, pp. 43–46 (2016)
8. Lavrič, P., Bohak, C., Marolt, M.: Collaborative view-aligned annotations in web-based 3D medical data visualization. In: Proceedings of 40th Jubilee International Convention, MIPRO 2017, 22–26 May 2017, Opatija, Croatia, pp. 276–280 (2017)
9. Žiga Lesar, Bohak, C., Marolt, M.: Real-time interactive platform-agnostic volumetric path tracing in webGL 2.0. In: Proceedings of Web3D 2018, pp. 1–7 (2018)
10. Lorensen, W.E., Cline, H.E.: Marching cubes: a high resolution 3D surface construction algorithm. SIGGRAPH Comput. Graph. **21**(4), 163–169 (1987)
11. Rannou, N., Bernal-Rusiel, J.L., Haehn, D., Grant., P.E., Pienaar, R.: Medical imaging in the browser with the A* Medical Imaging (AMI) toolkit. In: ESMRMB Annual Scientific Meeting 2017, p. 735 (2017)
12. Ollinger, J.M., Fessler, J.A.: Positron-emission tomography. IEEE Sig. Process. Mag. **14**(1), 43–55 (1997). https://doi.org/10.1109/79.560323
13. Rinck, P.A.: Magnetic Resonance in Medicine. The Basic Textbook of the European Magnetic Resonance Forum, 9th edn., vol. 9.1. TRTF, e-Version (2016)

Exploring the Benefits of the Virtual Reality Technologies for Assembly Retrieval Applications

Katia Lupinetti(✉) [iD], Brigida Bonino [iD], Franca Giannini [iD],
and Marina Monti [iD]

Istituto di Matematica Applicata e Tecnologie Informatiche "Enrico Magenes",
CNR Via De Marini 6, 16149 Genova, Italy
{katia.lupinetti,brigida.bonino,
franca.giannini,marina.monti}@ge.imati.cnr.it

Abstract. Virtual reality technologies offer several intuitive and natural interactions that can be used to enhance and increase awareness during the different phases of the product design, such as the virtual prototyping or the project reviews. In this work, we focus on the benefits of virtual reality environments for assembly retrieval applications, which usually are employed to reuse existing CAD models by operating minor modifications. The retrieval system adopted to evaluate the different similarity among CAD assemblies is able to detect three different similarity criteria (i.e. shape, joint and position) at different levels (globally and locally). Considering the complexity of assembly models, conveying these different types of information simultaneously is quite difficult, especially through the traditional desktop systems. Hence, this paper presents a system for the visualization and the inspection of the results returned by a retrieval system with the aim of easing the analysis especially when the retrieved objects to analyze are complex. The proposed tool exploits the 3D space to represent and communicate the different types of similarity and gestures as well as voice commands to interact in the VR environment. Finally, the usability of the system has been tested over a sample of interviewers, who used the system and the different functionalities provided and then expressed their judgment filling out a questionnaire. The results indicate that user are able to interpret the different similarity criteria and that the gestures help acting in a natural manner.

Keywords: CAD inspection · VR natural interaction ·
Assembly retrieval browsing

1 Introduction and Motivations

The application of immersive virtual reality (VR) environments has been proven beneficial in numerous applications [11,38], thanks also to the quality enhancement and cost reduction of technologies that allow interacting with the 3D space.

© Springer Nature Switzerland AG 2019
L. T. De Paolis and P. Bourdot (Eds.): AVR 2019, LNCS 11613, pp. 43–59, 2019.
https://doi.org/10.1007/978-3-030-25965-5_5

The large use of VR systems is due to the capabilities offered to have a better interaction and comprehension of the digital objects at their real size. Moreover, they also allow better visualization of multidimensional information than 3D screen. For instance, they are used in design and manufacturing applications to evaluate the final product at the real size, to facilitate engineer reviews [34] or to simulate and monitor complex systems [21]. These characteristics make VR an attractive choice to visualize in an efficient and communicative way the results of content-based retrieval of 3D assembly models. Experiences of using VR for content-based retrieval exist for images, e.g. [29], and 3D objects [18]. As for images, the difficulties behind CAD assembly model retrieval are due to the multiple criteria characterizing the similarity. Indeed, as explained in [27], the similarity concept is affected by the objective of the user, which may be interested in different assembly characteristics such as the shape of its components, their mutual relationships or the assembly structure. In addition, the ability to present properly the ranked results of a given query according to specific criteria becomes even more challenging when the retrieved objects to analyze are complex.

In this work, we propose a VR application exploiting the 3D space to collect and organize the results of the assemblies retrieval system proposed in [27] and to support their browsing and exploration. The evaluation of the 3D assembly similarity is facilitated by visual hints and by the possibility of inspecting the assembly components using voice commands and gestures that have metaphoric meanings and have become part of the daily life thanks to smartphones use.

The rest of the paper is organized as follows. Section 2 reviews the most pertinent related works. Section 3 illustrates the adopted methodology, while Sects. 4 and 5 report details on the proposed system. Finally, Sect. 6 shows the results of the experimentation carried out with end users using our prototype. Section 7 ends the paper providing conclusions and future steps.

2   Related Works

VR applications have received a notable amount of attention in manufacturing engineering and Sect. 2.1 provides an overview of related academic/industrial research. In addition, with the aim of understanding how VR could support designers and engineerings during product development, Sect. 2.2 presents some works that analyze human natural interaction.

2.1   VR in the Product Development Process

Examples of the use and the advantageous of adopting virtual reality applications in manufacturing processes are illustrated and discussed in several works (e.g. [7,10,22,28] and [33]). From their analysis, it follows that nowadays VR technologies are mature enough not only to visualize information or highlight problems over a digital model but also to interact with the environments modifying data related to a product and solving issues. Indeed, they find room in several phases of product development. Indeed, VR technologies enhance the virtual simulations [11] by replicating physical mock-ups, [6]; improve the collaborative design reviews on digital mock-ups [32] and the decision making on the manufacturing plan as well as on the assembly plan [22]; and support engineering education and training by reproducing learning situations [2]. In addition, they suggest that the VR visualization with the possibility of fulfilling interactive modification on the simulation input parameters can improve the decision-making capabilities of engineers thereby improving quality and reducing the development time for new products.

Conversely, design CAD models directly in VR application deserves a separate analysis. Although immersive CAD would convey many benefits, its achievement is tricky. A first complication arises from the different aims, CAD applications create detailed models and define product manufacturing plan, while VR responds to a need for visualization and uses simplified models. In addition, the complexity of the CAD models increases the difficulties of real-time interactions. A step forward the integration between CAD and VR systems has been performed by Bourdot et al. [8]. In this work, the authors propose a framework to combine VR and CAD allowing an intuitive and controlled modification of CAD models in VR environments. Hence, to overcome the bottleneck for the use of VR technologies in design activities, they suggest to label the B-Rep elements of a CAD model according to their design feature history; in this way, a parametrization of the CAD model is performed and it is used to guide the user in the possible modifications and render them in real-time. Jezernik and Hren [19], to share CAD models among different peoples without accessing a CAD workstation, propose a framework to import CAD models into VR and draw back to CAD the modification made to the model in VR. The authors explain the main weakness of the current VR systems and why so far it is not possible to modify CAD models in the VR system directly. Their proposition uses VRML format to represent the geometry data of the virtual models and an XML schema to store the assembly configuration. More recently, Okuya et al. [31] propose a system to modify native CAD data through a shape-based 3D interaction allowing the user to manipulate some parametric constraints of CAD parts by grabbing and deforming the part shape.

2.2 Natural Interaction

A selling point of VR applications is their ability to embed virtual objects in the 3D space replicating closely the physical environment. This performance can be even improved by allowing natural and intuitive interaction of the user with the virtual environment surrounding him. Hence, the development of devices to track and analyze hands movements has been encouraged in the last years. Two main types of devices can be identified: haptic or wearable based and vision based.

Wearable devices include data gloves and the Myo Gesture Control Armband. The latter is especially used in medical applications, but it still needs to be improved [4]. Instead, more established and widespread wearable devices are the data gloves. VR Glove by Manus, Cyber Glove System or Noitom Hi5 VR are some of the devices available on the market, but all of them are quite expensive. This kind of technology is very accurate and with fast reaction speed. Moreover, it avoids hands occlusion problems or loss of tracking, which are typical visual based devices issues. However, using gloves requires a calibration phase every time a different user starts and not always allows natural hand gestures because the device itself could constrain fingers motion [1,17,24].

Consequently, in the last few years, vision-based sensors have received increasing attention. These devices, in general, are easy to install and no calibrations are needed; even if in a limited range of space, they allow free-hand natural interactions. Kinect and Leap Motion Controller (LMC) are the two most commonly visual based devices used. However, LMC is more suitable for hand gesture recognition because it is explicitly targeted to hand and finger tracking, while Kinect is a depth sensor tracking full-body movement. Gunawardane et al. [17] point out the high repeatability and potential of Leap Motion compared to gloves devices in particular applications. In spite of its limitations, the controller has been evaluated with an overall better performance against other competing devices in various application domains [4].

LMC has been highly exploited in static and dynamic hand gestures recognition and 3D objects manipulation. The controller, used alone or together with a Head Mounted Display (HMD), in fact, offers the possibility of interacting with virtual objects by means of mid-air interactions. In the last few years, free-hands interactions and Virtual Reality (VR) have been involved in several research fields. For example, applications for medical purpose are described in [23] and [37]. An immersive archaeological environment is developed in [41] to simulate the creation of a statue. [25] provides an interactive application for virtual flower and plant manipulation. Beattie et al. [5] provide the manipulation and inspection of a CAD mechanic model and its individual parts first. However, in their work, interactions do not seem to be completely realistic since the only way to select parts is to break up by default the assembly.

Other works focus their attention precisely on studying and improving manipulation techniques. One of the challenges is to allow users to manipulate virtual objects in the most natural way and reaching a high level of accuracy. In [13] and [14], the authors investigate which mid-air gestures people prefer to complete different virtual manipulation and deforming tasks. Caggianese et al. [9] face the problem of canonical 3D manipulation (selection, positioning, and rotation) in VR proposing two different techniques: direct or constrained to a single dimension. The first is more natural, but the latter is heavily preferred for accomplishing complex tasks, such as the rotation. Also, Cui et al. [15] investigate the problem of deforming virtual objects' shape with natural free-hands interactions. They come up with a "steering wheel" visual metaphor to improve the precision of manipulation. Three handles are fixed along the three principal axes to guide the user during the deformation tasks. The experiment received overall positive feedbacks: users state that the process of manipulation is easy to learn and remember. Using virtual handles is intuitive but not really natural, however, it is a good way to increase the controllability of interactions, which is affected by Leap Motion's limited tracking. Later, the same authors propose in [16] a new "musical instrument" metaphor for 3D manipulation tasks, trying to achieve naturalness and avoid problems relative to hand tremor, jump release and occlusion. The idea is to distinguish between a non-dominant and a dominant hand. The first triggers events and controls precision, while the latter is tracked for the 3D object manipulation or deformation tasks. Tests' results point out that the proposed technique is easy to learn and apply, it effectively minimizes the problems relative to hands. However, mid-air interactions, compared with the use of a standard mouse, do not show significant advantages in time, learnability and comfort. Finally, also in this work, the naturalness has to be a little compromised to achieve reliable tracking.

3   Methodology and Usability

Assembly models are compared and retrieved through the system proposed by Lupinetti et al. in [27], where the user specifies a query model and sets the similarity criteria to be fulfilled. To take into account the multiple aspects characterizing an assembly, the retrieval system evaluates different criteria of similarity, as well as multiple levels of similarity (global, partial and local). In fact, assemblies can be similar in their whole or only in some parts. Specifically, the criteria of similarity include the similarity of components shape, their reciprocal position and the type of contacts with the allowed degree of freedom. Then, the similarity among assemblies is computed combining three different measures related to shape, position and joint, whose weights can be modified according to the user objectives. We refer to [26] for details on their computation.

Highlighting the diverse similarity levels between assemblies according to the different criteria clearly turn out to be tricky and challenging, especially when the assembly models are complex and made up of numerous parts. Rucco et al. [36] proposed a first effort in a desktop view to browse retrieved models according to their level of similarity and then to analyze 3D assemblies highlighting the matched parts in the query and in the target models and tuning part fade-out to allow the visualization of internal components. This approach is suitable for assemblies containing a relatively small number of parts (approximately 10 parts), anyhow in real industrial examples the results visualization is not able to provide the right level of information.

To overcome these limitations, we propose an immersive system to exploit the 3D space to increase the understanding of associated information by using metaphors. The methodology to visualize the results of the retrieval system [27] considers two important aspects: (i) how the information is communicated to the user and (ii) how the user can interact with the system and in particular with the 3D models. Each assembly model is located in a specific point of the 3D space and the more a model is close to the query, the more similar it is. In this manner, we take advantage of profundity and prospective perception of 3D space against a poorly communicative desktop representation. In addition, the proposed VR environment allows a realistic and natural interaction, where users do not need to know how CAD systems work to analyze models and their similarity. The interaction is as much natural as possible thanks to the use of gestures that have become part of the daily life because of the smartphones use. In addition, it provides a more realistic feeling by analyzing the models from the inside and in their real dimensions and evaluating also aesthetic and ergonomic proprieties. Finally, it is accessible easily and a CAD workstation is not required.

Considering existing works in literature, the proposed looks to be innovative. In fact, if on the one hand VR (and AR) has been exploited to allow the browsing and sorting of large collection of images [20,29,35], to support web pages content comparison and retrieval [38], or to provide virtual interfaces for retrieval typically associated with research in the physical books [12]; on the other hand the browsing and inspection of CAD models in an immersive virtual environment is a field little explored up to now.

An important difficulty to consider is that representing CAD in VR is not a standard so far. The integration of two different virtual words (CAD and VR) requires to manage heterogeneous data, such as the design history, semantic information, kinematic characteristics and all the elements defining the mobility of the assembly model. Unfortunately, most of this information is not embedded within CAD models [8] especially using standard formats for the exchange of CAD models. In the proposed application, the input CAD models are created with traditional CAD systems and stored in STEP format (AP 203–214). The models are then processed to obtain a standard graphics object format, more precisely an STL representation is stored for each part of the assembly model. Since this application does not require to draw back in the CAD files the potential editing performed on the models in the VR environment to inspect and analyze the different similarities, the proposed application recovers the hierarchical structure of an assembly, i.e. how the parts are gathered into sub-assemblies and the relative position of the single parts in the entire assembly model. The adaptation of the CAD models in a suitable format for VR applications is performed once for all the models present in the dataset used by the retrieval system of [27]. How these models are used in the VR environment and their possible interactions are described in the Sects. 4 and 5 respectively.

4   Virtual Environment Set Up

The proposed system adopts Unity 3D as graphical engine to visualize and perform model transformation operations. The user accesses the VR environment by wearing an HTC Vive head-mounted display (HMD), where, to allow as much natural as possible, gestures motions and vocal instructions are guaranteed.

To make more communicative the results rendering, we designed a *results browsing scene* (see Fig. 1), where each assembly is located in a specific point of the 3D space according to three similarity measures $\overrightarrow{\mu} = (\mu_{shape}, \mu_{joint}, \mu_{position})$. In this way, the Cartesian coordinates x, y and z of a general point will reflect respectively the three measures of similarity (shape, position and joint). Thus, the position $\vec{P}$ of the barycenter of each model is expressed in the Eq. (1), where ρ and ω indicate the radius of the sphere containing the model and the measurement accuracy respectively.

$$\vec{P} = 2\rho\omega\,\vec{\mu} \tag{1}$$

In this way, to avoid overlapping situations, two models whose similarity measures differ less than ω are collapsed into a sphere, on which the number of included models is drawn. The models represented with a sphere can be visualized either opening the sphere or increasing the value of the measurement accuracy ω.

Fig. 1. Results browsing scene. The query and the gazed models are ringed in yellow and in purple respectively. (Color figure online)

In addition, to help the understanding and to increase the communicability, the virtual room is furnished with several elements. Three arrows are set on the floor, under the query model, to indicate the directions along with the three measures of similarity decrease. Two groups of columns are projected on the walls and, once the user gazes an assembly, some data are displayed on them. In particular, the first group is made up of two gray columns and provides information relating to global/local similarity, showing the percentage of matched elements in the query and the target model. The second group consists of three colored columns, where the exact values of shape, position and joint similarities are visible.

This approach helps to have a first intuitive view of all results, where the VR system allows the user to move within the scene and explore them. To have a better understanding of the results, the user can select a model and then the comparison mode is automatically activated into the *inspection scene*.

Here, only the selected model and the query are visualized and the components identified as similar by the retrieval system are shaded with the same color in the two assemblies (see Fig. 2). This phase allows examining deeply the similarity between the 3D objects because the user can manipulate the entire assembly or disassembly it by manipulating part by part with realistic hands-free interaction. Indeed, it is possible to choose whether to manipulate the assembly or the single elements by means of a three buttons menu. The menu is available when the user opens the left hand with the palm facing his gaze. Notice that all the provided operations do not change the geometry of the object and can be always undone.

Fig. 2. Inspection scene. The user can manipulate the selected model; the matched parts are of same color. (Color figure online)

5   Interaction Behaviour

To comply as much as possible with the fundamentals of a Natural User Interface (NUI) [40], the interaction in the 3D space must be intuitive and simple and possibly without using controllers for input specification, such that user can act with the most natural behavior and the least training. In general, this involves two main problems: (i) specification of practical and natural object selection methods [3]; (ii) inaccuracies in hands tracking and difficulties in keeping hands motionless when using gestures without external controllers [30].

5.1   Selection

A suitable selection functionality in VR has to provide the selection of objects in dense environments and of occluded parts [39]. Many techniques have been studied [3].

To our aim, the selection by virtual hand is not recommended, even if it would be a very natural interaction. In fact, CAD assembly models are made up of many small nested parts and touching one involves colliding its neighbors too. An expensive disambiguation mechanism would be required to guess the object the user aims to select. Hence, we suggest the use of gaze: a ray is cast from the user's viewpoint and the first hit assembly/part is highlighted and becomes the selectable one. In this way, the user is guided during the gaze selection and disambiguation is easier. The voice command "select" confirms the choice. To reach the innermost occluded components, the user has to disassemble the model part by part using the gestures described in Sect. 5.2. This may seem tedious, but it is certainly the most realistic and natural method.

5.2   Gestures and Voice Commands

NUI can be achieved maintaining a connection to real-life, for instance, by allow-ing to choose operations using different gestures instead of selectable menus [15]. Thus, we propose a multi-modal system supporting gesture and voice interaction.

In the browsing scene, the user can move the whole assemblies' system in order to bring a result of interest near. This operation is achieved performing a uni-manual scrolling gesture constrained to one direction at a time (horizontally, vertically or in depth). Once selected a 3D object, it can be directly grasped and manipulated with the user's virtual hands. Three specific extra operations are allowed: translation, rotation and uniform scaling. Each of them is associated with a specific dynamic gesture: uni-manual drag, hands open with facing palms and bi-manual pinch. Table 1 shows all the gestures with a brief description. Before executing an operation, except the grasping one, the user has to maintain the hands in the correct position, until audio feedback occurs. This delay permits reliable gesture recognition and better disambiguation. These gestures have been chosen according to their use in VR [14] and in everyday life to achieve the required naturalness in a NUI.

Finally, few additional voice commands have been included as confirmation or shortcuts for specific actions. The commands are:

- *zoom (de-zoom)* to change the unit of measure in the results browsing scene,
- *select* to confirm gaze selection,
- *undo* to cancel the last performed transformation,
- *restore (restore all)* to reassemble the selected (all) disassembled model(s),
- *show matching parts* to highlight the matched parts between query and target models and put in transparency the not matched parts,
- *show assemblies* to undo the show matching parts voice command,
- *return* to come back to the previous scene.

6   Users Validation

To validate the approach experimentation has been carried out with users. In particular, the testing phase was conceived with the aim of evaluating the two

Table 1. Gestures description

Operation	Gesture	Description
Scroll the browsing scene in one direction		Keep one hand opened with the palm's normal parallel to the chosen direction. Scroll the scene by moving the hand in that direction.
Grasp the selected object		Grasp the selected object with the left or the right hand. Move the grasped object by moving the hand.
Translate the selected object in any direction		Keep the right hand with the index extended and the other fingers closed. Move the selected object by moving the hand.
Rotate the selected object in any direction		Keep the hands opened with the palms facing each other. Rotate the selected object by rotating the hands.
Scale uniformly the selected object		Keep the hands with the thumb and the index pinched. Scale the selected object by bringing your hands closer and further.

main aspects of our VR system: the naturalness and intuitiveness of interactions and the communicability of the proposed environment. In the following, the evaluation method and the achieved results are described.

6.1   The Validation Method

The evaluation process consists of three steps:

- *Explanation and training*
  First, we explain to the users what the proposed application is meant for: it is

a tool for browsing and inspecting in a virtual reality environment the results returned by a retrieval system for CAD assembly models. After summarizing the concepts of local and global similarity between assemblies and the three adopted similarity criteria, the challenges of this study are pointed out. Finally, brief training follows, in order to illustrate all the possible gestures and voice commands.

- *Use of application*
  Users are asked to wear the HTC Vive headset and they can explore the virtual environment freely, move around the results and select assemblies to inspect. We suggest them to think-aloud commenting on what they are observing and remarking the difficulties during the interaction with virtual objects. If necessary, we guide users during their experience to be sure all the functionalities are tested.

- *Questionnaire*
  Finally, users are required to fill in an assessment questionnaire. It consists of three sections: personal information, interaction evaluation and communication evaluation. Each question asks to evaluate a functionality or a feature of the virtual environment with a score from 1 (= Not at all) to 5 (= Totally) points.

16 participants (8 male, 8 female) were invited, with age between 21–60, distributed as shown in Fig. 3(a). All of them are scientific researchers (8 with a master degree, 8 with a Ph.D.) and the main subjects of their study are illustrated in Fig. 3(b). Most of the users have low or none experience with VR/AR environment and 3D CAD system; they are quite more familiar with Video Games. Few users are familiar with applications for 3D model retrieval and concepts of similarity between 3D models.

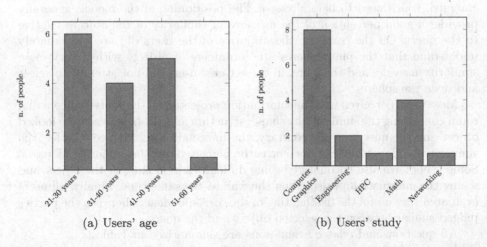

(a) Users' age   (b) Users' study

Fig. 3. Users' personal information

6.2   The Validation Results

Users have understood very promptly which objects were selectable and also the effective selection of assembly objects has been judged easy. Some difficulties have been encountered in a few cases when selecting individual parts, especially when there are small parts in big objects.

In general, the evaluation of the interaction commands is very positive for the majority of the users. The *grabbing* is judged easy and intuitive; only a few users have considered inconvenient to be obliged to approach the object to be able to take it. A few users have had troubles in distinguishing between the *translation* command (used to move the selected object in any direction) from the *scrolling* one (used to move the whole scene horizontally, vertically and deeply). Some users have encountered some difficulties in the *rotation* and *scaling* operations, mainly because the gestures leave the field of vision of the Leap Motion or because in certain positions of the hands the Leap Motion recognizes the gesture no longer.

A weakness perceived by different users is the slowness of the operation execution with respect to the command. All users agree that the voice commands are extremely useful. Some of them have recommended avoiding similar voice commands that can be difficult to remember and easily confused. Most users have assessed useful the activation of any gesture/operation by specific voice command; some users have suggested having the corresponding voice command for each gesture.

The ability to access a graphic menu showing the palm of the hand has been appreciated by most users, even if a certain difficulty arose in performing the operation of pressure on the buttons.

Regarding the communication of the information to convey to users through the virtual environment and the mutual position of the models, some issues emerged, which need to be addressed. The positioning of the models generally provides a good perception of the measure of similarity of the objects relative to the query. On the contrary, the majority of the users did not immediately understand that the pink spheres are containers of objects with a very close similarity measure and therefore, in these cases, users did not attempt to select and open the spheres.

Most users perceived well the information projected on the walls of the virtual room concerning the similarity by shape, structure and joints between the looked object and the query; on the contrary, the information relating to whether the similarity is global/local is not correctly perceived by the majority of users. Some users have also encountered some difficulties in looking at the object and seeing the information projected on the wall at the same time. Finally, all users evaluated very useful the highlighting of the correspondences between the parties judged similar between the selected object and the query.

All questions and related evaluations are summarized in Table 2.

In general, the experimental results confirm the validity of the approach to visualize and explore in an immersive environment the results of a 3D assembly models retrieval system. Most of the choices made in terms of both gesture and

Table 2. Summary of users' feedbacks

N.	Question	Evaluation				
		1	2	3	4	5
1	Was it easy to discern selectable and unselectable?	0%	0%	0%	44%	56%
2	Was it easy to select target assemblies?	0%	0%	0%	63%	37%
3	Was it easy to select target parts?	0%	12%	19%	44%	25%
4	Was the grabbing gesture natural and intuitive?	0%	0%	6%	56%	38%
5	Did the grabbing operation act as you expected?	0%	6%	25%	31%	38%
6	Was the translation gesture natural and intuitive?	0%	0%	31%	44%	25%
7	Did the translation operation act as you expected?	0%	0%	19%	56%	25%
8	Was the rotation gesture natural and intuitive?	0%	12%	25%	44%	19%
9	Did the rotation operation act as you expected?	0%	12%	13%	56%	19%
10	Was the scale gesture natural and intuitive?	0%	6%	13%	50%	31%
11	Did the scale operation act as you expected?	0%	0%	19%	56%	25%
12	Was the scrolling gesture natural and intuitive?	0%	12%	19%	50%	19%
13	Did the scrolling operation act as you expected?	0%	6%	44%	44%	6%
14	Were the voice commands useful?	0%	0%	6%	13%	81%
15	Were the adopted keywords natural and intuitive?	0%	0%	6%	13%	81%
16	Is the gesture to visualize the graphical menu natural and intuitive?	0%	0%	12%	25%	63%
17	Are the menu buttons easy to push?	0%	12%	12%	38%	38%
18	Is the meaning of the buttons easy to understand?	0%	0%	12%	19%	69%
19	To what extent does objects' position help in the perception of the object similarity with respect to the query according to the three similarity criteria?	0%	0%	31%	44%	25%
20	Is it easy to understand that the pink spheres represent a set of close objects?	19%	31%	31%	6%	13%
21	Does the information projected on the walls help you in understanding how much the object you are looking at is similar to the query for each criteria?	0%	6%	13%	31%	50%
22	Does the information projected on the walls help you in understanding how globally/locally the object you are looking at is similar to the query?	6%	31%	19%	38%	6%
23	Is it useful to visualize the matching parts among two similar assemblies(voice command *show matching parts*)?	0%	0%	0%	12%	88%

vocal commands and of communication strategy to the user have been appreciated with some exceptions. As far as the interaction with 3D objects is concerned, the perceived limits mainly derive from technological limits that cannot be completely overcome and require revising some interactions in order to reduce the negative effects perceived at least partially. As regards information that are not perceived correctly by most users, i.e. use of spheres as containers of several objects and communication of information on global/local similarity, it is necessary to study and propose alternatives solutions to make the communication more immediate and less ambiguous.

7   Conclusion

In this paper, we present a VR environment for the visualization and inspection of the results of a system for the retrieval of similar 3D assembly models. It exploits the capabilities offered by VR technologies for creating a 3D scene in which the retrieved 3D models are distributed according to their distance from the query model highlighting the characteristics that determine the similarity. Moreover, natural interaction capabilities are provided to explore the models and their correspondences with the query one.

The experimentation carried out with end users provided good indications confirming the validity of the approach and the usefulness and usability of the developed prototype. Moreover, it highlighted some issues to be faced to improve it. Future extension foresees the inclusion of functionality for the modification of assembly components to allow the updating of existing models and the possible combinations of existing components to create new assemblies.

References

1. Abraham, L., Urru, A., Normani, N., Wilk, M., Walsh, M., O'Flynn, B.: Hand tracking and gesture recognition using lensless smart sensors. Sensors **18**(9), 2834 (2018)
2. Abulrub, A.H.G., Attridge, A.N., Williams, M.A.: Virtual reality in engineering education: the future of creative learning. In: 2011 IEEE Global Engineering Education Conference (EDUCON), pp. 751–757. IEEE (2011)
3. Argelaguet, F., Andujar, C.: A survey of 3D object selection techniques for virtual environments. Comput. Graph. **37**(3), 121–136 (2013)
4. Bachmann, D., Weichert, F., Rinkenauer, G.: Review of three-dimensional human-computer interaction with focus on the leap motion controller. Sensors **18**(7), 2194 (2018)
5. Beattie, N., Horan, B., McKenzie, S.: Taking the leap with the oculus hmd and CAD-plucking at thin air? Proc. Technol. **20**, 149–154 (2015)
6. Bordegoni, M.: Product Virtualization: An Effective Method for the Evaluation of Concept Design of New Products, pp. 117–141. Springer, London (2011). https://doi.org/10.1007/978-0-85729-775-4_7
7. Bougaa, M., Bornhofen, S., Kadima, H., Rivière, A.: Virtual reality for manufacturing engineering in the factories of the future. Appl. Mech. Mater. **789**, 1275–1282 (2015)

8. Bourdot, P., Convard, T., Picon, F., Ammi, M., Touraine, D., Vézien, J.M.: VR-CAD integration: multimodal immersive interaction and advanced haptic paradigms for implicit edition of CAD models. Comput. Aided Des. **42**(5), 445–461 (2010)
9. Caggianese, G., Gallo, L., Neroni, P.: An investigation of leap motion based 3D manipulation techniques for use in egocentric viewpoint. In: De Paolis, L.T., Mongelli, A. (eds.) AVR 2016. LNCS, vol. 9769, pp. 318–330. Springer, Cham (2016). https://doi.org/10.1007/978-3-319-40651-0_26
10. Choi, S., Jung, K., Noh, S.D.: Virtual reality applications in manufacturing industries: past research, present findings, and future directions. Concur. Eng. **23**(1), 40–63 (2015)
11. Choi, S., Cheung, H.: A versatile virtual prototyping system for rapid product development. Comput. Ind. **59**(5), 477–488 (2008)
12. Cook, M.: Virtual serendipity: preserving embodied browsing activity in the 21st century research library. J. Acad. Librariansh. **44**(1), 145–149 (2018)
13. Cordeiro, E., Giannini, F., Monti, M., Mendes, D., Ferreira, A.: A study on natural 3D shape manipulation in VR. In: Livesu, M., Pintore, G., Signoroni, A. (eds.) Smart Tools and Apps for Graphics - Eurographics Italian Chapter Conference. The Eurographics Association (2018). https://doi.org/10.2312/stag.20181296
14. Cui, J., Kuijper, A., Fellner, D.W., Sourin, A.: Understanding people's mental models of mid-air interaction for virtual assembly and shape modeling. In: Proceedings of the 29th International Conference on Computer Animation and Social Agents, pp. 139–146. ACM (2016)
15. Cui, J., Kuijper, A., Sourin, A.: Exploration of natural free-hand interaction for shape modeling using leap motion controller. In: 2016 International Conference on Cyberworlds (CW), pp. 41–48. IEEE (2016)
16. Cui, J., Sourin, A.: Mid-air interaction with optical tracking for 3D modeling. Comput. Graph. **74**, 1–11 (2018)
17. Gunawardane, P., Medagedara, N.T.: Comparison of hand gesture inputs of leap motion controller & data glove in to a soft finger. In: 2017 IEEE International Symposium on Robotics and Intelligent Sensors (IRIS), pp. 62–68. IEEE (2017)
18. Henriques, D., Mendes, D., Pascoal, P., Trancoso, I., Ferreira, A.: Poster: evaluation of immersive visualization techniques for 3D object retrieval. In: 2014 IEEE Symposium on 3D User Interfaces (3DUI), pp. 145–146. IEEE (2014)
19. Jezernik, A., Hren, G.: A solution to integrate computer-aided design (CAD) and virtual reality (VR) databases in design and manufacturing processes. Int. J. Adv. Manuf. Technol. **22**(11–12), 768–774 (2003)
20. Koutsabasis, P., Domouzis, C.K.: Mid-air browsing and selection in image collections. In: Proceedings of the International Working Conference on Advanced Visual Interfaces, pp. 21–27. ACM (2016)
21. Kovar, J., Mouralova, K., Ksica, F., Kroupa, J., Andrs, O., Hadas, Z.: Virtual reality in context of industry 4.0 proposed projects at brno university of technology. In: 2016 17th International Conference on Mechatronics-Mechatronika (ME), pp. 1–7. IEEE (2016)
22. Lawson, G., Salanitri, D., Waterfield, B.: Future directions for the development of virtual reality within an automotive manufacturer. Appl. Ergon. **53**, 323–330 (2016)
23. Lin, J., Schulze, J.P.: Towards naturally grabbing and moving objects in VR. Electron. Imaging **2016**(4), 1–6 (2016)

24. Liu, H., Ju, Z., Ji, X., Chan, C.S., Khoury, M.: Human Motion Sensing and Recognition. SCI, vol. 675. Springer, Heidelberg (2017). https://doi.org/10.1007/978-3-662-53692-6

25. Liu, L., Huai, Y.: Dynamic hand gesture recognition using lmc for flower and plant interaction. Int. J. Pattern Recognit Artif Intell. **33**(01), 1950003 (2018)

26. Lupinetti, K., Giannini, F., Monti, M., Pernot, J.P.: Multi-criteria similarity assessment for CAD assembly models retrieval. IMATI Report Series (18–07), 21 (08/2018 2018). http://irs.imati.cnr.it/reports/irs18-07

27. Lupinetti, K., Giannini, F., Monti, M., Pernot, J.P.: Multi-criteria retrieval of CAD assembly models. J. Comput. Des. Eng. **5**(1), 41–53 (2018)

28. Mujber, T.S., Szecsi, T., Hashmi, M.S.: Virtual reality applications in manufacturing process simulation. J. Mater. Process. Technol. **155**, 1834–1838 (2004)

29. Nakazato, M., Huang, T.S.: 3D Mars: immersive virtual reality for content-based image retrieval. In: IEEE International Conference on Multimedia and Expo, 2001. ICME 2001. (ICME). pp. 44–47 (2001). https://doi.org/10.1109/ICME.2001.1237651

30. Nguyen, T.T.H., Duval, T., Pontonnier, C.: A new direct manipulation technique for immersive 3D virtual environments. In: ICAT-EGVE 2014: the 24th International Conference on Artificial Reality and Telexistence and the 19th Eurographics Symposium on Virtual Environments, p. 8 (2014)

31. Okuya, Y., Ladeveze, N., Fleury, C., Bourdot, P.: Shapeguide: shape-based 3D interaction for parameter modification of native CAD data. Front. Robot. AI **5**, 118 (2018)

32. Okuya, Y., Ladeveze, N., Gladin, O., Fleury, C., Bourdot, P.: Distributed architecture for remote collaborative modification of parametric CAD data. In: IEEE VR International Workshop on 3D Collaborative Virtual Environments (3DCVE 2018) (2018)

33. Ottosson, S.: Virtual reality in the product development process. J. Eng. Des. **13**(2), 159–172 (2002)

34. Park, H., Son, J.S., Lee, K.H.: Design evaluation of digital consumer products using virtual reality-based functional behaviour simulation. J. Eng. Des. **19**(4), 359–375 (2008)

35. Peng, C., Hansberger, J.T., Cao, L., Shanthakumar, V.A.: Hand gesture controls for image categorization in immersive virtual environments. In: 2017 IEEE Virtual Reality (VR), pp. 331–332. IEEE (2017)

36. Rucco, M., Lupinetti, K., Giannini, F., Monti, M., Pernot, J.-P.: CAD assembly retrieval and browsing. In: Ríos, J., Bernard, A., Bouras, A., Foufou, S. (eds.) PLM 2017. IAICT, vol. 517, pp. 499–508. Springer, Cham (2017). https://doi.org/10.1007/978-3-319-72905-3_44

37. Seif, M.A., Umeda, R., Higa, H.: An attempt to control a 3D object in medical training system using leap motion. In: 2017 International Conference on Intelligent Informatics and Biomedical Sciences (ICIIBMS), pp. 159–162. IEEE (2017)

38. Toyama, S., Al Sada, M., Nakajima, T.: VRowser: a virtual reality parallel web browser. In: Chen, J.Y.C., Fragomeni, G. (eds.) VAMR 2018. LNCS, vol. 10909, pp. 230–244. Springer, Cham (2018). https://doi.org/10.1007/978-3-319-91581-4_17

39. Vanacken, L., Grossman, T., Coninx, K.: Exploring the effects of environment density and target visibility on object selection in 3D virtual environments. In: IEEE Symposium on 3D User Interfaces, pp. 115–222. IEEE (2007)

40. Vitali, A., Regazzoni, D., Rizzi, C.: Guidelines to develop natural user interfaces for virtual reality solutions. In: International Conference on Innovative Design and Manufacturing: ICIDM2017, Milan, Italy, July 2017 (2017)

41. Vosinakis, S., Koutsabasis, P., Makris, D., Sagia, E.: A kinesthetic approach to digital heritage using leap motion: the cycladic sculpture application. In: 2016 8th International Conference on Games and Virtual Worlds for Serious Applications (VS-GAMES), pp. 1–8 (2016). https://doi.org/10.1109/VS-GAMES.2016.7590334

Metaphors for Software Visualization Systems Based on Virtual Reality

Vladimir Averbukh[1,2]([envelope]) [iD], Natalya Averbukh[2] [iD], Pavel Vasev[1] [iD],
Ilya Gvozdarev[2] [iD], Georgy Levchuk[2] [iD], Leonid Melkozerov[2] [iD],
and Igor Mikhaylov[1,2] [iD]

[1] Institute of Mathematics and Mechanics of the Ural Branch of the Russian
Academy of Sciences, Ekaterinburg, Russia
averbukh@imm.uran.ru
[2] Ural Federal University, Ekaterinburg, Russia

Abstract. The paper discusses research and development in the field
of software visualization based on virtual reality environments. Spatial
metaphors play an important role in such systems. A brief overview of
the projects of software visualization systems based on vir- tual reality is
provided. Among the systems developed over the past decades, one can
find systems both for program visualization and for visual programming.
Descriptions of prototypes of software visualization systems, software
objects visualization and supercomputer performance data visualization,
realized by the authors of the paper, are presented. These prototypes,
designed for virtual reality environments, were developed with the use
of several versions of a cosmic space metaphor and an extended city
metaphor. The paper also discusses psychological aspects of the human
factor in developing software visualization systems with the use of virtual
reality.

Keywords: Software visualization · Visual programming ·
Virtual reality · Visualization metaphors · Human factor · Presence

1 Introduction

Software visualization can be defined as an assembly of computer graphics
and human-computer interaction techniques employed for better explanation of
notions and efficient software maintenance, as well as for specification and software objects presentation in the process of program design and development.

The paper discusses the implementation of spatial metaphors in software
visualization systems using virtual reality. The success of visualization, to a
certain extent, depends on the choice of metaphors, although the factors of
implementation quality, the human factor and usability also play a significant
role. Provided below is the overview or existing solutions for systems based
on virtual reality, and several issues connected with the human factor in this
context.

© Springer Nature Switzerland AG 2019
L. T. De Paolis and P. Bourdot (Eds.): AVR 2019, LNCS 11613, pp. 60–70, 2019.
https://doi.org/10.1007/978-3-030-25965-5_6

2 Software Visualization Based on Virtual Reality Environments

Let us define a visualization metaphor as a mapping from concepts and objects of the simulated application domain to a system of similarities and analogies, and generating a set of views and a set of techniques for interaction with visual objects. Visualization metaphors form the basis for the views of specialized visualization systems. Designing such views is a crucial part of a thorough design of the human factor related aspects of these systems [1]. Spatial metaphors can be used in visualization systems based on virtual reality.

In early 90s an interesting software visualization system, Avatar, was developed [15], which makes extensive use of the means of virtual reality and a 3D version of a room building metaphor, and functions on the basis of a CAVE-type virtual reality environment. The Avatar system was designed to represent large volumes of data regarding the performance of parallel systems. In the process of work a user sort of finds themselves inside a 3D room, where a video image is projected on the walls.

A new wave of attention towards the use of virtual reality in software visualization systems rose in early 2000s [5,6]. As far as this decade, papers in this field have received a big boost. Papers [3,4] describes the prototype of a software visualization system based on a city metaphor. A city metaphor is rather popular among software visualization systems using the means of virtual reality. In this respect, two systems with similar names are worth mentioning VR City [20] and CityVR [10].

The VR City system employs a modified city metaphor to represent program systems and related analytical data. Static (metrics) and dynamic (traces) aspects of programs are visualized. Users can observe and interact with objects of the city in an immersive virtual reality environment. A program browsing function is available.

Note that publication [10] discusses also the approach of **gamification** of software engineering tasks. This approach presupposes creating tools that provide software engineers with an interface similar to that of computer games. **Gamification** in software visualization systems development based on virtual reality is also mentioned in paper [12]. Also worth mentioning are the publications of this research team, devoted to software visualization systems development based on virtual reality [12,14] (including the paper on visual programming with the use of virtual reality [11]). The paper [13] discusses the prospects of using immersion for software engineers into the program's structures, and paper [14] a virtual flight over software objects within various metaphors (including city metaphors and cosmic metaphors). Another paper deals with software system visualization based on virtual reality. This tool enables software engineers, project managers or clients to explore the architecture and get a first impression of the dimensions and inter-dependencies of components. The paper [16] discusses using mixed reality for code checkout. An interactive 3D visualization of the message flow is presupposed, combining traditional visualization techniques with augmented reality based on a VR-headset.

Paper [8] is devoted to the issues of usability assessment of software visual-ization systems. However, to a certain extent, it could be presented as a development of the CityVR project. The CityAR system was created on the basis of an immersive extended reality environment. Within a city metaphor, buildings are used to represent classes, and their sizes and colors code the software metrics. Paper [2] describes the tool for visual programming with the use of immersive virtual reality.

Worth mentioning is also the paper devoted to 3D software visualization systems performance evaluation [9]. For the purposes of the experiment, 3D visualization based on a city metaphor was realized on a standard computer screen, in a virtual reality environment and with a physical 3D printed model. A real-world island metaphor was presented in [17,18]. The entire software system is represented as an ocean with many islands on it. Also the approach for the exploration of software projects in VR and AR was described.

Even a brief overview shows that the agenda of using virtual reality means in software visualization arouses interest. At the same time, the process is currently on prototype development stage and the evaluation of virtual reality efficiency potential for software visualization. It appears that research and pilot projects should be extended with the use of various visualization metaphors. Performance evaluation should be carried out with the use of psychological techniques, in particular, those relying on the activity theory.

3 The Development of Prototypes of Software Visualization Systems Based on Virtual Reality Environments

Analysis of the examples of software visualization systems based on virtual reality environments shows that despite the 25 years of history a lot of questions remain as to the practical applicability of these environments. All this requires additional research and development of systems with account of specific software tasks. This section discusses projects and implementations of prototypes for software visualization systems based on several metaphors and with potential for using virtual reality environments.

Design and developments are proposed based on both the already familiar or modified metaphors, and new metaphors. At this stage, the main goal is to evaluate the use of metaphors in developing software visualization systems, in visual programming systems and in the visualization of supercomputer performance data. Further research, powered by these developments, involves the analysis of the human factor influence on the efficiency of using software visualization systems based on virtual reality media.

3.1 Software Visualization Based on a City Metaphor

A software visualization project is being developed for the needs of software engineers, programmers and testers, with possible use of virtual reality. The use of two forms of representation is presupposed: graphs and the one based on a city metaphor.

Let us consider a prototype of a software visualization system which is based on an extended city metaphor, through the addition of active agents. With the help of agents, one can conduct checkouts, testing and comparative analysis of the code with various types of data, logic and other features, responsible for the quality of the code. And due to the links, the user can monitor the interaction of various pieces of code, determine the type of connection, and have the opportunity to rationalize the logic of this code for smart use of quick or closest to quick connections that increase the speed of algorithm operation. Virtual reality makes it possible to construct presentations within objects, to place various transmitters in a 3d model and to see how the device functions under the program control.

A software visualization system prototype has been developed. Its interface is a menu with an option to choose a software engineer activity type (engineer, tester etc.), with level s of access for different kinds of work and for solving different tasks. There is opportunity to work in the framework of virtual reality with 3D visualization of the code structure based on a city metaphor. When choosing a task that is a part of the algorithm logic description, a corresponding visual presentation pops up with a window of the algorithm itself, its description, its versions revision and reset history, as well as various statistical data etc. Each type of activity presupposes its own data presentation. In the course of visual world realization in a virtual reality environment a corresponding scene is created, which reflects a city representing the project that the software engineer is planning to work on Fig. 1, Left.

The structure of a virtual city corresponds with the structure of a directory tree and file tree of the respective project: a block and a sub-block are a folder and a subfolder respectively, and a building represents a file. In this case, roads represent the borders of the first level of the projects subfolders. The next level of a software project will be represented by building interior. Other objects of a software package can be represented as trees, piping, power stations etc. According-ing to the scenarios of activity in VR, the user is supposed to walk along the streets of a schematic city entering buildings and premises in order to study visual representations of software. Active agents help study these objects. In the future, it is planned to conduct research into the possibilities of visual represen-tation of a code structure in virtual reality with enhanced quality of an image and addition of the ability to visualize codes of various programming languages on different platforms, media, to define the best interactive option.

3.2 Visual Programming System Based on a Cosmic Metaphor

A project for a programming system based on a visual language built on the principles of object-oriented programming is under consideration. The purpose of this project is to create a medium for visual programming with the possibility to disregard the rules and standards of a specific language, which makes it possible to focus on the problem at hand when the coding.

A cosmic metaphor in its modern sense, with a heliocentric worldview, is chosen as an idea for a visual programming medium. At the same time, part of the entities of the program is represented as planets, their satellites, rings (like the rings of Saturn), and other elements of outer space (see Fig. 1, Right).

Fig. 1. Left: presentation of a software project as a city in virtual reality and the inner part of the structure of the code with the figure of the active agent. Right: view of the visual programming environment

It is supposed to exercise interaction with the system by means of devices used in the environments of virtual reality, and a traditional keyboard is used for supplementary tasks, for example, for identifier coding. A task is set to realize the scalability (taking into account various levels of abstraction), to provide navigation along the code and easy transition between the levels of abstraction. The increase of software complexity should not amplify the complexity of graphical representation.

All user classes within this visual programming system are represented as planets. Each planet (class) has two types of views: a free one (traditional view of a planet with its satellites and rings) and an active one (when a class is chosen, it looks like a planets elevation drawing). The active view is a circle with a sector of rings on the left and ordered satellites on the right. The ring sector represents class methods. Each ring is a separate method. Outer rings are public methods; inner rings are private and protected ones. The satellites represent class fields. From top to bottom first go the satellites which are more remote from the planet (public fields), then less remote ones (static methods and fields), which belong to the class itself, and in the center there is a core, which keeps all class constructors within itself. The algorithm (a set of instructions) can be implemented as a context, that is as an area of space with a sequence of expressions that have their own visual representation. Each visual representation can be put in any place of this area. The order of implementation is supposed to be set manually.

The scenario presupposes writing a program represented by a set of planets. While working, a user moves from orbit to orbit, exploring the corresponding planets (classes). Then the user enters the planet in order to study additional information. Virtual reality provides visibility for a specific example, as well as the opportunity to keep all the classes, interfaces and their methods in sight, with the possibility of in-depth analysis of their structure by means of zooming. In the future, extension of visual software development medium is possible, for instance, by means of introducing graphic analogs of syntactic constructs from Java, so that one could create programs of any complexity.

3.3   Software Visualization Based on a Geocentric Metaphor

A two-dimensional and a three-dimensional option of this metaphor realization are discussed, as well as other possible modifications. Two prototypes of visualization systems are suggested in the geocentric metaphor: in 2D and 3D versions.

A 3D Option of the Geocentric Metaphor

The metaphor of a geocentric system in 3D space is used for representing the data on supercomputer performance at a certain moment in time. A parallel supercomputer consist of a number of processors. An IT administrator studies the data of the supercomputers work with the aim of increasing its efficiency, studying for situations when various delays occur in the work of the system. An IT administrators work scenario involves using a large number of representation views. Using representations based on the geocentric metaphor involving virtual reality is one of them. Visual representation within this metaphor consists of several layers: (0) In the center of a geocentric system, there is a supercomputers file subsystem. (1) The first sphere, surrounding the center, shows the supercomputers users, whose tasks are being accomplished at a given moment in time. (2) The second sphere contains the tasks being accomplished. The size of the tasks object means the number of supercomputers computation nodes dedicated for the task.

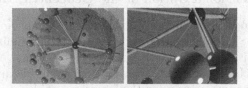

Fig. 2. Left: visualization of the load on the file system of a supercomputer using geocentric metaphor. Right: a fragment of the state of the supercomputer at some point in time, made from within a three-dimensional geocentric model. A view from within. (Color figure online)

An example of the above-described view work process is demonstrated in Fig. 2. In this case, one can see that the main load on the file system (the green sphere in the center) is carried only by several users. Also visible is the texture of supercomputer use: some users are working on one task, others are working on two or three tasks. However, there are exceptions: a user in the right part of the figure has launched about 30 small tasks.

It seems that in the case of virtual reality application the most archaic 3D version of a geocentric model may be more convenient, with the Earth represented as being at, and celestial bodies located on hemispheres covering the at

Earth. Left and right picture on Fig. 3 illustrate performance data representation, visualized as an archaic 3D geocentric metaphor.

Visualization program within the 3D geocentric system metaphor is built on the basis of the Viewlang.ru web library of 3D graphic. This library relies on WebGL standard. It also provides support for WebVR virtual reality technologies. If there is a compatible device (for example, Oculus Rift), the visualization programs user has at their disposal the switch to VR mode button. When pressed, it initiates the graphic transmission onto the VR device. It also reacts to head tilts and users movements. At the same time, the user also has traditional navigation through mouse at their disposal. An example of an image that can be observed by a user in VR mode is shown at the left side of Fig. 3.

Fig. 3. Left: visualization of the load on the file system of a supercomputer using an archaic geocentric metaphor. Right: a fragment of the state of the supercomputer at some point in time, made from within a three-dimensional geocentric model. A view from within.

3.4   A Two-Dimensional Option of the Geocentric Metaphor

A two-dimensional option of the geocentric metaphor can be used to represent programs within the paradigm of object-oriented programming. This methodology is based on visualizing a program as a set of objects, each of which is a sample of a certain class, and classes form an inheritance hierarchy. Based on the assumption that software design using the languages of object-oriented programming is conducted in class-files, the observers viewpoint (the Earth) can be the current work class. The interface system can be represented as a set of satellites of the current planet. In accordance with the perception of stars being far away and immutable objects, third-party dependencies and libraries used in projects can be represented by the metaphor of stars or constellations. Closest celestial bodies-planets represent other classes of the project, their size and appearance depending on various features of the class, its size, used language and file extension. A system of cascaded placement of the code according to frequency of use or level of connectedness with the current class can serve to interpret numerous transparent celestial spheres. See Fig. 4, Left.

A Python module is chosen on entry and used as a starting point, the Earth. Planets and stars are built based on the imported modules. Planets represent dependencies from the same project, and stars represent external libraries, including standard ones. By choosing a planet one can review the contents of the represented module. Currently a version has been released accepting files with the number of dependencies of up to 20 (out of accuracy and legibility concerns). Expansion of the systems possibilities is planned, in particular, it is intended to add planet assortment based on coupling (how many imported language units are used in the source file) and, possibly, visual representation of the sizes of planet-modules. Visual representation of object-oriented programs in 3D view is possible for visual programming systems. In this case, the use of virtual reality means is presupposed.

3.5   Visualization Based on a Starry Sky Metaphor

Apart from the above mentioned cosmic metaphors, a visualization idea emerged based on a starry sky metaphor. This metaphor was inspired by the images in old engravings. The example shown in Fig. 4, Right uses stylized images of stars and constellations. Visualization of up to three levels of data is possible, with the use of sky color and the position of constellations and stars for visualization. The stars correspond directly with the objects themselves, constellations with groups or clusters of objects, and the color of the sky can correspond with both super-groups and various features of objects in this part of the sky. For the purpose of visualizing different object features size, shape and color of stars can be used. For increased expressivity (like constellation images in old sky atlases of the ages of Renaissance and Enlightenment) an image corresponding to the given cluster is depicted atop constellations. The objects of city public transport system are depicted. This system is suitable for visualization of recurrent or long-term processes by means of moving the objects across the sky with their dynamic development. For the purposes of filtering objects, images of clouds covering parts of the sky are very suitable. For a more extended and efficient use of this idea one can use virtual reality systems.

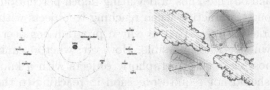

Fig. 4. Left: a two-dimensional geocentric representation of the program within the object-oriented paradigm. Right: example of visualization based on a starry sky metaphor.

4   The Human Factor

Software visualization, as any other field of knowledge connected with IT, bumps into the human factor. Software visualization based on virtual reality in particular requires taking the human factor into account and studying it. Virtual reality as an environment for human activity brings specific states inherent only to it.

The most psychological state is the state of presence. The notion of presence is discussed by a number of researchers extensively. The sense of presence is characterized as a basic state of consciousness a conscious feeling of being located in an external world, at the present time. This applies to the physical world, in which our bodies are located, to virtual worlds created through technological mediation, and to a blended mixture of the two: the physical and the virtual (see [21]). However, within the framework of this paper, we shall focus on the notion of presence in virtual reality as a special state that is different from presence in the real world, 'presence as the extent to which participants in a VE respond to virtual objects and events as if these were real [7]. We consider presence in virtual reality as a special state that can influence performance. We are talking about using virtual reality for software visualization. This paper describes system prototypes that employ metaphors, which present abstract data as familiar objects. The program engineer faces the challenge of simultaneously operating familiar images and abstract notions. Will presence occur in this situation? How is it going to influence performance? For instance, when using a city metaphor, the actor is supposed to roam the streets, enter buildings and, at the same time, sort out the structure of a software code and data. Can we talk about presence in such an environment, and what exactly is going to happen with this environment? Are we sure it will not be perceived as a city instead of software data presentation?

Speaking about the human factor in software visualization, one cannot but mention the issue of control and navigation. It is most convenient to look to 3D controllers or to use motion capture technology; however, the latter raises questions connected with precision and alignment. This partially depends on the used technologies and programs, but there is still the question of human factor, of distorted perception of distances in a virtual environment. The issue is solved to a degree, as is shown in paper [19]: familiar size is taken into account when reaching to a virtual object. Thus, a deviating depth perception might be less a problem for 3D touch interaction when reaching to objects with known sizes as compared to abstract objects. That is why a program engineer should familiar objects with known size when possible. However, when organizing interaction it is not always possible to use metaphoric objects with a known size. In this case, interaction should not require precision in reaching to the objects of the environment.

5 Conclusion

The development of prototypes of systems based on virtual reality is just the first stage of research. It is necessary to search for the criteria of visualization metaphors and approaches to developing views. It is not clear as to how effective the virtual reality environments are in solving various software visualization tasks. Numerous questions remain concerning the adaptation of such systems with regard to user peculiarities, including individual perception limitations and restrains on the time spent in virtual reality environments. The gamification idea is surely extremely interesting, but it cannot always be employed in major development. New research and pilot projects are required in the field of virtual reality application for software visualization systems. Software design for real-life tasks requires specialization, and even personalization of visual systems. It is necessary to extend the psychological knowledge regarding interaction with virtual reality and the processes occurring in the course of a professionals activity while working on software visualization systems.

References

1. Averbukh, V.: Visualization metaphors. Program. Comput. Softw. **27**, 227–237 (2001)
2. Ens, B., Anderson, F., Grossman, T., Annett, M., Irani, P., Fitzmaurice, G.: Ivy: exploring spatially situated visual programming for authoring and understanding intelligent environments. In: Proceedings of the 43rd Graphics Interface Conference GI 2017, pp. 156–162 (2017)
3. Fittkau, F., Koppenhagen, E., Hasselbring, W.: Research perspective on supporting software engineering via physical 3D models. In: 2015 IEEE 3rd Working Conference on Software Visualization (VISSOFT), pp. 125–129 (2015)
4. Fittkau, F., Krause, A., Hasselbring, W.: Exploring software cities in virtual reality. In: 2015 IEEE 3rd Working Conference on Software Visualization (VISSOFT), pp. 130–134 (2015)
5. Glander, T., Döllner, J.: Abstract representations for interactive visualization of virtual 3D city models. Comput. Environ. Urban Syst. **33**(5), 375–387 (2009)
6. Glander, T., Döllner, J.: Automated cell-based generalization of virtual 3D city models with dynamic landmark highlighting. In: The 11th ICA workshop on generalization and multiple representation, 20–21 June 2008, Montpellier, France, 14 p. The International Cartographic Association (2008)
7. Khanna, P., Yu, I., Mortensen, J., Slater, M.: Presence in response to dynamic visual realism: a preliminary report of an experiment study. In: Proceedings of the ACM Symposium on Virtual Reality Software and Technology, pp. 364–367. ACM (2006)
8. Merino, L., Bergel, A., Nierstrasz, O.: Overcoming issues of 3D software visualization through immersive augmented reality. In: 2018 IEEE Working Conference on Software Visualization (VISSOFT), pp. 54–64 (2018)
9. Merino, L., et al.: On the impact of the medium in the effectiveness of 3D software visualizations. In: Proceedings on 2017 IEEE Working Conference on Software Visualization (VISSOFT), pp. 11–21 (2017)
10. Merino, L., Ghafari, M., Anslow, C., Nierstrasz, O.: CityVR: gameful software visualization. In: IEEE International Conference on Software Maintenance and Evolution (ICSME TD Track), pp. 633–637 (2017)
11. Oberhauser, R.: Immersive coding: a virtual and mixed reality environment for programmers. In: Proceedings of The Twelfth International Conference on Software Engineering Advances (ICSEA 2017), pp. 250–255 (2017)
12. Oberhauser, R., Carsten, L.: Gamified virtual reality for program code structure comprehension. Int. J. Virtual Reality **17**(02), 79–88 (2017)
13. Oberhauser, R., Lecon, C.: Immersed in software structures: a virtual reality approach. In: The Tenth International Conference on Advances in Computer-Human Interactions, ACHI 2017, pp. 181–186 (2017)

14. Oberhauser, R., Lecon, C.: Virtual reality flythrough of program code structures. In: Proceedings of the 19th ACM Virtual Reality International Conference (VRIC 2017), article No. 10. ACM (2017)

15. Reed, D., Scullin, W., Tavera, L., Shields, K., Elford, C.: Virtual reality and parallel systems performance analysis. IEEE Comput. **28**(11), 57–67 (1995)

16. Reipschläger, P., Gumhold, S., Ozkan, B.K., Majumdar, R., Mathur, A.Sh., Dachselt, R.: DebugAR: mixed dimensional displays for immersive debugging of distributed systems. In: Extended Abstracts of the 2018 CHI Conference on Human Factors in Computing Systems, 6 p. (2018)

17. Schreiber, A., Misiak, M.: Visualizing software architectures in virtual reality with an island metaphor. In: Chen, J.Y.C., Fragomeni, G. (eds.) VAMR 2018. LNCS, vol. 10909, pp. 168–182. Springer, Cham (2018). https://doi.org/10.1007/978-3-319-91581-4_13

18. Schreiber, A., Misiak, M., Seipel, P., Baranowski, A., Nafeie, L.: Visualization of software architectures in virtual reality and augmented reality. In: IEEE Aerospace Conference Proceedings. IEEE Aerospace Conference 2019, 02–09 March 2019. Big Sky Montana, USA (2019)

19. Schubert, R.S., Müller, M., Pannasch, S., Helmert, J.R.: Depth information from binocular disparity and familiar size is combined when reaching towards virtual objects. In: In Proceedings of the 22nd ACM Conference on Virtual Reality Software and Technology, pp. 233–236 (2016)

20. Vincur, J., Navrat, P., Polasek, I.: VR city: software analysis in virtual reality environment. In: 2017 IEEE International Conference on Software Quality. Reliability and Security Companion, pp. 509–516 (2017)

21. Waterworth, J., Riva, G.: Feeling Present in the Physical World and in Computer-Mediated Environments. Palgrave Macmillan, London (2014)

Design and Architecture of an Affordable Optical Routing - Multi-user VR System with Lenticular Lenses

Juan Sebastian Munoz-Arango[⊠], Dirk Reiners, and Carolina Cruz-Neira

Emerging Analytics Center, UA Little Rock, Little Rock, AR 72204, USA
eacinfo@ualr.edu
http://eac-ualr.org/index.php

Abstract. One of the all time issues with Virtual Reality systems regardless if they are head-mounted or projection based is that they can only provide perspective correctness to one user. This limitation affects collaborative work which is the standard in any industry. Several approaches have been proposed to generate perspective correct images for different users but not only are they highly complex but also require lots of custom circuitry. On this paper we present the design, architecture and the mathematical background of an affordable optical routing multi user VR system that uses lenticular lenses for separating users and providing perspective correct images.

Keywords: Multi-user VR · Lenticular lenses · Computer graphics

1 Introduction

Lately, Virtual Reality (VR) has experienced a massive appropiation by the mass consumer market. Although VR is not a new topic, different systems have emerged across the past years like CAVEs, projection based VR systems, head mounted displays (HMDs), trackers, wands, etc. Such systems used to be only affordable by laboratories or industries with high amount of funds. Today, these systems can be accessed by the consumer market with a relative affordable pricepoint.

In many industries, Virtual Reality is now the norm for prototyping, and fast iteration processes for reducing costs when designing new products. When the person who is interacting in VR is working alone, HMDs work quite well to some extent. But when groups of individuals interact and exchange ideas, HMDs become short for the task. Even though there have been several attempts to work collaboratively with HMDs [1], the weight and eyesight exhaustion in prolonged sessions affect the productivity on a working environment.

When groups of experts gather together to discuss, interact and gesture around a common dataset, they hope to achieve a consensus between them. Such behavior is consistent across disciplines and begs for a system that can

© Springer Nature Switzerland AG 2019
L. T. De Paolis and P. Bourdot (Eds.): AVR 2019, LNCS 11613, pp. 71–90, 2019.
https://doi.org/10.1007/978-3-030-25965-5_7

accomodate small groups of people working together with perspective correct point of views for each interacting user.

Different studies have been presented in the past [2,3], such studies back up the idea that collaboration times get significantly reduced when users have correct perspective point of views when working in VR.

Several approaches have been presented throughout the past years for generating multiple images from a single screen to different users. Bolas in [4] introduces a "Solution Framework" where he classifies current existing solutions within four categories: Spatial barriers, optical filtering, optical routing and time multiplexing.

The optical routing category encompasses all the solutions that use angle-sensitive optical characteristics of certain materials to direct or occlude images based on the user's position.

Lenticular lenses are commonly used for lenticular printing, corrective lenses for enhacing vision, autostereoscopic 3D TVs and lenticular screens for increasing perceived brightness on projection television systems [5]. Such lenses can also be used for multiplexing views from a single screen by taking advantage of the physical properties of the lens of bending the light that comes from each pixel and routing it depending on the users's physical positions.

With the affordability of VR equipment nowadays and the simplicity of lenticular lenses; on this paper we are going to present the mathematics, design and architecture of an optical routing multi-user VR system that uses lenticular lenses for generating perspective corrected images to users depending on where they are located in the physical space.

2   Background

2.1   Parts of a Virtual Reality System

For the purposes of clarity in this paper, we are going to adopt the term proposed by Huxor in [6] which considers VR to be "A computer-based representation of a space in which users can move their viewpoint freely in real time". This definition incorporates several components inherent in a VR system like input devices, output devices and the program itself (VR Engine).

These subsystems have to work in complete harmony in order to produce a smooth experience for the user(s). If one of the subsystem fails/gets delayed. The experience gets affected.

As show in Fig. 1, in a VR system, all its parts are strongly tied. The input has to react as soon as possible so the VR Engine can process the world and produce imagery/sounds/haptic reactions to the user which in return reacts and acts again on the input. If any of the subsystem gets delayed; the experience is noticeably damaged.

2.2   Lenticular Lenses

A lenticular lens is basically a set of cylinders (lenticules) placed next to each other overlapping with one side flat and the other side with the protubing lenses

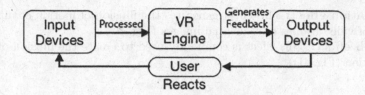

Fig. 1. Parts of a VR system.

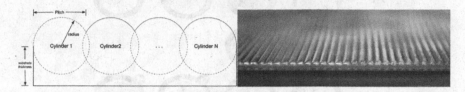

Fig. 2. Left: diagram of a lenticular lens, right: real lenticular lens

(Fig. 2), These lenses are designed so that when viewed from different angles different slices of an image are displayed.

The most common usage of lenticular lenses is in lenticular printing. Here, the technology is used to present images that appear to change as the image is viewed from different angles [5].

In the industry, lenticular lenses are classified by the lenses per inch (LPI) a lens sheet has. The higher the LPI the smaller each lenticule is. Manufacturers offer lenses in different LPI ranges like 10, 15, 20, 30, 50, 60, 75, 100LPI; however. With enough funds one can manufacture a lens with custom specifications.

Depending on how deep the lenticules are engraved in the sheet, different effects can be achieved (Fig. 3). Such effects can be just about anything one can do with video; some of the most common effects include morphing, animations, flipping and 3D effects [7,8].

Fig. 3. Flip to 3D lenticular lenses.

1. **3D effect:** This effect provides an illusion of depth and perspective by layering objects within an image (Fig. 4A).
2. **Flip effect:** In this effect, a dramatic swap of two images occurs vanishing and then reappearing from one to another (Fig. 4B).

3. **Animation effect:** This effect generates the illusion of motion coming from a set of video frames or sequential images (Fig. 4C).
4. **Morph effect:** This effect is commonly used to create the illusion of transformation (Fig. 4D).

Fig. 4. Lenticular lens effects. A: 3D, B: Flip, C: Animation, D: Morph

With the same principle on how lenticular printing works, lenticular lenses have also been used for multiplexing images to different users in VR as an optical routing approach.

3   Previous Work

Optical routing uses the angle-sensitive optical characteristics of certain materials to direct or occlude images based on the user's position. [4].

In 1994, Little et al. talk about a design for an autostereoscopic, multiperspective raster-filled display [9]. Here, they propose a time multiplexed approach and an optical routing approach. The optical routing approach features video cameras and LCTV projectors. They use an array of video cameras to capture multiple perspective views of the scene and then they fed these to an array of LCTVs and simultaneously project the images to a special pupil-forming viewing screen. The viewing screen is fabricated by either a holographic optical element or a Fresnel lens and a pair of crossed lenticular arrays. Their resolution is limited by the LCTV projectors and they use a lot of projectors/cameras to provide multiple views. For the time-multiplexed approach they investigated the use of a DMD projector that provided 20 sequentially perspective views, each one separated by one degree in viewing angle by a spinning disk.

Van Berkel et al. in [10,11] built a prototype display using a LCD and a lenticular lens from Philips Optics to display 3D images; they slanted the lenticular lens with respect to the LCD panel in order to reduce the "picket fence" effect.

Later, Matsumoto et al. in [12] proposes a system that consists of combination of cylindrical lenses with different focal lengths, a diffuser screen and several projectors to create a 3D image.

Omura presents a system that uses double lenticular lenses with moving projectors that move according to the tracked user's position to extend the viewable area [13], their system needs a pair of projectors per person and their projectors move to adjust each user's position. Their system suffers from latency due to the mechanical movement.

Lipton proposed the Synthagram [14], a system that consists of an LCD Screen with a lenticular screen that overlays the LCD display. They angled the lenticular screen in order to reduce the moiré pattern and their system uses nine progressive perspective views from a single image. They sample these views into a program called the Interzig where they process the images and assign each pixel to a specific position in the screen.

Matusik proposes a system that consists of an array of cameras, clusters of network connected PCs and a multi-projector 3D display with the purpose to transmit autostereoscopic realistic 3D TV [15]. They record the imagery with a small cluster of cameras that are connected to PCs. The PCs broadcast the recorded video which later on is decoded by another cluster of consumer PCs and projectors. Their 3D Display consists of 16 NEC LT-170 projectors that are used for front or rear projection. The rear projection approach consists for two lenticular sheets mounted back to back with an optical diffuser material in the center and the front projection system uses one lenticular sheet with a retro reflective front projection screen material.

Another way of optical routing approach use is the display proposed by Nguyen et al. [16,17]. Here, they propose a special display which consists of a screen with 3 layers that has directional reflections for projectors so each participant sees a customized image from their perspective; their system supports up to 5 viewing zones and needs a projector per participant.

Takaki et al. proposes a system that can produce 72 views [18]. Their system consists of a light source array, a micro lens and a vertical diffuser (a lenticular sheet). They mention that as the horizontal positions of all light sources are different, rays from different light sources proceed to different horizontal directions after passing through the micro lenses thus generating different views.

Later on. In [19,20] followed by [21], Takaki discusses a multiple projection system that is modified to work as a super multiview display. Here, they attach a lens to the display screen of a HDD projector and by combining the screen lens and the common lens, they project an aperture array. This aperture array is placed on the focal plane of the common lens, and the display screen (a vertical diffuser) is placed on the other focal plane. Hence, the image of the aperture array is produced on the focal plane of the screen lens. With this, the image of an aperture array gets enlarged generating enlarged images that become viewpoints.

In 2009. Takaki and his team introduce a prototype panel that can produce 16 views [22]. They do this by building a LCD with slanted subpixels and a lenticular screen. They place a diffusion material between the lenticular sheet and the LCD screen in order to defocus the moiré pattern but increase the crosstalk among viewpoints. They mention that by slanting the subpixel arrangement instead of the lenticular sheet, they can increase the number of views but the optical transmittance of the display decreases. They conclude that by slanting the subpixels in the screen instead in the lenticular sheet, they can reduce significantly the crosstalk and moire compared to the normal approaches.

Finally, in 2010 Takaki and his team combine several 16-view flat-panels that have slanted subpixels [22] and creates a system with 256 views [23]. They superimpose the different projected output of the panels to a single vertical diffuser. The multiple viewing zones for each flat panel are generated on an incident pupil plane of its corresponding projection lens. Each projection lens projects to the display surface of its corresponding flat panel system on the common screen and finally a screen lens is located on the common screen so the lens generates viewing zones for observers.

Another system that takes advantage of the optical routing approach is the Free2C display, a project proposed by Surman in [24]. Here, they created a single viewer autostereoscopic display using a head tracker. The display accommodates the head movement of the viewer by continually re-adjusting the position of the lenticular lens in relation to the LCD to steer the stereoscopic views onto the eyes of the viewer. Their display resolution is 1200×1600, the viewing distance goes from 40 cm to 110 cm and side to side movements range of approximately $\pm 25°$ from the center of the screen. They also attempted a multi-user display that steers the LCD instead of the lenses to produce image regions for the users but they mention the display performance was really poor.

Similarly to Free2C, Brar et al. use image recognition to track users' heads to produce multiple steerable exit pupils for left and right eyes [25,26]. Here, they describe the design and construction of a stereoscopic display that doesnt require wearing special eye wear. A stereo par is produced on a single LCD by simultaneously displaying left and right images on alternate rows of pixels. They propose steering optics controlled by the output the aforementioned head tracker to direct regions, referred as exit pupils to the appropriate viewers' eyes.

Kooima et al. [27] uses 24″ and 42″ 3DHD Alioscopy displays which come with integrated lenticular lenses. They propose a system that consists of scalable tiled displays for large field of views and use a generalization of a GPU based autostereoscopic algorithm for rendering in lenticular barriers. They tried different methods for rendering but they had issues where they perceived repeated discontinuities, exaggerated perspectives and as the displays pixels cannot be moved smoothly but in discrete steps. The tracked viewer moves into transition between channels, the user begins to see the adjacent view before the channel perspective is updated to follow the user's head.

Zang et al. propose a frontal multi-projection autostereoscopic display [28]. Their approach consists of 8 staggered projectors and a 3D image guided screen. The 3D image screen is mainly composed of a single lenticular sheet, a retro-reflective diffusion screen and a transparent layer that is filled between them to control the pitch of the rearranged pixel stripe in interlaced images. Their system is space efficient compared to previous approaches that produce light from the back of the screen, but the loss of intensity and crosstalk are seriously increased out of the system boundaries.

We have presented here some research that has been done throughout the past years that uses an optical routing approach; specifically, employing lenticular lenses. Still, none of these projects cover the mathematics, design and architecture of a system that separates users by generating an interlaced image that gets accommodated depending on where the users are located in the physical space.

4 System Design

This system mainly consists of three parts. A *tracking subsystem* which is "aware" all the time where every user is located in relation to the screen, a *display/lens subsystem* which outputs an interlaced image that gets generated depending on where each user is located and the *software subsystem* that runs all the logic.

Having in mind the affordability of the system, SteamVR® would be used for the tracking subsystem, an off the shelf 4K/8K (Preferably) TV that supports stereo for the screen and the only thing that could cost would be the lens that matches the screen. A general overview of the system can be seen in Fig. 5.

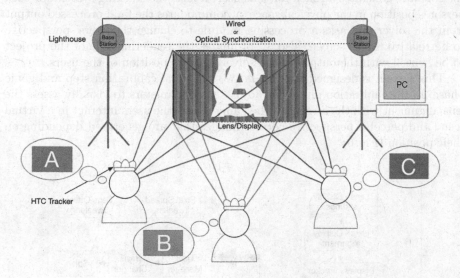

Fig. 5. Hardware overview of the system.

4.1 Tracking Subsystem

One of the core parts for doing multi-view user VR is tracking. The system needs to know all the time each user position and rotation, not only to provide feedback to the system to selectively render the relevant parts of the screen for each user depending on their location but also to provide a perspective correct view to each person engaging in the experience.

SteamVR® tracking system is a relatively easy to choose option. It is a system that is easy to get, cheap compared to other alternatives and reliable enough to provide a room tracking experience for several users.

For this system, we are interested in using a subset of this tracking system; the HTC VIVE™Trackers (as trackable objects) and base stations only as they provide exactly what we need, tracking.

4.2   Display Lens Subsystem

The display/lens subsystem consists of a lenticular lens attached to a screen. These parts (lens and display) are further calibrated and aligned through software so the calculations of the refraction for each pixel are correct. For correct functioning of the system, the lens attached to the TV needs to be completely flat so no gaps of air affect the refraction properties of the pixels unevenly.

4.3   Software Subsystem

The software subsystem manages each user; when an user joins the experience, the software generates a token for that user, it asks the tracking system for each person's position in the physical space, accommodates the final processed output from the solver and asks a processing module to change the users' perspective to be relative to the screen, also. Coordinates the rendering side of the project to be synced with the output of the solver and the position of the users.

This system is designed to work in two phases, a calibration step and game phase. In the calibration phase one adjusts the parameters to visually assess the lens alignment with the screen. In the game phase, the users interact in a virtual scene and perceive perspective correct images that are generated depending on their position (Fig. 6).

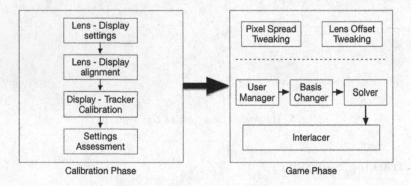

Fig. 6. System phases overview workflow.

4.4   Game Phase

Three sub modules work together in the game phase to process a final texture that tells the GPU how to render an interlaced image depending on the number of users that are interacting in the scene. These modules are: the User Manager, the change of basis component and the solver module.

The user manager detects when a tracker gets connected, creates, associates and register users to VIVE trackers. With the information at hand of each user's position and orientation in tracker space, a change of basis is performed to convert these positions relative to the tracker to solver-space positions. Afterwards, With the user's positions in lenticular space and the data from the calibration phase, the solver module processes all the mathematics and decides which pixels are seen by which users. Finally, a gray-scale texture is generated by the solver and passed to the interlacer in the GPU to render the screen (Fig. 7).

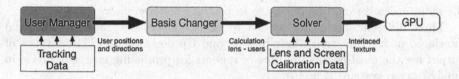

Fig. 7. Game phase components.

4.5   Rendering

When the solver finishes calculating which pixels belong to which users depending on their positions, it generates a grayscale texture that identifies pixel by pixel the user owner. This texture tells the GPU for the pixel i, j in the screen, which user camera should be rendered.

To achieve this a simple shader receives the render textures from the engaged users and with the grayscale texture generated by the solver, it generates a final interlaced image that gets rendered to the main camera in the screen. This interlaced image is driven by the resulting texture from the lens simulation in the Solver module (Figs. 8 and 9).

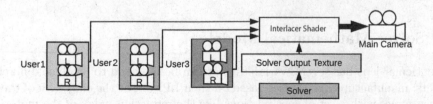

Fig. 8. Overview of how the interlacer works.

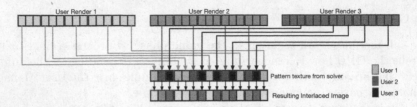

Fig. 9. Interlacer output dependent on the solver output texture.

4.6 Stereoscopy

Last but not least, one of the most important parts of a VR system is the perception of depth. Stereoscopy can be achieved by using the already built in stereo capabilities from the monitor the system runs into. With this the system just needs to output the image compensated by the screen's required format and the monitor itself would be in charge to produce a stereo view.

Three different modes are needed to be developed for producing the final interlaced image: No stereo, Side by side and Up down images. These types of output are the most commonly type of renders for producing user separation in passive stereo systems (Fig. 10).

Fig. 10. Supported stereo modes for one user.

The Up/Down stereo would the preferred type of stereo; this is to be able to use the full horizontal resolution and lose half of the vertical resolution. With this less resolution will be lost when separating users horizontally due to the nature of lenticular lenses.

5   Solver Mathematical Model

A lenticular lens has several variables that can be modified to produce different effects; manufacturers sell lenses based on their LPI but at the end, each of these lenses comes with a set of specifications like the refraction index of the material the lens is made of, substrate thickness, viewing angle, lenticule radius, etc. (Fig. 11) [5]. To simulate how a lens works; a discretization of the light emitted from each pixel is represented with several rays that start from each subpixel

along the lens substrate that get refracted to the air from the lenticular lens.
To achieve this, we generate a number of rays for each pixel and calculate ray
trajectories from when they start in the pixel until they get refracted by the lens
in three steps: Substrate contact, lens contact and lens refraction.

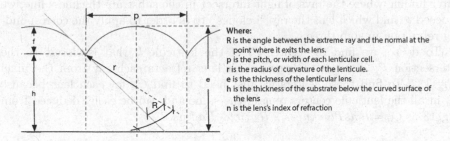

Fig. 11. Detailed lenticular lens

5.1 Step 1: Substrate Contact

In this phase n number of rays are calculated for each pixel with a spread S (in
deg) in order to get n contact points $P_0, P_1 \ldots P_n$ from a horizontal line (parallel
to the screen) that defines the substrate thickness of the lens (Fig. 12). To find
the points $P_0, P_1 \ldots P_n$ where the rays intersect the end of the lens substrate we
represent the rays from the pixel and the substrate with line equations and find
their respective intersection points.

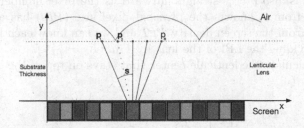

Fig. 12. Step 1: Find where rays intersect substrate line.

Given the points R_1 and R_2 from the line L_1 that represents the ray that
gets generated from the pixel and points S_1 and S_2 from the line L_2 that defines
the substrate of the lens, we can generate two standard line equations with the
form $y = mx + b$, make them equal on the y axis (as it is the substrate thickness
that the manufacturer gives) and find the intersection point P_i on x (Fig. 12).

$$L_1 \rightarrow y = m_1 x + b_1 \qquad\qquad L_2 \rightarrow y = m_2 x + b_2$$

$$\boxed{x = \frac{b_2 - b_1}{m_1 - m_2}} \qquad \text{intersection X axis } (P_x).$$

Again, finding the P_y becomes trivial as is given by the lens manufacturer and is the lens substrate thickness.

5.2   Step 2: Lens Contact

After finding where the rays of light intersect in the substrate thickness line, we proceed to find which lens the ray "belongs" to in order to apply the corresponding refraction in step three.

To do so, we find the center C_i of the lenticule l_i that is closest to the intersection P_j in order to know which lens refracts each ray from the pixel (Fig. 13). To find these centers we just need to find C_x for each lens because C_y in all the lenticule centers remain the same and can be easily deduced from Fig. 13 as $C_y = lensThickness - lenticuleRadius$.

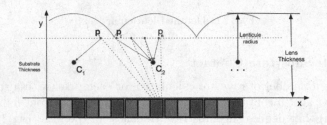

Fig. 13. Step 2: Find points P_j closest to lens center C_i.

Getting C_x is also pretty straight forward as the pixel number ($PixNum$) the ray comes from is known, the physical pixel size (Pps) has already been pre-calculated from the screen density PPI and we can know each lenticule size (Ls) by just dividing the LPI of the lens by 1 in. ($Ls = \frac{1}{LPI}$).

First, we calculate the lenticule center that stays on top of the current pixel (Lct) with:

$$Lct = \left(\left\lfloor \frac{PixNum * Pps}{Ls} \right\rfloor * Ls \right) + \left(\frac{Ls}{2} \right) \tag{1}$$

After calculating Lct (Eq. 1), we proceed to check if the distance with P_x i is lesser than half of an individual lenticule size ($|Lct - P_x| < \frac{Ls}{2}$), if it is we have found the lenticule center C_x the ray belongs to, else we carry checking for neighbour lenticule centers with $C_x \pm \frac{Ls}{2}$ until the condition gets satisfied.

In conclusion, the algorithm for finding the lenticule center C_j that belongs to the ray intersection P_i we are calculating in this phase can be seen in Algorithm 1.

5.3   Step 3: Lens Refraction

After finding the closest lenticule center C_j from a given ray intersection P_i we can continue with the ray direction r and finally find the intersection point Q_i with the lenticule L_j where we can calculate the ray refraction (Fig. 14). rf.

Algorithm 1. Get closest lens center from ray intersection P_i

---

1: **procedure** CLOSESTCENTER($PixNum, Pps, Ls, P$)   ▷ Return closest C to P
2:     $C_y \leftarrow lensThickness - lenticuleRadius$
3:     $Lct \leftarrow \left(\left\lfloor \dfrac{PixNum * Pps}{Ls} \right\rfloor * Ls \right) + \left(\dfrac{Ls}{2} \right)$
4:     **if** $|Lct - P_x| < Ls/2$ **then**
5:         **return** $C(Lct, C_y)$
6:     **end if**
7:     $counter \leftarrow 1$
8:     **while** true **do**   ▷ Search neighboor lens centers
9:         $centerL \leftarrow Lct - (counter * Ls)$
10:        **if** $|centerL - P_x| < Ls/2$ **then**
11:            **return** $C(centerL, C_y)$
12:        **end if**
13:        $centerR \leftarrow Lct + (counter * Ls)$
14:        **if** $|centerR - P_x| < Ls/2$ **then**
15:            **return** $C(centerR, C_y)$
16:        **end if**
17:        $counter \leftarrow counter + 1$
18:    **end while**
19: **end procedure**

Finding Lens Intersection Point: To find Q_i (Fig. 14) we can treat each lenticule as a circle and the rays that come from each pixel as lines and then the lens-ray intersection point can be treated as a line-circle intersection as follows [29,30]:

Given a circle with center (c_x, c_y) with radius r representing the lenticule with center C_j and a line representing the ray of light that comes from a given pixel:

$$Ray \rightarrow y = mx + b \qquad\qquad Lens \rightarrow (x - c_x)^2 + (y - c_y)^2 = r^2$$

$$0 = (x - c_x)^2 + (mx + (b - c_y))^2 - r^2 \; Replace\ ray\ in\ lens\ equation$$

$$0 = x^2(1 + m^2) + x(2mb - 2c_x - 2mc_y) + (c_x^2 + b^2 - 2bc_y + c_y^2 - r^2)$$

Solving the quadratic form of the resulting equation we end up with:

$$x_{1,2} = \frac{-mb + c_x + mc_y \pm \sqrt{-2mbc_x + 2mc_xc_y - b^2 + 2bc_y - c_y^2 + r^2 + r^2m^2 - m^2c_x^2}}{1 + m^2} \tag{2}$$

After finding the x component in Eq. (2), we can see we have three possible values that the quadratic equation gives us under the square root. Lets call this D.

$$D = -2mbc_x + 2mc_xc_y - b^2 + 2bc_y - c_y^2 + r^2 + r^2m^2 - m^2c_x^2$$

If $D < 0$ there is no intersection point, when $D = 0$ the line touches the lens tangentially and finally if $D > 0$ there are two intersection points (as each lenticule is at the end a circle.) We are only interested in the positive value as the lenses point toward the positive Y axis so on this case, the x component of the intersection point Q_i where the ray touches the lens ends up being:

$$x = \frac{-mb + c_x + mc_y + \sqrt{-2mbc_x + 2mc_xc_y - b^2 + 2bc_y - c_y^2 + r^2 + r^2m^2 - m^2c_x^2}}{1 + m^2} \qquad (3)$$

Finally, just by replacing this value (Eq. 3) on the line equation from the ray one can get the y component of Q_i.

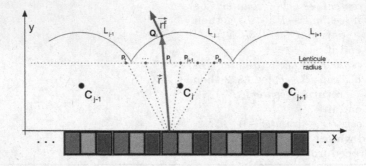

Fig. 14. Step 3: Ray intersection with lens and refraction.

Generating Refracted Ray from the Lens: After finding the point of intersection where the ray (coming from the pixel) touches the lens (Q_i). One can finally calculate the refracted pixel ray (rf) (Fig. 14) using Snell's law [31].

Snell's Law states that the products of the index of refraction and sines of the angles must be equal (Eq. 4).

$$n_1 \sin(\theta_1) = n_2 \sin(\theta_2) \qquad (4)$$

Snell's equation (Eq. 4) can be re-written as:

$$\sin(\theta_2) = \frac{n_1}{n_2} \sin(\theta_1) \qquad (5)$$

One can immediately see a problem here, and is that if $\sin(\theta_1) > \frac{n_2}{n_1}$ then $\sin(\theta_2)$ has to be bigger than 1 which is impossible. So when this happens, we have a TIR (*Total Internal Reflection*), TIR only happens if you go from a denser material (lens) to a less dense material (air). When TIR happens, we just ignore that ray and do nothing about it. So Eq. 5 can be written like this:

$$\sin(\theta_2) = \frac{n_1}{n_2} \sin(\theta_1) \longleftrightarrow \sin(\theta_1) \leq \frac{n_2}{n_1} \qquad (6)$$

To find rf, lets begin by splitting it up in a tangent and a normal part:

$$rf = rf_\parallel + rf_\perp \qquad (7)$$

As all the vectors are normalized and any vector v can be decomposed in its tangent and parallel parts, and its parts are perpendicular to each other ($v_\parallel \perp v_\perp$), with basic trigonometry, the following rules apply:

$$\sin(\theta) = \frac{|v_\parallel|}{|v|} = |v_\parallel| \qquad \cos(\theta) = \frac{|v_\perp|}{|v|} = |v_\perp| \tag{8}$$

Since Snell's law talks about sines (Eq. 6), we can use Eq. 8 and rewrite Eq. 6 as:

$$|rf_\parallel| = \frac{n_1}{n_2}|r_\parallel| \tag{9}$$

Since $rf_\parallel$ and $r_\parallel$ are parallel and point in the same direction, Eq. 9 becomes:

$$rf_\parallel = \frac{n_1}{n_2}r_\parallel = \frac{n_1}{n_2}(1 - \cos(\theta_r)n) \tag{10}$$

To find $rf_\perp$ one can simply use pythagoras ($|v|^2 = |v_\parallel|^2 + |v_\perp|^2$) and get:

$$rf_\perp = -\sqrt{1 - |rf_\parallel|^2}n \tag{11}$$

Replacing Eq. 9 and Eq. 11 in Eq. 7 we get:

$$rf = \frac{n_1}{n_2}r - \left(\frac{n_1}{n_2}\cos(\theta_r) + \sqrt{1 - |rf_\parallel|^2}\right)n \qquad \text{Replacing 8}$$

$$\boxed{rf = \frac{n_1}{n_2}r - \left(\frac{n_1}{n_2}\cos(\theta_r) + \sqrt{1 - \sin^2(\theta_{rf})}\right)n}$$

Finally, we need to find $\sin^2(\theta_{rf})$ in this last equation, but this can be easily deduced it using Snell's law in Eq. 9.

$$\boxed{\sin^2(\theta_{rf}) = \left(\frac{n_1}{n_2}\right)^2 \sin^2(\theta_r) = \left(\frac{n1}{n2}\right)^2 (1 - \cos^2(\theta_r))}$$

With these two equations one can finally obtain the refracted vector rf.

6   Basis Changer Mathematical Model

SteamVR® when installed for the first time requires the user to calibrate the system to know where the floor is and where the monitor of the computer the system is running in is pointing to (forward). With these two values, an origin and a coordinate system is generated for the system (Fig. 15 left).

On the solver side, the calculations for the pixel rays and lens refractions are performed in a plane where $0, 0, 0$ is located at the bottom left of the screen. $+X$ points to the right, $+Y$ towards outside of the screen and $+Z$ up (Fig. 15 right).

As one can see in Fig. 15, a way to communicate these two coordinate systems is needed. To achieve this, a change of basis is performed. With this, the system can convert coordinate systems from the tracker and translate them to screen coordinates and vice versa.

This is important for the solver so it knows where the users are relative to where the simulation is being calculated. After the change of basis is performed the solver can work with the users positions in screen space and generate the final texture to pass it to the interlacer in the GPU.

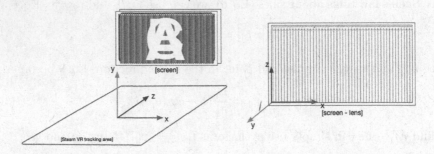

Fig. 15. Coordinate systems for SteamVR® (left) and the screen (right)

As presented in Fig. 15, this design relies on two coordinate systems, one for the tracker space where all the users interact and the other for the screen space where all the lens calculations are done. To be able to change between coordinate systems (bases) we set up both systems to share the same origin.

Let the tracker space base be S_B (also called standard base) composed by axes X, Y, Z. And let the lens base be L_B, composed by axes X', Y', Z'. (Fig. 16) The standard coordinate system (S_B) is defined by the basis [32,33]:

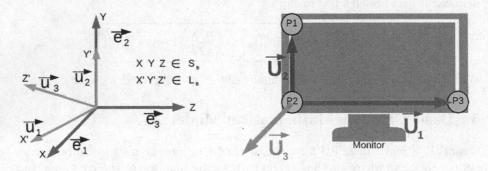

Fig. 16. Left: standard base (S_B) and Lens Base (L_B), right: screen points for getting L_B basis

$$S_B = (e_1, e_2, e_3) = \begin{bmatrix} 1 & 0 & 0 & 0 \\ 0 & 1 & 0 & 0 \\ 0 & 0 & 1 & 0 \\ 0 & 0 & 0 & 1 \end{bmatrix} \tag{12}$$

The lens space base (L_B) is defined as:

$$L_B = (U_1, U_2, U_3) = \begin{bmatrix} U_{1x} & U_{2x} & U_{3x} & 0 \\ U_{1y} & U_{2y} & U_{3y} & 0 \\ U_{1z} & U_{2z} & U_{3z} & 0 \\ 0 & 0 & 0 & 1 \end{bmatrix} \tag{13}$$

To find a coordinate vector $[a_B]$ of the vector a with respect to the coordinate system defined by the basis B, we need to express a as a linear combination:

$$a = X'U_1 + Y'U_2 + Z'U_3 \Leftrightarrow [a]_B = \begin{bmatrix} X' \\ Y' \\ Z' \end{bmatrix} \tag{14}$$

Noting that every vector is equal to its coordinate vector in the standard basis. In order to convert from S_B (Tracker space) to L_B (Lens space), let the vector V in tracker space (S_B) be denoted as V_{SB}, from Eq. 14, V can be rewritten as a matrix equation:

$$V = \begin{bmatrix} | & | & | \\ U_1 & U_2 & U_3 \\ | & | & | \end{bmatrix} \begin{bmatrix} X' \\ Y' \\ Z' \end{bmatrix} \tag{15}$$

Given that $V = [V]_{SB}$ and $[V]_{LB} = (X', Y', Z')$ one can write:

$$[V]_{SB} = \begin{bmatrix} | & | & | \\ U_1 & U_2 & U_3 \\ | & | & | \end{bmatrix} [V]_{LB} \tag{16}$$

The matrix composed by $U_1 U_2 U_3$ Is called the "Change of basis matrix" from the standard basis S_B, Lets call this matrix P; Rewriting Eq. 16 with P, will look like:

$$[V]_{SB} = P[V]_{LB}$$
$$P^{-1}[V]_{SB} = P^{-1}P[V]_{LB} \qquad \text{Multiplying by } P^{-1} \text{ on the left}$$
$$P^{-1}[V]_{SB} = I_3[V]_{LB}$$
$$\boxed{[V]_{LB} = P^{-1}[V]_{SB}}$$

With this equation, we can simply multiply any given position in tracker space (S_B) and the result will return the respective position in lens space (L_B).

Now, the reader might ask. How are the axes for lens space generated?, its pretty straight forward. In the calibration phase, one sets three points (from tracker space) that denote the physical position of the screen; These three points are $P1, P2, P3$ as shown in Fig. 16 right.

With these points, the system subtracts $P1 - P2$ to generate U_2, $P3 - P2$ to generate U_1 and the cross product from $U_2 \times U_1$ to generate the final axis U_3 (Fig. 16 right). With this we can perform the change of basis between the two coordinate systems.

7   Conclusions

The design, architecture and mathematical background of a multi-user VR system was presented. Such system is designed to work with SteamVR and commercially available 4K/8K TVs that come with stereo.

This system is designed to work with small groups up to three persons due to the amount of pixels that screens currently have. Still, this design can be scaled to support as many users as needed as far as the physical resolution of the screens afford it.

A general overview of the workflow of the data from the tracker and the lenses was presented and an easy three step process was also shown on how to calculate rays that come from pixels to assess if they belong to specific interacting users.

The software design presented here consists of two phases; a calibration phase where the system aligns the lenses with the screen pixels and a game phase where users interact in VR.

Three main submodules compose the game phase system. The user manager which controls user positions, connection and disconnections, the basis changer module which converts user coordinates to solver coordinates and the solver module which generates a grayscale texture that tells the GPU how to assemble the final interlaced image.

Two main coordinate systems interact in this design; the tracker coordinate system where users move and the solver coordinate system where the mathematics of the lenses are calculated. The mathematics behind for changing axes between the tracker space and simulation space where also presented.

An interlacer subsystem was also depicted in this paper where we show that by generating a grayscale texture that identifies each grayscale color to a user we can generate an interlaced image from the output of the solver and the render textures that come from each user.

8   Future Work

Even though the design and mathematics of a multi-user VR system where covered here, some external factors where not taken into account that would need more research.

A Frame for holding the screen to the lens needs to be designed; such frame needs to hold tight the whole lens to the screen and needs to not allow air gaps between the lens and the screen, else the mathematics of the system would not converge and images would not look correct.

Ghosting between users, crosstalk from the stereo system and moire patterns where not discussed in this paper; such topics need further research to assess to what extent they affect the performance of the system.

In this paper the mathematics of the lenticular lens where presented and explained but no optimum values that minimize ghosting and maximize unique pixels perceived for such lenses where mentioned. This is a topic that is important to maximize the performance of the system.

Pixel spread is also another important aspect that needs to be taken into account as this directly affects ghosting between users. Collimated light is known to minimize ghosting but we dont know to what extent the pixel spread can be pushed to get more unique pixels per user.

Finally color banding produced by lenses is a topic that needs also to be researched on as depending on the lens used and the subpixel layout arrangement of the screens it gets magnified or reduced.

References

1. Szalavari, Z., Schmalstieg, D., Fuhrmann, A., Gervautz, M.: Studierstube: an environment for collaboration in augmented reality. Virtual Reality 3(1), 37–48 (1998)
2. Pollock, B., Burton, M., Kelly, J.W., Gilbert, S., Winer, E.: The right view from the wrong location: depth perception in stereoscopic multi-user virtual environments. IEEE Transact. Visualization Comput. Graphics 18(4), 581–588 (2012)
3. Chen, K.B., Kimmel, R.A., Bartholomew, A., Ponto, K., Gleicher, M.L., Radwin, R.G.: Manually locating physical and virtual reality objects. Hum. Factors 56(6), 1163–1176 (2014)
4. Bolas, M., McDowall, I., Corr, D.: New research and explorations into multiuser immersive display systems. IEEE Comput. Graphics Appl. 24(1), 18–21 (2004)
5. Lenticular lens. https://en.wikipedia.org/wiki/Lenticular_Lens. Accessed 7 Feb 2019
6. Huxor, A., Lansdown, J.: The design of virtual environments with particular reference to VRML
7. Choosing the correct lenticular lens sheet. http://www.microlens.com/pages/choosing_right_lens.htm. Accessed 5 Feb 2019
8. Lenticular Effects. https://www.lenstarlenticular.com/lenticular-effects/. Accessed 20 Jan 2019
9. Little, G.R., Gustafson, S.C., Nikolaou, V.E.: Multiperspective autostereoscopic display. In Proceedings SPIE, vol. 2219, pp. 388–394 (1994)
10. van Berkel, C., Clarke, J.A.: Characterization and optimization of 3D-LCD Module design. In: Proceedings of SPIE, 3012, 179-186 (1997)
11. Van Berkel, C.: Image preparation for 3D-LCD. In: Proceedings of SPIE, vol. 3639(1), pp. 84–91 (1999)
12. Matsumoto, K., Honda, T.: Research of 3D display using anamorphic optic. Proc. SPIE 3012, 199–207 (1997)
13. Omura, K., Shiwa, S., Miyasato, T.: Lenticular autostereoscopic display system: multiple images for multiple viewers. J. Soc. Inf. Disp. 6(4), 313–324 (1998)
14. Lipton, L., Feldman, M.: A new autostereoscopic display technology: the SynthaGram. In: Proceedings of SPIE, vol. 4660, pp. 229–235 (2002)
15. Matusik, W., Pfister, H.: 3D TV: a scalable system for real-time acquisition, transmission, and autostereoscopic display of dynamic scenes. ACM Transact. Graph. (TOG) 23(3), 814–824 (2004)
16. Nguyen, D., Canny, J.: MultiView: spatially faithful group video conferencing. In: Proceedings of the SIGCHI Conference on Human Factors in Computing Systems, pp. 799–808. ACM (2005)
17. Nguyen, D.T., Canny, J.: Multiview: improving trust in group video conferencing through spatial faithfulness. In: Proceedings of the SIGCHI conference on Human factors in computing systems, pp. 1465–1474. ACM (2007)

90   J. S. Munoz-Arango et al.

18. Takaki, Y.: Thin-type natural three-dimensional display with 72 directional images. In: Proceedings of SPIE, vol. 5664, pp. 56–63 (2005)
19. Nakanuma, H., Kamei, H., Takaki, Y.: Natural 3D display with 128 directional images used for human-engineering evaluation. eye, 3, 3D (2005)
20. Kikuta, K., Takaki, Y.: Development of SVGA resolution 128-directional display. In: Proceedings of SPIE, vol. 6490, p. 64900U (2007)
21. Takaki, Y.: Super multi-view display with 128 viewpoints and viewpoint formation. In: Proceedings of SPIE, vol. 7237, p. 72371T (2009)
22. Takaki, Y., Yokoyama, O., Hamagishi, G.: Flat panel display with slanted pixel arrangement for 16-view display. In: Proceedings of SPIE, vol. 7237(723708), pp. 1–8 (2009)
23. Takaki, Y., Nago, N.: Multi-projection of lenticular displays to construct a 256-view super multi-view display. Opt. Express **18**(9), 8824–8835 (2010)
24. Surman, P., et al.: Head tracked single and multi-user autostereoscopic displays. In: 3rd European Conference on Visual Media Production, 2006. CVMP 2006. IET (2006)
25. Brar, R.S., et al.: Laser-based head-tracked 3D display research. J. Dis. Technol. **6**(10), 531–543 (2010)
26. Brar, R.S., et al.: Multi-user glasses free 3D display using an optical array. In: 2010 3DTV-Conference: The True Vision-Capture, Transmission and Display of 3D Video. IEEE (2010)
27. Kooima, R., et al.: A multi-viewer tiled autostereoscopic virtual reality display. In: Proceedings of the 17th ACM Symposium on Virtual Reality Software and Technology. ACM (2010)
28. Zang, S.F., et al.: A frontal multi-projection autostereoscopic 3D display based on a 3D-image-guided screen. J. Dis. Technol. **10**(10), 882–886 (2014)
29. AmBrSoft: Intersection of a circle and a line (2018). http://www.ambrsoft.com/TrigoCalc/Circles2/circlrLine_.htm, Accessed 20 Nov 2018
30. Attila's Projects: Circle and Line Intersection (2018). http://apetrilla.blogspot.com/2011/12/circle-and-line-intersection.html. Accessed 20 Nov 2018
31. De Greve, B.: Reflections and refractions in ray tracing (2006). Accessed 16 Oct 2014
32. Change of basis. http://www.math.tamu.edu/~julia/Teaching/change_basis_Narcowich.pdf. Accessed 12 Nov 2018
33. Harvey Mudd College Math Tutorial: Change of basis. https://www.math.hmc.edu/calculus/tutorials/changebasis/changebasis.pdf. Accessed 6 Nov 2018

Training Virtual Environment
for Teaching Simulation and Control
of Pneumatic Systems

Carlos A. Garcia[1], Jose E. Naranjo[1], Edison Alvarez-M.[1],
and Marcelo V. Garcia[1,2]

[1] Universidad Tecnica de Ambato, UTA, 180103 Ambato, Ecuador
{ca.garcia,jnaranjo0463,ealvarez,mv.garcia}@uta.edu.ec
[2] University of Basque Country, UPV/EHU, 48013 Bilbao, Spain
mgarcia294@ehu.eus

Abstract. It is clear that in recent years have emerged countless simulators in all possible work areas for reaching the goals of Industry 4.0. Unity Pro software focuses on studying the options and possibilities offered virtual environments for the teaching in schools and universities. The aim of this paper is intended to create a practical training environment for the simulation and control of a pneumatic process for the students of Industrial Process Engineering of Automation. This virtual environment uses MQTT protocol into a Raspberry Pi to transmit instruction for control a real FESTO pneumatic lab based on what students develop during their virtual classes, granting and facilitating the student and teacher a new method that accelerates the learning process.

Keywords: Virtual learning environment · Pneumatic systems · Unity Pro · MQTT protocol

1 Introduction

As technology has advanced over time, computers have increasingly been used as spaces for education. With the advent of virtual environments in 3D, they have created new forms of learning and teaching through the simulation of work environments and experiences that facilitate education. The fact that users can design their own virtual representation, in 3 dimensions, are just some of the peculiarities that are provided to the users of these worlds, an experience different from the traditional learning spaces, better known as physical learning spaces [1].

A virtual learning environment admits technical aspects like building virtual objects, promotes communication and understanding between people and emphasizing social learning, it designs simulations about new educational models both in primary, secondary and university education systems. In fact, the first simulators emerged in the decades of the 60s with the aim of reducing the level of human error [2].

© Springer Nature Switzerland AG 2019
L. T. De Paolis and P. Bourdot (Eds.): AVR 2019, LNCS 11613, pp. 91–104, 2019.
https://doi.org/10.1007/978-3-030-25965-5_8

The use of technological elements, whether they are software for simulation, has to be seen as a resource that allows a better understanding to the apprentices in the construction of knowledge. Such that, the simulation must be seen and understood as a technical resource that facilitates the execution of educational methodologies and, consequently, of learning procedures [3].

The simulation environments are structured events that embody causal relationships between the element and event that represents a real or physical world situation and that, in turn, allows the manipulation of the variables that intervene in the simulation. As already mentioned, this virtual environment can be a very advantageous instrument when it comes to learning to solve problems and acquire knowledge, by allowing the student, user or teacher to witness and experience situations in a controlled manner [4].

In the field of teaching, the simulations fulfill a fundamental role since they generate in the students a more responsible attitude of their own learning and their motivation consists in the accomplishment of objectives. The benefits of a simulation are: use by the teacher to illustrate a specific process or procedure, use by the student without a teacher's guide and, finally, supervised or guided use by the teacher, in order for the student to acquire the enough domain and understanding of the virtual environment [5].

In industry 4.0 field, a MQTT (Message Queue Telemetry Transport) protocol is used because its was development for single board devices (SBC) with few memory resources [6]. This protocol has a huge use in Industrial Internet of Things (IIoT) because is based on publications and subscriptions to the so-called "topics", whose low consumption of resources and speed of response make it suitable for the communication needs of the IIoT. The methodologies of this protocol allow the data of an industrial system to be available to all the devices from the system and to the entire enterprise, without straining data bandwidth.

The aim of this paper is to achieve a virtual training environment for teach pneumatic systems at university students, the Unity Pro^{TM} software is used in this research. The main advantage of this training environment is to encourage creativity where the student can not only simulate pneumatic control models already made, but can build their own. This virtual environment can test the pneumatic circuits development in a real FESTO^{TM} lab using a MQTT communication into a Raspberry Pi board, which transmit the instruction of pneumatic model designed in virtual environment. In addition, it saves time and money since we substitute training equipment, laboratories and test plants for a virtual environment, granting and facilitating the student and teacher a new method that accelerates the learning process.

This article is divided into 7 sections including the introduction. In Sect. 2 the related works are detailed. In Sect. 3 the state of technology is analyzed. Study Case of this research is shown in Sect. 4. The proposed system Architecture is presented in Sect. 5. The analysis of the results of this investigation states in Sect. 6. Finally, Sect. 7 develops the conclusions and future work.

2   Related Works

In this section is analyzed the research and work directly related to the areas in which Virtual Reality (VR) has been used for virtual training environments with pneumatic systems. However, with regard to these antecedents, it is necessary to state that there are not currently investigations that have focused on simulating training environments in pneumatic systems, with which it is evident that this research that is being developed is about a completely innovative theme.

During the last few years, some researches have been published about virtual environments in Unity pro, as is the case of the project presented by Naranjo et al. [7], which deals with a design of a bilateral tele-operation system that allows the operator to carry out maintenance or remote inspection activities in the Well-Pads in Oil & Gas industry using a mobile robot. For the development of this system, the MQTT protocol is selected over other IIoT protocols because of its high speed of response in real time as well as its low consumption of resources. In the local site a virtual reality environment developed in a 3D graphic engine is implemented. Unlike our research, the controller used is a low cost SBC as Raspberry Pi, which is use to control pneumatic circuits designed in a virtual environment.

In [8], authors present a related research work for the development of a remote automation laboratory, allows students to interact and perform practices of automation and process control by using the Internet in a Modbus TCP/IP network. In this remote lab users can manipulate the whole process through a simulator, and then physically implement it to the food manufacturing process, with the possibility of remote monitoring between the system and the process, uses a set of JAVA applications for the simulation environment.

The fields of application of virtual simulation for student training, is in medicine, as published by the authors Gallagher et al. [9], make a comparative research of commercial and open use virtual environments for students to training laparoscopic surgery. These simulators use interfaces for the sensitivity to the touch, hearing and sight of the practitioners, increasing the learning capacity and the realism of the surgical procedures. Being a field in which this research work wants to be involved.

In the industrial field, virtual reality is a means to develop practical skills in the management of machinery, as described by the authors Di Gironimo et al. [10], in his scientific research. They presents the development and testing procedure of a Virtual Reality application for the training of practical skills, corresponding to the operation of forklifts. This paper explains the aim for the application, as well as basic ideas, concepts of knowledge forms and organization of the virtual course, being more efficient in time compared to the traditional and perceived with enthusiasm by the students. The difference with our training environment proposed is that this research paper teach different kinds of pneumatics circuits used in different industrial machinery.

In the scientific article Virtual Technologies Trends in Education mentions that educational institutions will benefit from better accessibility to virtual technologies [11]. This will allow teaching in virtual environments that are impossible to visualize in physical classrooms, such as access to virtual laboratories, visualization machines, industrial plants or even medical scenarios. The enormous possibilities of accessible virtual technologies will allow breaking the limits of formal education. However, this research is purely theoretical, but did not implement a virtual environment, on the contrary, in our research work was executed a virtual environment in the laboratory of pneumatic for the use of students.

3   State of Technology

Nowadays, we can find simulation environments in almost all industrial areas, which allows us to face real life situations from a different perspective, allowing a level of learning and user experience in a fast and dynamic way at the same time [12].

A virtual training consists of generating a virtual environment similar to a real-world event, representing the behavior of a process through models whose parameters and variables are the reproduction of the processes studied. In the world there are a variety of three-dimensional virtual environments, where most of these are to apply knowledge in different subjects, encourage design, engineering and computer-aided simulation [13,14].

3.1   Virtual Reality (VR) in Education Process

With the aim of achieving better learning in students today new teaching methods are proposed, this is how virtual reality occupies an important role when it is desired to impart knowledge in a didactic, practical and safe way to students. It is one of the best options that teachers currently have, with the possibility of updating and optimizing teaching methods, reducing in a certain way the excessive expenditure on acquisition of physical tools and materials, for this reason, students are they feel limited to learn in group and many times the knowledge varies between students. Today we can find many pages on the Internet with virtual simulation environments in different areas of teaching, motivating students to investigate, practice and analyze with their own judgment the activities practiced in virtual environments, acquiring useful experience for their academic training [15].

The simulation that the virtual reality develops can refer to virtual scenes, creating a world that only exists inside the computer. Additionally, it allows working in a completely virtual world, disconnecting us from reality and introducing us into the virtual world that had been created. Applications found nowadays are activities form everyday life, reconstruction of cultural heritage, medicine, crowd simulation and presence sensation [16,17].

3.2   Raspberry Pi

Raspberry Pi is a low-cost Single Board Computer (SBC), which has a processor of up to 1 GHz and 512 Mb RAM. Its purpose of use involves the stimulation of computer education in the academic field of vocational training and in some low-level industrial applications. Raspberry Pi are able to work with many Linux distributions making them more versatile. Additionally, have a considerable number of general purpose I/O ports (GPIO): 26 ports in the RPI that allow interaction with the physical world. One of the disadvantages of this cards is that their kernel does not allow direct manipulation of their I/O ports, but this can easily be offset by the wide variety of libraries developed to manage these ports [18]. Table 1 shows specific electric characteristic of this card.

Table 1. RPI3 B+ electrical characteristics.

	Raspberry PI3 B+
System-on-a-chip (SoC)	Broadcom BCM2835
Central processing unit (CPU)	ARM 1176JZFS at 700 MHz
RAM memory	512 MB
GPIOs	8 x GPIO, SPI, I2C, UART
Networks	Ethernet 10/100

3.3 Unity Pro™

Is a software tool developed by Unity Technologies in 2005 [19]. It is a completely integrated development engine that provides a complete functionality create both three-dimensional and two-dimensional games as well as simulations for its many platforms. Developers, students, designers, corporations, researchers, etc., use this platform because it allows reduction in time, effort and costs, in many branches. Unity Pro 3D™is one the existing versions that eases software development for a large range of platforms, while being very attractive for its benefits for developers and researchers. User benefit from cross-platform native C++ performance with the Unity-developed backend IL2CPP (Intermediate Language To C++) scripting [20].

This software utilize the multicore processors available, without heavy programming. This software enabling high-performance which is made up of three sub-systems: the C# Job System, which gives you a safe and easy sandbox for writing parallel code; the Entity Component System (ECS), a model for writing high-performance code by default, and the Burst Compiler, which produces highly-optimized native code [21,22].

3.4 MQTT Protocol

It is a protocol based on a subscription publishing architecture with a message broker to communicate to both the publisher and the subscriber. Due to its low bandwidth consumption, this protocol is oriented to the communication of sensors. It can also be implemented on most integrated devices with few resources. The MQTT architecture follows a star topology, in which the intermediary has

a capacity of up to 10,000 clients. The visualization of the messages in this protocol are of JSON type allowing a comfortable and clean reading. Its design
principles make it ideal for emerging paradigms such as the Internet of Things
(IoT), and therefore Smart Factories [23]. MQTT is a very useful protocol for
wireless systems which experience fluctuating levels of latency due to unstable
connections. If the subscription connection is lost, the broker saves the message
packet and forwards it to the subscriber when the connection is re-established.
On the other hand, if the publication link becomes unstable, the broker can
terminate the connection [24] (See Fig. 1).

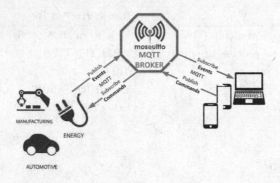

Fig. 1. MQTT protocol architecture.

4   Case Study

The aim of study case is creating a virtual environment of an electro-pneumatic
system laboratory in order to teach students techniques of control and simulation. This laboratory will be used for learning of designing principles deals
about creation of electro-pneumatic control system and its simulation. This type
of pneumatic circuits are used in the field of mechanization and automation in
the industrial context.

The virtual laboratory was development using Unity Pro 3D software and
consists: (i) With the help of RV Oculus glasses students visualize and can move
in the designed environment in a realistically way with a immersive sensation
allowing interaction with the system through the recognition of manual gestures. The teaching process begin when students receive an circuit pneumatic
problem about the use of two pneumatic cylinders. (ii) Once students solve the
proposed problem, their use the virtual laboratory to begin to model the proposed pneumatic solution. Here students interact with virtual environment in
order to connect in right way cylinders, end-of-stroke sensors and pneumatic
supply. In many cases the mechanism is not working exactly, so student have to
rebuild the exercise, what leads to the increase of manual skills and this activity
increase failure searching skill. (iii) When the pneumatic problem is solved and
simulated the student transmit the control instructions using MQTT protocol

to a Raspberry PI board who is connected to a real FESTO electro-pneumatic system to show its function in a practical form (See Fig. 2).

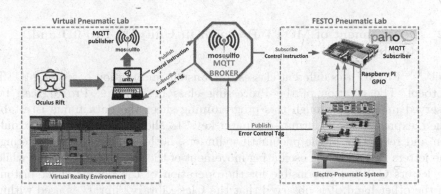

Fig. 2. Study case architecture.

This virtual laboratory teach advanced calculation technique that allows simulating and exercising on the different pre-designed pneumatic and electropneumatic (logical) circuits associated to different specific subjects such as Mechanization and automation, Theory of systems and automatic machines and Assembly devices and machines.

Creation of such laboratory will lead to the quality increase of teaching process of some subjects which belongs to our study branch called manufacturing devices and systems. Student which will go through courses in these kind of laboratory will get needed professional skills and competences. These competences and skills will strengthen their skills which are very important for designing process of such devices. The students will also know the methodology how to create several types of systems, which will lead to their labour market value increase.

5   Proposed System Architecture

5.1   Methodology

An iterative and incremental development methodology has been used. This procedure allows to plan a project in different time blocks called iteration. So, this methodology will allow us to divide our project at different stages, which will be validated at the end of each. You can verify that all the results are the expected ones without the need to have finished the software completely, and you can identify situations of error (or simply apply improvements) in the phases already developed.

The analysis phase is about what to do, how to do it, with what tools and to outline the limitations and objectives to be achieved. In the design phase, a sketch of the application is made, both of the scenes and the Users interfaces. The implementation phase: the logic of the application is developed. Then comes the tests: phase in which the correct functioning of the implemented functionalities is checked.

5.2 Development of MQTT Protocol in Unity™ Pro 3D and Raspberry PI

Unity™ Pro 3D uses different classes to transmit information based on MQTT protocol. The creation of the Processing class in Unity™ Pro 3D can be observed in Fig. 3. Through this programming sheet the data frame is established, expressed in the form of computer keys, to allow the movement of expulsion and retraction of the pneumatic cylinders, both single and double acting. The letters F-R allow the oscillating movement of the simple effect piston, while the letters G-T are responsible for the operation of the double effect piston. On the other hand, it is observed that the ClassLibrary2 library is used within which the communication stack of the MQTT protocol is located. The class called Class1 is used in order to establish the IP of the subscriber device (Raspberry Pi3) in addition to the topic to which it must subscribe.

```
public static class Processing
{

    static StringBuilder keyBuffer;
    public static string[] Teclas = new string[] { "R", "F", "G", "T" };
    static ClassLibrary2.Class1 cn;

    public static void initConection(String ip, String topic)
    {
        cn = new ClassLibrary2.Class1(ip,topic);
    }

    public static void Trama(int valor, int posicion)
    {
        int[] cadena = new int[] { 0, 0 };
        cadena[posicion] = valor;
        cn.envio(CadenaToString(cadena));
    }
    public static string CadenaToString(int[] cadena)
    {
        string aux = "";
        for (int i = 0; i < cadena.Length; i++)
        {
            aux += cadena[i];
            if (i < (cadena.Length - 1))
                aux += ",";
        }
        return aux;
    }

}
```

Fig. 3. MQTT class used in Unity™ Pro 3D.

The RPi board incorporates the C++ library corresponding to Eclipse Paho which implements the MQTT protocol versions 3.1 and 3.1.1. Paho is a lightweight alternative and is suitable for use on devices from low power single board computers to full servers. The Python library incorporates the methods to: initialize and clean up the library, create and eliminate clients, specify authentication and encryption methods, configure wills, connect the created clients to the broker, publish messages, subscribe to topic, among others. In Fig. 4 shows MQTT initialization where Raspberry PI subscribe to a topic in Unity^{TM} Pro 3D.

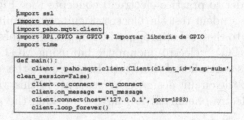

```
import ssl
import sys
import paho.mqtt.client
import RPi.GPIO as GPIO # Importar libreria de GPIO
import time

def main():
    client = paho.mqtt.client.Client(client_id='rasp-subs',
clean_session=False)
    client.on_connect = on_connect
    client.on_message = on_message
    client.connect(host='127.0.0.1', port=1883)
    client.loop_forever()
```

Fig. 4. MQTT class used in Raspberry Pi.

5.3   Virtual Reality Environment Development

Virtual reality is a technology used most frequently to generate learning environments that improve the level of knowledge of students in their professional instruction. Many educational institutions do not have the economic or physical resources to carry out a correct teaching to their students, for this reason the technological resources help in the process of knowledge acquisition.

When the student begins to interact with the proposed virtual training system by using the Oculus Rift^{TM} device, sends data through coordinated movements of user's hand to the programmed environment and can perform actions on the pneumatic system, observing the behavior and functioning of the elements present in the circuit, previously counting on a training environment and tests for the student.

Figure 5(a) shows the main screen where the user or the student visualizes the options to become familiar with the instruments and to start with the virtual training. Likewise, a cylinder is present in three dimensions, where with the Oculus Rift^{TM} device can interact with the screen.

In Fig. 5(b) presents a screen where the instruments that our virtual system has, in which user can see cylinders of single and double effect, optical and capacitive sensors, 24 V source, maintenance unit, among others. On this screen you can go through the whole environment, Oculus Rift^{TM} allows to the user to move around the environment with hands movements without touching the computer or any key of the same.

The next step of this virtual environment is show the electro-pneumatic circuit that the student is going to be solved (See Fig. 5(c)). Furthermore, the proposed circuit solved is presented in the next screen in order to the student watches the correct operating of pneumatic cylinders and electric circuit (See Fig. 5(d)).

Once the student know the aim of proposed electro-pneumatic circuit, begin to manipulate the pneumatic cylinders and sensor in the virtual environment.

In this step the student test the operation of pneumatic equipment to be use. Student use the virtual environment in order to practice how the pneumatic elements should be connected because it has the pneumatic supply and pneumatic hoses. When the pneumatic circuit is connected, user connect sensors and cables to 24 V supply in order to practice electrical concepts (See Fig. 5(e)).

At final step, the student test the electro-pneumatic circuit and if the test is successful according to the goal of proposed example all of these instructions are sent to the real FESTO™ electro-pneumatic lab using MQTT protocol. Raspberry Pi actuating as controller of this Festo Lab receives the control instruction and begin the test of the circuit designed by student. In this way the student see in the FESTO™ electro-pneumatic lab how is working the proposed example by teacher. (See Fig. 5(f)).

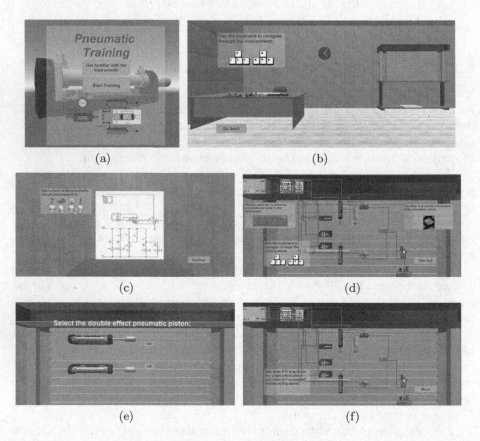

Fig. 5. VR user interface. (a) Main screen of training electro-pneumatic lab. (b) Virtual environment to begin with the electrical and pneumatic equipment to be used (c) Virtual screen where the proposed electro-pneumatic circuit is presented. (d) Screen where is showing the right function of proposed electro-pneumatic circuit, (e) Virtual environment where user design and test the circuit, (f) Test of pneumatic circuit finished.

6   Discussion of Research Findings

The application is evaluated by an anonymous survey provided to the students (See Fig. 6). The results show that the user interface was considered easy to understand by most of the students and, additionally, that they required a reasonable period of time to learn how to use the program. More than half of the interviewers said that the application helped them to better understand the operation of the pneumatic components.

Fig. 6. The virtual training pneumatic system evaluated by students.

Figure 7 shows the network traffic when a client subscriber communication is implemented. The communication packets are measured in bytes/seconds. It can also be appreciated that the bandwidth reaches a maximum of 500 bytessec and a minimum of 1000 bytes/sec. The red line represents the errors in the communication, however, it seemed that the error in the transmission of data is zero.

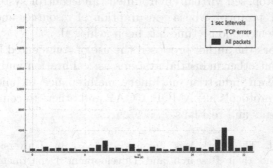

Fig. 7. MQTT bandwidth. (Color figure online)

In Fig. 8, the load of the CPU of the Raspberry pi 3 can be observed. A variation in the load is observed which corresponds to the processing performed by the MQTT subscriber. This happens when RPI receives a message activating or deactivating the pneumatic pistons. Although an industrial process is being carried out, no alteration or overload of processing is observed in the controller CPUs.

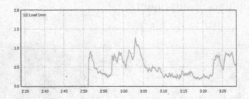

Fig. 8. RPI memory overhead.

7 Conclusion and Future Works

An educational application that has been designed for a virtual training environment for pneumatic systems is presented. The application acts as a virtual laboratory that helps students understand the operation of complex systems. The real implementation of this kind of laboratories by universities is expensive and sometimes does not have all equipment for personalize student education. This research has achieved an efficient immersion of the user easing education of different electro-pneumatic control algorithm avoiding traditional educational methodologies.

With this technology, the vision of custom the training of students and developing more efficient methodologies, it becomes a conceivable goal for any educational unit. When talking about the interaction of real lab with virtual reality interfaces, it is necessary to take into account the response time and delays inherent to the proposed virtual environment and control system. Due to the use of the MQTT protocol, whose consumption of resources and bandwidth is very low, an efficient response time has been achieved.

As a future work the author proposed the use of Augmented Reality (AR) to custom learn about other industrial systems as: calibration and commissioning of instruments, learn industrial machinery maintenance techniques and using different kinds of protocols like AMQP, COAP and others to communication the virtual environments and real labs.

Acknowledgment. This work was financed in part by Universidad Tecnica de Ambato (UTA) and their Research and Development Department under project CONIN-P-0167-2017.

References

1. Anokhin, A., Alontseva, E.: Implementation of human–machine interface design principles to prevent errors committed by NPP operators. In: Bagnara, S., Tartaglia, R., Albolino, S., Alexander, T., Fujita, Y. (eds.) IEA 2018. AISC, vol. 822, pp. 743–753. Springer, Cham (2019). https://doi.org/10.1007/978-3-319-96077-7_81
2. Daniel, G.: Skill training in multimodal virtual environments. Work **41**, 2284–2287 (2012). https://doi.org/10.3233/WOR-2012-0452-2284
3. Cardoso, A., et al.: VRCEMIG: a virtual reality system for real time control of electric substations. In: 2013 IEEE Virtual Reality (VR). Presented at the 2013 IEEE Virtual Reality (VR), pp. 165–166. IEEE, Lake Buena Vista (2013). https://doi.org/10.1109/VR.2013.6549414
4. Chessa, M., Maiello, G., Borsari, A., Bex, P.J.: The perceptual quality of the oculus rift for immersive virtual reality. Hum.-Comput. Interact. **34**, 51–82 (2019). https://doi.org/10.1080/07370024.2016.1243478
5. Dave, I.R., Chaudhary, V., Upla, K.P.: Simulation of analytical chemistry experiments on augmented reality platform. In: Panigrahi, C.R., Pujari, A.K., Misra, S., Pati, B., Li, K.-C. (eds.) Progress in Advanced Computing and Intelligent Engineering. AISC, vol. 714, pp. 393–403. Springer, Singapore (2019). https://doi.org/10.1007/978-981-13-0224-4_35
6. MQTT Org: MQTT standard (2018). http://mqtt.org/
7. Naranjo, J.E., Ayala, P.X., Altamirano, S., Brito, G., Garcia, M.V.: Intelligent oil field approach using virtual reality and mobile anthropomorphic robots. In: De Paolis, L.T., Bourdot, P. (eds.) AVR 2018. LNCS, vol. 10851, pp. 467–478. Springer, Cham (2018). https://doi.org/10.1007/978-3-319-95282-6_34
8. Mujber, T.S., Szecsi, T., Hashmi, M.S.J.: Virtual reality applications in manufacturing process simulation. J. Mater. Process. Technol. **155–156**, 1834–1838 (2004). https://doi.org/10.1016/j.jmatprotec.2004.04.401
9. Gallagher, A.G., et al.: Virtual reality simulation for the operating room: proficiency-based training as a paradigm shift in surgical skills training. Ann. Surg. **241**, 364–372 (2005). https://doi.org/10.1097/01.sla.0000151982.85062.80
10. Di Gironimo, G., Matrone, G., Tarallo, A., Trotta, M., Lanzotti, A.: A virtual reality approach for usability assessment: case study on a wheelchair-mounted robot manipulator. Eng. Comput. **29**, 359–373 (2013). https://doi.org/10.1007/s00366-012-0274-x
11. Martín-Gutiérrez, J.: Virtual technologies trends in education. EURASIA J. Math. Sci. Technol. Educ. **13** (2017). https://doi.org/10.12973/eurasia.2017.00626a
12. Davies, R.C.: Adapting virtual reality for the participatory design of work environments. Comput. Support. Coop. Work. (CSCW) **13**, 1–33 (2004). https://doi.org/10.1023/B:COSU.0000014985.12045.9c
13. Merchant, Z., Goetz, E.T., Keeney-Kennicutt, W., Cifuentes, L., Kwok, O., Davis, T.J.: Exploring 3-D virtual reality technology for spatial ability and chemistry achievement: exploring 3-D virtual reality technology. J. Comput. Assist. Learn. **29**, 579–590 (2013). https://doi.org/10.1111/jcal.12018
14. Merchant, Z., Goetz, E.T., Keeney-Kennicutt, W., Kwok, O., Cifuentes, L., Davis, T.J.: The learner characteristics, features of desktop 3D virtual reality environments, and college chemistry instruction: a structural equation modeling analysis. Comput. Educ. **59**, 551–568 (2012). https://doi.org/10.1016/j.compedu.2012.02.004

15. Grajewski, D., Górski, F., Zawadzki, P., Hamrol, A.: Application of virtual reality techniques in design of ergonomic manufacturing workplaces. Procedia Comput. Sci. **25**, 289–301 (2013). https://doi.org/10.1016/j.procs.2013.11.035

16. Hedberg, J., Alexander, S.: Virtual reality in education: defining researchable issues. Educ. Media Int. **31**, 214–220 (1994). https://doi.org/10.1080/0952398940310402

17. Langley, A., et al.: Establishing the usability of a virtual training system for assembly operations within the automotive industry. Hum. Factors Ergon. Manuf. Serv. Ind. **26**, 667–679 (2016). https://doi.org/10.1002/hfm.20406

18. Garcia, M.V., Perez, F., Calvo, I., Moran, G.: Developing CPPS within IEC-61499 based on low cost devices. In: Proceedings of the IEEE International Workshop on Factory Communication Systems, WFCS (2015)

19. Unity [WWW Document] (2018). https://unity3d.com/es/unity. Accessed 23 Sept 2018

20. Lanzotti, A., et al.: Interactive tools for safety 4.0: virtual ergonomics and serious games in tower automotive. In: Bagnara, S., Tartaglia, R., Albolino, S., Alexander, T., Fujita, Y. (eds.) IEA 2018. AISC, vol. 822, pp. 270–280. Springer, Cham (2019). https://doi.org/10.1007/978-3-319-96077-7_28

21. Limniou, M., Roberts, D., Papadopoulos, N.: Full immersive virtual environment CAVETM in chemistry education. Comput. Educ. **51**, 584–593 (2008). https://doi.org/10.1016/j.compedu.2007.06.014

22. Sacks, R., Perlman, A., Barak, R.: Construction safety training using immersive virtual reality. Constr. Manag. Econ. **31**, 1005–1017 (2013). https://doi.org/10.1080/01446193.2013.828844

23. Naik, N.: Choice of effective messaging protocols for IoT systems: MQTT, CoAP, AMQP and HTTP, pp. 1–7. IEEE (2017). https://doi.org/10.1109/SysEng.2017.8088251

24. Asaad, M., Ahmad, F., Alam, M.S., Rafat, Y.: IoT enabled monitoring of an optimized electric vehicle's battery system. Mob. Netw. Appl. **23**, 994–1005 (2018). https://doi.org/10.1007/s11036-017-0957-z

Real Time Simulation and Visualization of Particle Systems on GPU

Bruno Ježek$^{(\boxtimes)}$, Jiří Borecký, and Antonín Slabý

University of Hradec Králové,
Rokitanského 62, 50002 Hradec Králové, Czech Republic
{bruno.jezek, jiri.borecky, antonin.slaby}@uhk.cz

Abstract. The article deals with simulation of fluids in the digital environment. It provides a brief overview of the use and various ways of simulation and displaying results and stresses the possibility of particle representation through particle systems and describes the particle method Smoothed Particle Hydrodynamics simulation. It focusses on description of the implementation of this method using the possibilities of a modern programmable graphics processing unit (GPU) in both the simulation and the visualisation process. For implementation of multi pass GPU algorithms the special data representations of simulation and rendering properties are designed.

Keywords: Fluid simulation · Particle systems · Real time rendering · GPU · GPGPU

1 Introduction

The article summarizes our own approach to real time simulation of fluids. It briefly introduces the selected elements of fluid dynamics and the possibilities of its representation in the digital environment. It presents elements of Smoothed Particle Hydrodynamics (SPH) simulation method and the technique for optimizing the entire method. Then it focuses on implementation of simulation and rendering using GPU. This is the principal part of the article and describes authors own approach to real time solving the problem of simulation and rendering of fluids. Two multi pass GPU algorithms for the implementation of simulation and rendering task are designed together with appropriate data representation to store required properties. Then the article brings some other details of the solution and shows samples of results and summarizes our testing outputs concerning time requirements of our solution.

2 Simulation of Fluids

The process of simulation of liquid substances is controlled by many parameters and is also significantly influenced by the overall design of the domain. Wrong setting of material properties, particle counting, smoothing functions and other parameters can lead to unwanted behaviour or even collapse of the entire simulation. It is necessary, among other things, to pay close attention to these details for good functioning of the

L. T. De Paolis and P. Bourdot (Eds.): AVR 2019, LNCS 11613, pp. 105–119, 2019.
https://doi.org/10.1007/978-3-030-25965-5_9

whole system. A lot of calculations and demanding operations need to be solved during the process of simulation. Proper optimization can greatly speed up the process. In addition to proper selection of the closest particles for interaction, we have to select also suitable smoothing functions that are not complicated for processing. When calculation of the forces is involved, it is necessary to know the density at the point of the particle, which also depends on its surroundings. Consequently it is necessary to iterate over all adjacent particles twice in each simulation cycle. The density is determined for each particle in the first iteration, and sum of the forces and the subsequent time integration is established during the second iteration. This results in new position and velocity of particles for the next cycle. When using a single particle list, time integration is only necessary after calculating the forces involved for each particle.

2.1 Smoothed Particle Hydrodynamics Simulation Method

The oldest, most well-known and most widely used method for particle fluid simulation is the Smoothed Particle Hydrodynamics (SPH) method [1, 2]. SPH is the interpolation method using approximation kernels and based on calculation of parameters called properties. The properties of a substance as density of substance, internal forces (pressure and viscosity), external forces (gravity, surface tension) and others define its state and dynamics at a given moment, but in more complex systems, the resulting motion is often influenced by collisions and the behaviour of the environment. Ultimate property of the substance at the point of space is given by the weighted sum of the given property of all surrounding particles that serve as interpolation points of the properties of the substance. Particles are influenced by the environment, but they are not logically connected to the nearest particles, so they can move independently and the closest neighbours change over time. The magnitude of their effect on adjacent particles is the result of the kernel function. The fluid property at any point is calculated as the sum of contributions of surrounding particles, for each property separately. This is where the kernels are used.

2.2 The Kernel

The kernel, sometimes called smoothing function or weighing function, is an approximation function that calculates the value of the contribution of surrounding particles depending on their distance. Thanks to these functions, it is possible to accurately estimate the values of functions defining the behaviour of the fluid and its derivation anywhere in the space. Each kernel must necessarily satisfy basic conditions for simulation see more [1, 2].

2.3 SPH-Based Simulation Procedure

The SPH based simulation process can be briefly described as follows:

1. Initialization of the system, this stage includes preparation of data structures, calculation of constant values, generation of particles
2. The process of simulation repeats the following items

a. For each particle
   (1) Find all the surrounding particles
   (2) Calculate the total density by surrounding particles
b. For each particle
   (1) Calculate the pressure at the particle point (point characterizing the position of the particle)
   (2) Find all the surrounding particles
   (3) Calculate the total compressive force using the surrounding particles
   (4) Count the total viscosity force with the surrounding particles
   (5) Calculate the total surface tension by the surrounding particles (for liquids)
   (6) Add the influence of gravitational acceleration (for liquids)
   (7) Calculate sum all the forces and find the acceleration
c. For each particle
   (1) According to the integration scheme, update its position and velocity
   (2) Check and resolve collisions with objects

2.4 Rendering Methods

The principal problem (after the simulation is finished) is appropriate visualization of the acquired data. It is sufficient to represent each particle as a sphere of suitably chosen size for simulation purposes. It is important to represent the visible surface and maintain as much details as possible for detailed viewing (smooth surfaces, sharp gradients, small splashes of droplets separated from the rest of the substance, minor changes in the surface, such as capillary waves, and even thin volumes that can be during the display process be completely lost) [1]. Another problem arises if the role of transparency should be included. Then it is not enough to display only the surface itself but it is necessary to include into considerations depth and visibility too. The main methods of surface reconstruction include Marching Cubes (MC) [3], Volume imaging [4], Screen Space methods [5, 6].

3 Implementation of Simulation and Rendering on GPU

Our solution of the problem is based on application of advanced display methods based on the Screen Space. The freely available OpenGL graphical interface has been used to access to GPU. The essence of our application is to use textures as source data (instead of image data) and in addition to it as objects to store the results of the calculations. This approach is known as GPGPU - General Purpose Graphics Processing Unit. Figure 1 shows schematically the main data structures, elements and relationships of the process. The calculation is based on processing a single quadrangle (quad) that is rasterized, and its each rasterized pixel corresponds to one texture position in the result into which the fragment shader writes the results of the calculations.

The OpenGL interface in version 4.3 introduced for this purpose the compute shader (CS). Compute shader is a program running on the GPU that is not tied to a visualization pipeline, and still has many graphics processor sources available. Ability to be run at any point of the application independently of the visualization pipeline is its great advantage. The CPU only passes the GPU request to execute it. Compute shader program, however, has no inputs and outputs and it is therefore necessary to preload the necessary data into the GPU memory and determine the location of the inputs and outputs in the GPU memory. The conversion of outputs is required for subsequent use on the CPU. This often results in considerable bus load and it is desirable to avoid these operations as much as possible. One possible partial solution of this problem is to divide the whole calculation task into multiple shader programs and leave the individual results in the form of a texture stored on the GPU until all needy operations are made via the GPU.

3.1 Simulation Procedure Using Data Storage on GPU

All the data needed for the simulation are stored on the GPU in the form of textures. All the SPH property calculation formulas use position and the velocity of the particles as parameters. These are three-dimensional quantities that can be easily placed into RGB channels in the texture. The value stored in each pixel therefore represents the position XYZ or the velocity Vxyz of one particle, see Fig. 1. Other properties are constant or counted during the algorithm, and do not need to be saved in textures. Two textures are created for calculating SPH property to preserve data integrity - one for reading and the other for writing in one simulation cycle. This fact also allows to do time integration and collision control immediately after computing forces without additional iterations, because writing new values is done in independent textures and consequently causes not overwriting of source values. The target texture contains updated information at the end of the cycle, and so data in the source texture can be overwritten in the next cycle. Each simulation cycle is therefore terminated by changing the source and target texture.

The second important part of simulation is searching the surrounding particles on the GPU. The nearest neighbour search (NNS) solution proposed in [2, 7] is very efficient and simple, but unsuitable for GPU implementations. Sophisticated data structures do not allow simple texture representation, although they are optimal for the original CPU implementation.

A different way of representation of the environment is therefore necessary to choose. One possibility is based on the division of the space into imaginary 3D cells with a size of maximum searching distance that corresponds to the kernel radius. It is then obvious that for each particle all neighboring particles at a given distance must be located in the 27 adjacent cells, including the central one. Each cell and all the particles contained therein can be easily identified by a unique spatial identifier ID composed of cell coordinates. When searching for surrounding particles, it is necessary to generate uv coordinates stored in the new IDX texture (Fig. 1) referring to the original location of the particle having the position XYZ and velocity Vxyz textures together with the spatial particle identifier ID. This texture is then sorted in ascending order according to ID. The particles of one cell are then stacked in a sorted texture and such arrangements enables easily iterate through the surrounding particles of one cell. The actual position and velocity of the particle in the textures are achieved from uv coordinates. This process is performed by the CPU and the sorted result is converted to the GPU at each step to avoid parallelization implementation problems.

It is not possible to admit in the process of simulation of the fluid an error, which would significantly affect the dynamics of the substance. Consequently the additional NNS texture was created for this purpose. It provides detailed information about the entire area in which the substance is located. This texture is one-dimensional and its size is equal to the number of cell identifiers found. Each pixel contains three data: cell identifier ID, number of all particles with a lower identifier obtained in sorting (prev), and number of particles in the given cell (count). Values are placed in bytes and coded in RGB channels, alpha channel remains unused. It is sufficient to convert the number of particles with the lower identifier for each 3D cell (stored in NNS) into the index to the IDX texture, and in such a way generate the coordinates of the first particle of the cell in the sorted texture. Information of particle number allows to iterate through all the 3D cell particles in the loop. The use of this auxiliary texture is more time consuming, but ensures maximum accuracy in calculating particle interactions. Relations among textures shows the following figure Fig. 1.

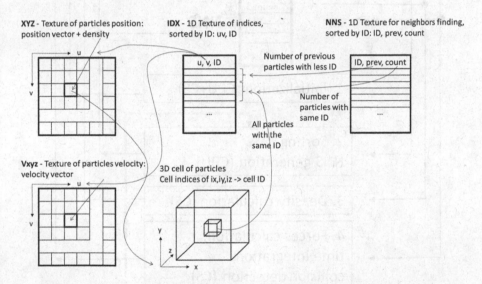

Fig. 1. Properties stored in textures used for calculation of simulation process

Six textures, four of them for both source and target particle position and velocity storage and the two others used for sorted and unsorted uv coordinates and identifiers, are created during the system initialization, and the source texture containing the positions is filled with the primary data. Process of the initialization includes starting of the first CS program, which generates the uv coordinates and identifiers for the first sorting cycle. This phase takes place on the CPU and its result is converted to the GPU memory where it is used in the following two steps by running compute shaders. Each CS is run in several GPU workgroups that process continuous but distinct parts of the texture with particles arranged according to the workspace definition. Each work item in a group (by its identifier) performs calculations for one single particle i.e. pixel in texture (Fig. 2).

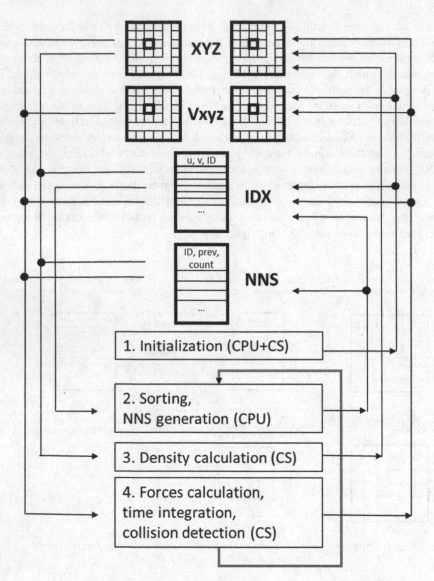

Fig. 2. Data structures and steps of simulation process performed on CPU and GPU.

The second step of the loop utilizes texture with positions and textures for particle search. Density calculation of each particle requires the information of all 27 surrounding cells. Their identifier is calculated from the position of the particle shifted by the cell size in all directions. Each surrounding cell is searched in the NNS texture. The adjacent cells without any particles in the auxiliary texture cannot be found and are labelled to prevent unwanted computations with unknown data in the memory. The CS further ensures cyclical processing of the obtained values. The labelled cells are omitted and information about the remained particles is obtained from the auxiliary texture. The contribution of surrounding particles to density at the point of the particulate currently processed particle is calculated. The resulting value is stored in the free alpha channel of the pixel for use in the next step.

The third step of the loop provides all the remaining calculations. It includes search for the surrounding particles (just like the previous shader) and therefore it uses both the same textures (information on particle position and particle velocity in textures) and the same search algorithm. The texture containing positions is used to store density for pressure value calculation. After identifying information about surrounding cells, the SPH method ideas are used to calculate the interactions between particle pairs. The last step on this shader updates particle position and velocity to the output textures along with updating the spatial identifier of the new position into the unsorted texture for the next sorting cycle.

3.2   Particle Rendering Implementation Details

The starting point for the particle rendering procedure is the Screen Space method presented in [5]. The resulting simulation solution is aimed at visualization of a liquid substance. The essence of this rendering method is to plot the necessary attributes into several image buffers and their subsequent combinations. The possibility to render into multiple image buffers in one running shader program at the same time is used, to speed up the process. The first step determines the spatial location of the substance (depth of the scene), to obtain the rough properties of the visible surface (normal vectors) and to determine the overall distribution of the volume of the substance, thickness of the substance denoting the mass, amount of particles visible through the given pixel. This phase, when passing through the visualization pipeline, uses the vertex shaders (VS) to transform particle positions from the world coordinate system (WCS) into the camera coordinate system (CCS), geometry shaders (GS) to create a spatial object representing a single particle, and to complete volume transformations and fragment shaders (FS) to calculate and store raster pixel values in buffers that contain the resulting attributes. Required operations are performed by two passes through the visualization pipeline. The results obtained must be processed by auxiliary filters for better results. Filtering is a post-processing operation, consisting in simple texture mapping on a quad transformed by VS to cover the entire target buffer and texture used to read the data. The filtering calculations are performed by FS at the pixel level in the local area. For the application of optical phenomena, it is also necessary to find the parameters of the substance in the direction from a light source by the same procedure. The depth of the scene and the mass of the substance that can be gained in one pass. Once the rough content of a scene has been rendered, the final appearance of the substance is reached by the common way of rendering objects. It is achieved by creating a simple quad geometry (as well as by filtering), and subsequent composing the content of the obtained textures and placing them in the resulting image.

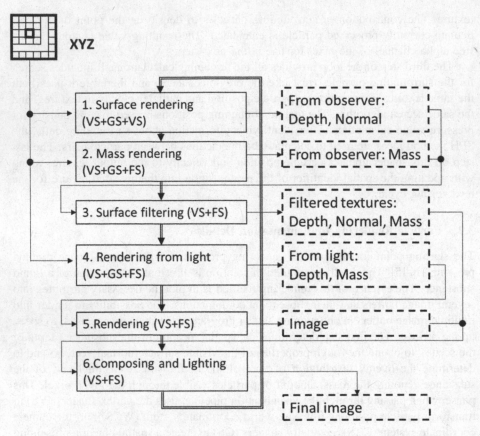

Fig. 3. Rendering steps of visualization process performed on GPU

3.3 Special Rendering Problems and Related Procedures

The result of each simulation step is a texture of the x, y, and z coordinates of particles positions coded in the RGB channels. Each particle represents a certain elemental volume of matter, but this element is not the geometric point in the space. It is therefore necessary to choose a suitable method of representing the volume, based on representing of one particle. Spherical surface representation appears to be the best choice. Displaying a large number of spheres with a detailed description of the surface using a number of triangles in order to prevent the surface from appearing a bit angular may be overly complex for real time display. The billboarding method for rendering particles, which creates single-point quads with their surface facing to the viewer is more effective and adequate. Pixels of a rasterized quad are then modified to form a spherical surface. This method was used in our approach.

The amount of fluid is to some degree transparent, depending on the thickness of the volume, representing the mass of the substance through which the light passes. Thickness does not represent precise visible volume, it only approximates its representation. It determines the effect of a substance to the light passing through it and is required to determine the intensity of surface colour, light refraction and overall transparency of the substance [8]. Thickness is obtained by the process based on additive colour blending.

The blending is active for all output image buffers of the pass, but different functions can be set separately for each buffer, when an output pixel of FS is combined with a pixel of the same position in the image buffer. The blending cannot be performed simultaneously with the hardware Z-buffer test, which is required in the previous step to save the normal vectors of the visible surface correctly (step 1 in Fig. 3). To obtain information about the mass of the displayed substance, therefore, the rendering of all the particles must be re-performed (step 2 in Fig. 3). VS and GS ensure transformations and quads generation. FS clips pixels out of the circular area of the rasterized quad, generated in GS, and sets a low value for the other pixels representing the particle thickness, respectively the amount of substance represented by the particle at the output. The blending function set to add pixel values will ensure the accumulation of values in the image buffer. This determines number of particles and the mass of substance is displayed per pixel.

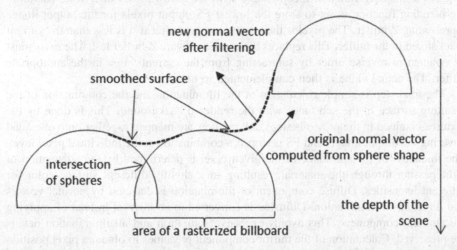

Fig. 4. Image buffer filtering generates a smooth surface and new normal vectors are created from original distances and normal vectors.

The resulting image buffers at the points of intersection of the individual spheres produce sharp edges, which need to be smoothed to make smooth surface (Fig. 4). This is achieved by the subsequent application of a bilateral filter to buffers containing depth, normal vectors and mass rendered from the viewpoint of the camera (step 3 in Fig. 3). The essence of filtering is to calculate the new pixel value in the image depending on its local environment. The bilateral filter is ideally indivisible for application in the horizontal and vertical plane separately. Therefore, accurate filtering results are achieved if the whole square area is processed for each pixel. This operation is very demanding for a large filter radius, so it may be more appropriate to split the filtering into horizontal and vertical processing even though the artefacts are created. The bilateral filter for smoothing uses two scales - the distance of the individual pixels in the image and their individual values in the color scale. Pixel with a very different color has minimal weight when filtering the image, although it may be neighboring. With increasing distance in the image its weight decreases. By using a bilateral filter, the silhouettes are maintained, smoothing the intersections near the surface is achieved, and in addition to it the amount of spatial detail in the image remains (Fig. 5).

Gaussian filter is used for comparison, which gives different graphical results when finalizing the image, especially in low thickness areas (Fig. 6). The Gaussian filter, unlike the bi-directional, can be applied sequentially in the horizontal and vertical planes with two passes through the display chain without any inaccuracies.

The addition of the shadows that cast the substance onto the background pad is provided by the shadow mapping method described in details in [9]. The process of displaying the particles is repeated from the viewpoint of a light source (step 4 in Fig. 3). In this case, the depth of the scene is sufficient to find the shaded area and accumulated mass of the substance to distinguish the shadow intensity. Obtaining these two results can be done in one rendering process, in spite the accumulation of thickness requires a blending function incompatible with the hardware Z-buffer test. Therefore the blending function is set to store the lowest FS output pixels for the output image representing Z buffer. The pixel is then written to the buffer if it is less than the current pixel stored in the buffer. This replaces the usual hardware Z-buffer test. The mass must be counted in reverse order by subtracting from the current value in the appropriate buffer. The actual value is then complemented to one.

The final step is simple calculation of the illumination and the combination of the resulting surface of the substance with the rendered background. This is done by the textures obtained in the previous steps. All textures are mapped together onto one quad covering the output image, and FS performs a combination at the individual pixel level. The thickness of the substance of the given pixel is determined by the attenuation of light passing through the material, resulting in a slightly different surface color for different intensities. Diffuse component of illumination is detected by normal vectors and direction of illumination. Diffusion is converted to an interval instead of applying the ambient component. This avoids excessive saturation and all illumination details are preserved. Calculation of the mirror component presumes to obtain a pixel position in the CCS from the Z-buffer by a reverse order to determine the vector directed to the camera and to determine the resulting light rays.

The surface of the substance also reflects and refracts rays of surrounding light. This phenomenon defining the ratio of light reflected back to the environment and refracted into a substance describes the Fresnel equations. Although they are physically correct, their calculation may be unnecessarily demanding and often replaced by a simpler Schlick approximation [10]. Using Schlick approximation enables to see how much light is reflected from the visible surface of the substance to the camera and how much is refracted into the substance volume. The light passing through the substance to the camera refracted at the distant boundary of the substance and its surroundings is determined by the modified background texture sampling. The xy component of the normalized normal vector is scaled by the material thickness and added to the pixel's uv coordinate, and the result is used as coordinates to sample the background texture.

The physically accurate representation of the refracted light is impossible in the given situation, and the described process gives a graphically highly facefull approximation of light refraction.

Fig. 5. Result of filtering of normal vectors. The left side contains the image before filtering, the right shows the result of the operation. New normal vectors credibly show true surface curvature and allow correct application of optical phenomena.

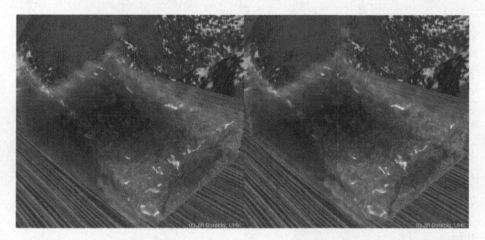

Fig. 6. Comparison of Bilateral and Gaussian filter are used for smoothing. They yield different graphical results in the final composition of the image, especially in low thickness areas

4   Computational Time Measurement Results

It is essential for the real-time interactive rendering that each part of the simulation and rendering process takes up as little time as possible. The following measurements are specifically focused to simulation and rendering time. The purpose is to monitor the computational demands of individual operations in order to determine their processing time and to determine the average load over the entire simulation cycle and rendering cycle. The OpenGL interface provides functionality for measuring the time of calculations processed on the GPU used for these purposes. This results in the real time passing through the visualization pipeline, or compute shader processing. The particle sorting time is measured by system time. No overhead due to program control and data transfer between CPU and GPU is counted. Its overall effect results from a comparison of the individual times and the total loop time.

The implementation is tested using the nVidia GeForce GTX 660 hardware and the Intel® Core ™ i5-3470 processor. The measurements use for simplicity, the default settings - three predefined particle numbers: 4 096, 16 384 and 65 536. The camera is set to be as close as possible to the liquid substance volume and at the same time could absorb during the simulation maximum possible number of particles. This ensures the same conditions for the rasterization of particle pixels, and rendering times can be compared. The obtained measurement results are summarized in Tables 1 and 2. Table 3 shows the course of the whole algorithm and gives the ratio of simulation time and display time compared to the entire program cycle. The image has the height and the width of 500 pixels. The average simulation and rendering times are determined from the first 500 cycles of a simulation run without any user interaction. The dynamism of the scene is as follows: The examined substance in the scene strikes the pad, its motion stabilizes and finally ends in the state of calm. Simulation time can be considered as a credible picture of implemented situation.

Table 1. Overview of the individual simulation steps separately. Values are the average processing time of the first 500 simulation cycles.

Number of particles	4 096		16 384		65 536	
Computational load	Time (s)	%	Time (s)	%	Time (s)	%
Sorting of particles	0.001368	11.57	0.006272	9.82	0.095146	12.73
Density calculation	0.004476	37.86	0.024150	37.83	0.289725	38.76
Calculation of forces, collisions, integration	0.005366	45.38	0.032387	50.73	0.359494	48.09
Overhead of simulation	0.000614	5.19	0.001037	1.62	0.003135	0.42
Total simulation time	0.011824	100.00	0.063846	100.00	0.747500	100.00

Tables 1 and 2 provide the detailed analysis of simulation and rendering phases. Particle sorting takes about 11% of the simulation time. Minor deviations may be caused by the operating system load. The total time increases about 5 times and 15 times with increasing the number of particles. Sorting does not take a lot of processing time, and this process can be accelerated using parallel sorting. The difficulty phase for the implementation is then generation of an NNS auxiliary texture. The course of the

Table 2. Overview of individual rendering steps separately. Values are the average processing time of the first 500 images.

Number of particles	4 096		16 384		65 536	
Computational load	Time (s)	%	Time (s)	%	Time (s)	%
Rendering from the perspective of light	0.000296	0.86	0.001161	3.13	0.004612	9.68
Rendering attributes of substance	0.000514	1.49	0.002019	5.44	0.008273	17.37
Scene rendering	0.000164	0.48	0.000168	0.45	0.000173	0.36
Application of a bilateral filter	0.031650	91.97	0.031827	85.79	0.032367	67.94
Gauss filter application	0.000157	0.46	0.000159	0.43	0.000160	0.34
Composition of the final image	0.000247	0.72	0.000256	0.69	0.000286	0.60
Rendering overhead	0.001384	4.02	0.001510	4.07	0.001769	3.71
Total rendering time	0.034412	100.00	0.037100	100.00	0.047640	100.00

Table 3. Overall algorithm overview. Values are the average processing time of simulation and rendering of the first 500 images.

Number of particles	4 096		16 384		65 536	
Simulation time (s)	0.011824	25.36%	0.063846	62.84%	0.747500	93.86%
Display time (s)	0.034412	73.79%	0.037100	36.51%	0.047640	5.98%
Overhead of the processing (s)	0.000396	0.85%	0.000656	0.65%	0.001266	0.16%
One loop time (s)	0.046632		0.101602		0.796406	
Image/frame rate (FPS)	21.44		9.84		1.26	
Processing time per 500 frames (s)	23.3		50.8		398.2	

first CS for density calculation occupies the same part of the cycle, approximately 38%. This phase can be considered stable and is only affected by the number of particles used. The processing time is increased almost 6 times and then 12 times. The remaining calculation of forces, collisions and data updates are the most demanding operation. They occupy almost half of the cycle. The use of 16 384 particles instead of 4 096 leads to an increased number of collisions with the container, which results in increasing of the processing time. If we use 65 536 particles, most of the additional particles fill the container volume, and the rest of the container does not cause significant load. The time amount required is extended approximately 6 times and then 12 times depending on the increasing number of particles. The simulation overhead is caused by program and

GPU control and, in particular, by the transfer of NNS auxiliary texture into the GPU memory. This texture is getting larger and its transfer takes longer depending on the decreasing number of particles but from the point of view of the entire simulation cycle its effect is decreasing. The four times increase of the number of particles results in increasing the processing time four times when rendering particles as substance attributes. The remaining operations process of preparation of images, and their load is almost constant. The most prominent part of the rendering cycle is application of the bilateral filter, because it is implemented for a better graphical result by a single pass with the filtering of the entire square pixel environment. It is however just a fraction of the load compared with the simulation time.

Table 3 provides comparison of simulation and rendering time with the average full- program loop. The rendering time takes up a minimum of total time and does not increase much depending on the increasing number of particles. The simulation time definitely causes the highest load and should be the first place to optimize or modify the design solution.

5   Conclusion

The article describes the original way of implementation and visualization of liquid substances using parallel GPU calculations with OpenGL graphical interface and summarizes the results of the implemented solution. The simulation phase utilized advanced graphics adapter functions for non-graphical computing using compute shaders. Special data structures represented by 1D and 2D textures were designed to compute position, velocity and density in multi pass GPU algorithm. Particle rendering was performed using the Screen Space method that rasterizes each particle as a quad object into image buffers, which represent the individual parameters needed for final image rendering. The individual buffers are additionally filtered to obtain a smoother surface appearance. The final result was achieved by combining the parameters of the liquid and the background image. Properly set rendering and simulation parameters resulted in high quality graphical results. The presented solution is a foundation for possible follow-up work and still offers plenty of room for expansion. Implementation can be generalized to simulate both fluids and gases or multiple different substances at the same time. The calculation is fast enough to allow simulation in real-time and can be further improved and accelerated.

Acknowledgements.   This work and the contribution were supported by a project of Students Grant Agency (SPEV) - FIM, University of Hradec Kralove, Czech Republic. The authors of this paper would like to thank Milan Košťák, a PhD student of Applied Informatics at the University of Hradec Kralove, for help with GPU implementation.

References

1. Ihmsen, M., Orthmann, J., Solenthaler, B., Kolb, A., Teschner, M.: SPH fluids in computer graphics (2014)
2. Kelager, M.: Lagrangian Fluid Dynamics Using Smoothed Particle Hydrodynamics. University of Copenhagen, Department of Computer Science (2006)
3. Akinci, G., Ihmsen, M., Akinci, N., Teschner, M.: Parallel surface reconstruction for particle-based fluids. Comput. Graph. Forum **31**, 1797–1809 (2012)
4. Fraedrich, R., Auer, S., Westermann, R.: Efficient high-quality volume rendering of SPH data. IEEE Transact. Visual. Comput. Graph **16**, 1533–1540 (2010)
5. Green, S.: Screen space fluid rendering for games. In: Proceedings for the Game Developers Conference., Moscone Center, San Francisco (2010)
6. Harada, T., Koshizuka, S., Kawaguchi, Y.: Smoothed particle hydrodynamics on GPUs. In: Computer Graphics International, pp. 63–70. SBC Petropolis (2007)
7. de Berg, M., van Kreveld, M., Overmars, M., Schwarzkopf, O.: Algorithms and Applications, pp. 1–17. Springer, Heidelberg (1997)
8. van der Laan, W.J., Green, S., Sainz, M.: Screen space fluid rendering with curvature flow. In: Proceedings of the 2009 Symposium on Interactive 3D Graphics and Games, pp. 91–98. ACM, New York (2009)
9. Williams, L.: Casting curved shadows on curved surfaces. In: Proceedings of the 5th Annual Conference on Computer Graphics and Interactive Techniques, pp. 270–274. ACM, New York (1978)
10. Schlick, C.: An inexpensive BRDF model for physically-based rendering. Comput. Graph. Forum **13**, 233–246 (1994)

A Proof of Concept Integrated Multi-systems Approach for Large Scale Tactile Feedback in VR

Daniel Brice[✉], Thomas McRoberts[✉], and Karen Rafferty[✉]

Queen's University Belfast, Belfast, UK
{dbrice01,wmcroberts02}@qub.ac.uk, k.rafferty@ee.qub.ac.uk

Abstract. It is well understood that multi-sensory stimulation can be used to enhance immersion in virtual environments such as Virtual Reality (VR) and Augmented Reality (AR). State of the art VR technologies have enhanced visual stimulation in aspects such as pixel density in recent years, whilst the area of haptics has remained less developed. The Ultrahaptics Evaluation Kit is a relatively new technology that consists of a 16×16 array of ultrasound transducers used to create ultrasound haptic sensations. We have developed a proof of concept large scale haptic system by integrating this device with an HTC VIVE, Leap Motion and Rethink Robotics' Baxter Robot to provide ultrasound haptic feedback in a volume greater than $1.5\,\mathrm{m}^3$ for VR users. The system was evaluated through a user study with 19 participants. The study focused on users' assessments of the location of the haptics produced by the system. The results of the study offer a means of validating the system, as well as providing comparisons in accuracy for a haptic perception task vs a visual perception task. There have been opportunities recognised for improving accuracy of the system. However, the system has been deemed suitable in creating haptic feedback for low fidelity models within large volumes in VR.

Keywords: Haptics · Ultrasound · Virtual Reality · Immersion · System integration

1 Introduction

In recent years there has been a rise in the popularity of Virtual Reality (VR) for both leisure and work. Modern releases of VR hardware bring advances in tracking area size, increasing user immersion through more spacious interactions. Visual stimulation is improved through the increase of pixel density and reduction of latency to the Head Mounted Displays (HMD). Multi-sensory feedback on the other hand is developing at a slower rate. Two of the most ubiquitous headsets on the market are the HTC Vive and the Oculus Rift, both of which provide hand-held controllers. Through the use of small motors these controllers

© Springer Nature Switzerland AG 2019
L. T. De Paolis and P. Bourdot (Eds.): AVR 2019, LNCS 11613, pp. 120–137, 2019.
https://doi.org/10.1007/978-3-030-25965-5_10

are capable of providing vibrational feedback with ranging magnitude and frequency. Though this can provide feedback over the full VR volume it is still limited and is not capable of offering high fidelity tactile feedback to the user's hands.

There are many applications for VR within professional occupations such as surgical training where the impact of AR and VR is already anticipated to be large [24]. Consequently there have been haptic systems developed for medical purposes [11,22]. VR training is not limited to medical use. Training in VR has also been investigated in areas such as sports [30] and hazardous environments such as mining [32]. Haptics have been shown to be capable of providing an alternative method to sight for spatial awareness in VR. This can be seen in design of HMDs with haptics [16] and other wearable haptic devices [8].

A recent technology brought to market is the Evaluation Kit from Ultrahaptics, developed by Bristol University [3], a 16 × 16 array of ultrasound transducers capable of providing tactile feedback through a phenomenon called acoustic radiation force [15]. The Evaluation Kit has been shown to be capable of rendering distinguishable 3D shapes during user trials [18]. The ultrasound array can generate haptic sensations with a beam width of 60° at a range of 15–80 cm [31].

Another area that has made great progress in recent years is that of collaborative robotics. Within the manufacturing industry there has been a drive for robots capable of working alongside humans. This has led to the development of robots that are considered safe to physically interact with during operation, with safety features such as torque sensing and elastic actuation. One such example of this is Baxter from Rethink Robotics, a robot with two large 7-Degree of Freedom (DoF) arms used in manufacturing pick and place tasks [25]. Applications for these robots are not limited to manufacturing, they can also provide safe platforms for haptic interactions in VR.

VR has moved over recent years from desk based experiences to larger room scale interactions. However, many of the state of art haptic technologies are still limited to small volumes. Those technologies that operate over larger volumes are limited to vibrations across handheld controllers and vibrotactile gloves which require the user to be tethered to equipment. In an attempt to investigate the capabilities of haptics for large volumes in VR we have developed a proof of concept system, providing ultrasound haptic feedback in a volume greater than $1.5\,m^3$. The system combines the Ultrahaptics Evaluation Kit with a Baxter robot's 7-DoF arm, a Leap Motion tracker [21] and an HTC Vive HMD in order to produce haptics sensations over a large volume for VR users. A primary user study has been conducted as a means of validating the system, as well as investigating comparisons between a haptic sensory/visual sensory task.

We have developed a proof of concept system integrating existing technologies to deliver haptic feedback over a large volume in VR. There are solutions for providing tactile feedback to users in VR, but only few are capable of delivering feedback over such a large volume ($1.5\,m^3$). In this paper the development will be reviewed. The system prototype will be introduced, followed by technical details of the implementation. Challenges identified during the development process will be discussed. A user study carried out in order to provide a means of validating the large mid-air haptic system will be presented. Finally conclusions will be drawn followed by intentions for further work.

The key contributions of our work are:

1. For the first time we have developed a system that enables ultrasonic feedback to be used to deliver haptics over a large volume in VR (1.5 m^3), enabling richer multi-sensorial experiences.
2. We have utilised the human in the loop as a means of validating the capabilities of the system through user perception.

2   Related Work

2.1   Force Feedback Haptics

One popular haptic platform capable of rendering 3D shapes is the professional grade Omni Phantom [10]. The device is capable of providing 3.3 N force feedback through a stylus interface through a series of articulating linkages. The Phantom is most suitable for high precision tasks involving small movements, due to it's limited 160 mm × 120 mm × 70 mm interaction volume. The system has been shown to be effective in applications such as surgical training and is a popular platform for haptics in small volumes [23]. For larger interaction volumes the Omni is not as well suited.

When it comes to wearable haptic interfaces for VR there have been many demonstrations of haptic glove technologies. These can come in the form of vibrotactile prototypes [20], capable of delivering tactile sensations to users. They also exist in the form of exoskeleton gloves, such as the commercially available Dexmo glove from Dexta Robotics [7]. These exoskeleton gloves are typically used to provide force feedback over large volumes in VR [1,19]. Haptic glove technologies are naturally suitable for large volume haptics, however they require that the user always be tethered to the device.

A haptic device capable of replicating curved geometry in VR was created in the work of Covarrubias and Bordegoni [5]. Through the use of a servo-actuated metallic strip different curvatures can be created and controlled in runtime from Unity 3D. The method is effective in recreating geometry for haptics at a fixed location in VR, but is limited to a smaller interaction volume.

Knopp et al. have also looked at utilising collaborative robots as a haptic interface where they have integrated a Kuka LBR iiwa robot with VR in order to provide haptic feedback during hip replacement VR training surgery [17]. In their work they too recognised the potential large collaborative robot arms offer in haptics. For their specific application of VR surgery training contact haptics, holding the robot end effector, was best suited in order to replicate large forces.

In similar work a system was designed to provide force feedback haptics over a large area to users [2]. This system was demonstrated through a VR exposure therapy application for acrophobia, where a user could climb up a wall in VR by grabbing a number of physical ledges held by the Baxter Robot. This system relied on creating custom end effectors to replicate geometry of objects in VR, taking preparation time and limiting the range of the haptic sensations plausible during a single interaction.

2.2 Mid-Air Haptics

When it comes to Augmented Reality (AR) many of the existing haptic interfaces fall short in contributing to immersion due to the user's being able to see them. In this field there is a natural synergy with mid-air haptic technologies. These technologies come in a few forms. There are air jet based technologies which have been shown to lack accuracy [28] and are incapable of producing 3D feedback. Air vortices can be used to produce sensations for users, however there are latency issues and fidelity issues present [26]. Focal points capable of being detected by users have been created using ultrasound arrays [13]. Work has been carried out demonstrating an ultrasound array capable of rendering 3D shapes with high fidelity by controlling focal points of emitted ultrasound waves [18]. This work in modulating ultrasound from an array of transducers has led to the product design of the Evaluation Kit from Ultrahaptics. Our research has sought to utilise the board to produce tactile feedback over a large volume to users in, VR.

3   A Tactile Feedback System for VR

3.1   Concept

The goal of the research was to integrate existing technologies to develop a system capable of delivering tactile feedback to users over a large volume for 3D shapes in VR. The system demonstrator consists of users being in a space themed Virtual Environment (VE) where they are able to reach out and interact with celestial objects through the addition of ultrasound haptic feedback, as shown in Fig. 1. During development we have employed techniques to deal with practical challenges in the alignment of physical and virtual systems. We have also determined limitations in the technologies employed, with problem solving techniques suggested.

3.2   Applications

The main function of the system is to enhance immersion of VR users through the addition of tactile feedback. It can therefore be considered as a platform for many VR applications.

The contribution of haptics from a system such as ours can benefit in immersive experience. Haptics in magical VR experiences have been shown to be effective [12]. This has provided motivation to use Space as a theme for a haptics demonstrator.

VR has also proven to be effective in exposure therapy for those with phobias [4,27,29]. In such therapies haptics normally come in the form of manually

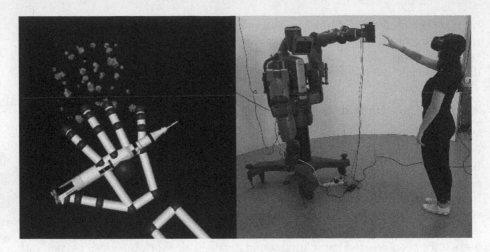

Fig. 1. A user interacting with the Saturn V model.

applying contact to the patient at set points during VR interactions. Introducing an automated haptics system for VR exposure therapy could be very effective in immersing participants with their fears.

3.3   Capabilities

The system is capable of rendering tactile feedback for moving 3D shapes over a volume greater than $1.5\,m^3$. These 3D haptic sensations can been synced and scaled with visual meshes in VR. The haptic feedback provided for these sensations comes in one direction, the reverse of the Baxter robot's forward direction.

Within the system users are capable of seeing their hands at all times whilst still being able to encounter haptic feedback at different locations.

3.4   System Implementation

Communication. The ubiquitous Unity 3D is used to bring together incorporated technologies to provide the large volume haptic feedback system. The Ultrahaptics, Leap Motion and HTC Vive communicate with one another through written C# scripts during runtime. There is also a bi-directional Web-Socket server, RosBridge, which is used for communicating between Unity and ROS running on a Virtual Machine(VM) to control the Baxter robot.

Within ROS there is a nodal publish/subscribe message protocol. This is used for publishing the desired arm position (1–5) from Unity ROSbridge server, where it is redirected to a custom c++ ROSNode on a seperate VM. The ROSNode's role is to receive positions and move the arms to preset IK positions. A high level overview of the data flow can be seen in Fig. 2.

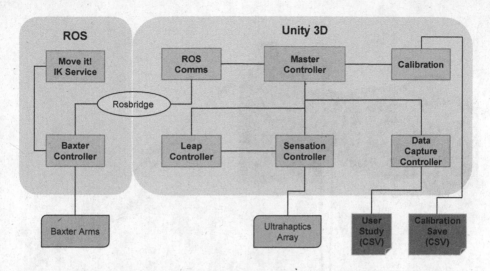

Fig. 2. An overview of the system nodes and scripts.

Robot Arm Control. The Baxter robot, operating on the Robot Operating System (ROS) framework, was chosen as a safe robot to position the board in space during runtime. The robot has been used in many applications and though its precision is not the best for a collaborative robot [6] it is still accurate enough for our application. Each arm has a reach of over 1.2 m and is capable of carrying a payload up to 2.2 kg. The arm is safe for humans to be in contact with during operation, whilst providing an end effector tolerance precision of 5 mm. In our work we have chosen to use the MoveIT IK Library [9] for solving Inverse Kinematics (IK) for the robot arms.

When using the system there is an emphasis on safety of participants as they cannot see the robots in front of them due to the VR HMD. The board positions were always far enough from Baxter that the outreached arms could not extend further and make contact with users. The Baxter robot arms also have two torque sensing features on them, one whereby a crushing force (slowly incrementing) can be measured from the arms, and another where a sudden impact force if measured. In both cases the arms deactivate.

The additional 520 g mass from the Ultrahaptics board is added to the URDF file on ROS. This is so that the arm can compensate in torque applied to the motors to reach end effector positions.

Leap Motion. The Ultrahaptics product comes with a Leap Motion tracker integrated above the Board. Initial versions of our system utilised this tracker with poor results. The tracker has a Field Of View (FoV) of 150° wide and 120° deep. When testing the board at varying heights the users hand would move to the edge of FoV when close to the board at higher positions on the robot. The user's palm orientation would often make it difficult for the Leap to track

Fig. 3. Example of positioning of hand relative to board causing tracking issues. The red line indicating Leap Motion line of sight. (Color figure online)

based on the Leap's position, see Fig. 3. The result of this was the VR hand models twisting and jerking unnaturally. A decision was made to instead mount a separate leap motion to the head for hand tracking. The orientation of the hand better suited tracking from the HMD position, where users could see their hands at all times with fewer tracking issues.

In order to render 3D shapes with the ultrasound board the distance between the board and the hand must be known. The Leap tracker tracks positions using a right-handed coordinate system. Unity on the other hand uses a left-handed coordinate system, as does the transform representing the Leap in VR. To map between the coordinate systems a number of transformations are performed.

One issue with the proposed technique is determining the position of the Leap tracker in space. When the Leap is attached to the board it is situated at a known offset from the array center point. When the Leap tracker is mounted to the front of the HTC Vive HMD the offset from the 6-DoF tracked point for the HMD in VR is unknown. There is no technique to attain the value of this offset and instead it is assessed visually through observing hand models in VR.

A second challenge with the proposed Leap positioning is when the haptic sensation requires hand tracking, as is the case for rendering 3D haptic shapes. This means that the user cannot look away from their hands during haptic interactions. If they look away hand tracking is lost and there is no known location to target the haptic sensation in space.

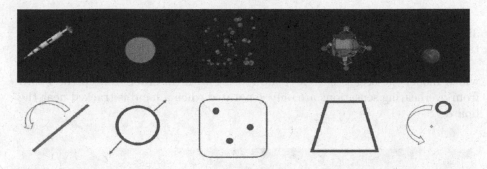

Fig. 4. Concept drawing for haptic sensations. Order from left to right; Saturn V, Star, Asteroids, Lunar Lander and Earth.

Haptic Sensations. The Ultrahaptics Evaluation Kit was chosen as a platform for producing the haptics due to its ability to produce high fidelity haptics at low latency. Ultrahaptics recently released a Unity Core Asset (UCA) in closed beta for Unity developers to use for creating sensations. Using the Ultrahaptics API library, and the UCA, 2D lines and circles are capable of being generated in the form of moving ultrasound focal points.

3D shapes can then be rendered by manipulating the focal points on 2D planes based on distance from the hand to the board and the virtual object, as previously shown to be successful [18].

A total number of 5 haptic sensations were developed using the UCA, each corresponding to a different space-themed object. A conceptual drawing for these sensations and their model counterparts can be seen in Fig. 4.

For the Saturn V NASA rocket a 2D line sensation was generated, scaled to roughly match the visual mesh. The model pivots about its centerpoint clockwise, the focal points for the line sensation change position in a relationship with the visual mesh. This corresponds in the haptic line remaining on the rocket while it rotates.

The star sensation is a 3D sphere scaled to fit the mesh, which varies in size over time with a sine function. The sphere radius for the 2D haptic circle emitted by the array was related to the mesh scale, resulting in the sensation staying attached to the model. Position of the palm is then used to produce the a 2D haptic circle of a radius varying based on palm distance from the board. The overall effect of this is the haptic rendering of a 3D sphere.

The Asteroids are a thin cuboid of small asteroid models. Small 2D haptic circles targeting the hand were constructed. This was done to immerse the user with the idea the asteroids were colliding with them.

For the Lunar Lander a conical frustum 3D shape was placed over the visual mesh. Where the radius of a 2D haptic circle would vary size as the user moved hand from one side to the other.

The Earth model orbits a central point. To create a matching sensation a 3D sphere, scaled to match the Earth model, was positioned with an orbit over time matching the Earth's.

Only one sensation can be generated at a time with the array, depending on which model the user is currently interacting with. In order to keep the board from overheating sensations are only generated when a hand is tracked near the board.

Fig. 5. Custom 3D printed parts. On the left side is the array grip, on the right side is the Baxter adapter.

Attaching the Board. In order to mount the board to the robot arm we designed our own 3D printed custom end effectors for the Baxter robot. The custom design has two functional parts, see Fig. 5. The first part is used to provide a flush surface with the end of the robot arm's cuff. The second part is designed to grip the Ultrahaptics array by its circuit board. The mass of the board is 520 g and is within the strength capabilities of the nylon printed gripper, as well as within the 2.2 kg payload specification for the Baxter robot. There is a slot for a 180 mm PC fan in the 3D print in order to provide additional cooling for the Ultrahaptics board between interactions.

Mapping the Virtual Environment to the Real World. In order to align ultrasound haptic feedback with virtual objects in VR we developed a calibration technique for position and orientation between real world and VR. The first purpose of the technique was to identify the position of the physical Ultrahaptics board in VR at 5 different pre-set arm positions on the Baxter Robot. The second purpose was to determine the forwards direction of the board in VR space, as the haptic sensations are generated relative to this.

5 positions were set for the virtual objects by moving the Baxter robot arm into 5 different configurations across an area of $1.5\,m^2$ and recording the end effector Cartesian coordinates using Forward Kinematics (FK). These positions were determined to be reachable for users ranging in height. Once the positions were set it was time to determine where these positions existed within the VR world, and where haptics placed in front of the array would occur in VR.

The calibration tool was programmed in C# for Unity. The technique employed was to have the Baxter robot move through the 5 different positions sequentially, each time waiting at a position for a set time before moving on to the next. The robot arm held the Ultrahaptics board with the 6-DoF HTC Vive Tracker [14] on the board center point and moved through each position by a set number of cycles. The spatial, (X, Y, Z), information acquired through the tracker was then averaged over each position to determine board position in VR.

The 5 positions are cycled 5 times with the positions recorded. The maximum change in repeatability of any position in any one VR World dimension was $14.2\,mm$. All other ranges in repeatability of (X, Y, Z) were less than $5\,mm$, giving validation in the positioning of the array in VR space.

With the average position of the board (1–5) now known there is still the issue of determining the forwards direction of the end effector in VR so that the haptic sensations generated in front of the board are in the correct position. All 5 board positions are on the same plane $0.9\,m$ in front of Baxter. Ignoring height dimension from the positions presents a set of 2D data points. Then using the Math.net library regression analysis can be undertaken to determine a line of best fit for the 2D points. This equation can then be used to generate 3 positional vectors on the line with different heights. These generated points are then used within Unity libraries to generate a plane between them, with a normal vector representing Baxter's forwards in VR.

4 Haptic Validation User Study

4.1 Purpose

There is a challenge in validating a system which produces ultrasound haptic feedback in VR. Traditional force feedback haptics typically have physical geometries which can be measured with different instrumentation to assess positioning. When it comes to ultrasound haptic feedback physical experimentation can be undertaken by observing the effects of the ultrasound on a liquid such as oil [18].

As our system is to be used by humans it was decided that the positioning of haptics should therefore be validated through user testing. The intended outcomes were to find the differences between where we were attempting to place ultrasound focal points and where users perceived these points to be.

A similar visual condition was carried out in to provide a baseline of results free from error in misalignment of the ultrasound board position between real and virtual environments.

4.2 Participants

There were a total of 19 participants involved in the Haptic Validation study. The majority of these participants were of an engineering background; however, only a couple had previously experienced ultrasound haptic feedback. All participants were between the age of 20 and 60. The participants ranged in height from approx 180 cm down to 167 cm. Out of the 19 participants 14 were male (74%) and 5 were female (26%). When asked, none of the participants claimed to have had any problems with mobility in their arms or sensitivity in their hands. Those who wore glasses were able to keep them on during studies.

4.3 Apparatus and Setup

The user study was conducted on the developed system. The equipment involved consisted of the HTC Vive, Leap Motion tracker, Ultrahaptics board and Baxter robot. The study was run using C# scripts in Unity 3D. The study was carried out over the course of a week, resulting in different instances of Steam VR being used with an initial calibration performed on the first instance.

Hand positions were tracked via a Leap tracker mounted on the VR HMD, these were then transformed to find the position relative to the Ultrahaptics board in its own coordinate frame.

The role of the Baxter robot in the study was to move the Ultrahaptics board to different positions in space. These positions were set to be within a $1\,m^2$ area. It was decided to not use the extremities ($1.5\,m^2$ volume) of the system as this could potentially cause implications in additional movement of users around the robot and ergonomic challenges.

4.4 Methodology and Protocol

Haptic Condition. The task for the participants was to find the strongest tactile feedback point around a location in VR. In order to locate the haptic point in space the participant could see only a small 1 cm radius blue sphere in space. They were told that the haptic point would be within 30 cm of this, this was done to give a starting point for the user to feel around in space.

Each participant was given a training attempt at finding the point in space so that they would understand what they were searching for. Once they understood this we would indicate, by touching, a central position on their palm. They were then informed that they would need to move their hands in space until they were satisfied that the point felt strongest in the center of their palm.

The haptic sensation created for the user study was a small 2D circle, of radius 2.5 mm. This gave the effect of creating a small haptic point in space. The intensity for the sensation was set to 1, and draw frequency set at 70 Hz.

Once the participants demonstrated they understood the concept the process begun. The process was to place the haptic sensation at 7 different positions relative to the board. These positions were the top, bottom, right, left, back, front and center of a sphere offset 20 cm from the board unknown by the participants. These 7 assessments of position were repeated 5 times with different Baxter arm configurations. The virtual sphere the haptic points were generated on was set to a random radius varying from 2 cm to 6 cm, in increments of 1 cm, at each IK position.

The Leap Motion tracker was used to provide hand tracking. Participants were informed of this and told to ensure they could see a graphic model for their hand when they were happy with their palm position.

When a participant felt they had situated the haptic point in the center of their palm they gave a verbal cue, the person conducting the study would then press a key on the PC resulting in the position of the palm being recorded. The sphere indicating approximate location of the haptic point would then change color, so as to indicate registry of the position and that the point had now moved. The time taken to assess each point was recorded.

In the interest of safety the robot was programmed to not move to a new position until the person conducting the study had informed the user to move their hand back to resting position.

Fig. 6. The participant view during user study visual condition. (Color figure online)

Visual Condition. Each participant would also complete a visual condition, where a blue sphere was presented to the user in space with an adjacent floating green cuboid, see Fig. 6. Participants were asked to line up the center point of the palm position sphere, seen in the VR model for their hands, with the point on the sphere closest to the green cuboid. Each participant was given a training run and opportunities for questions to demonstrate they understood the process.

The rest of process for the visual condition was identical to the haptic condition. Each participant would assess 7 points on a sphere that they could now see, the only point they could not see was the center point of the sphere they were assessing. Each time the participant was satisfied with their palm position they gave a verbal cue and palm position was recorded, the cuboid would then move to indicate a new target position on the sphere. The virtual spheres, were randomised in size as in the haptic condition and situated at each of the same positions the haptic ones were, i.e. in front of different robot arm configurations in space.

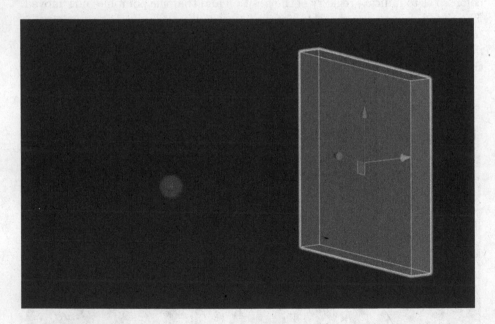

Fig. 7. The array coordinate system; X is the red arrow, Y is the green arrow and Z is the blue arrow. Blue sphere indicates region of haptics. (Color figure online)

4.5   Results

Palm Accuracy. Each study involved the assessment of 7 points on a virtual sphere at each of the 5 sphere positions. With a total of 19 participants there were a total of 665 points recorded for both visual and haptic conditions. 6 results were void as a result of the hand not being tracked during recording.

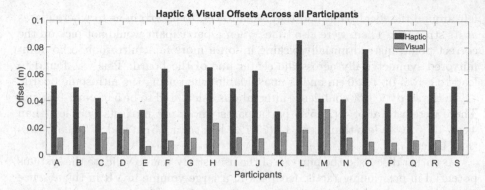

Fig. 8. A bar graph showing the accuracy in haptic and visual conditions for each participant.

The results from the study were recorded as the absolute difference between the theoretical position of the target point and the recorded palm position for visual & haptic conditions. These results were recorded in the form of Δ X, Δ Y and Δ Z in the array's frame of reference, see Fig. 7.

The mean values for the haptic condition Δ X, Δ Y and Δ Z were 1.32 cm, 2.91 cm and 2.87 cm respectively. The Δ X, Δ Y and Δ Z standard deviations were 1.51 cm, 2.25 cm and 2.47 cm. Combining the Δ values for each participant into single distance vectors and taking a mean for the haptic condition results in an offset of 5.08 cm.

For the visual condition the Δ X, Δ Y and Δ Z mean values were 0.65 cm, 0.83 cm, 0.66 cm respectively. Their standard deviations were 1.08 cm, 1.39 cm and 1.42 cm. The combined mean offset distance derived from the mean of each Δ value is 1.50 cm.

A distance between intended position and the position chosen by participants for visual & haptic conditions was derived from respective vector components for each trial. A mean distance offset, derived from distances for each robot arm position, for each participant can be seen in Fig. 8.

4.6   Discussion

The visual mean offset, 1.50 cm, is accurate as expected. This offset is formed from a combination of human error and the offset of the Leap from the HMD position in VR. Without being able to physically measure this, instead relying on visual inspection, the Leap offset error cannot be avoided.

The mean offset found from the haptic condition was 5.08 cm, determined from the Δ vector components. This offset is contributed to by the same errors as with the visual condition, human error and the leap-HMD offset error. There is also the error resulting from the position of the virtual and real Ultrahaptics boards being slightly offset from one another. Participants found it harder to pinpoint the haptic focal point with confidence as there would be a haptic force felt around the focal point and they would need to ascertain as to where it was at its strongest. There were also times when a participant would not pick up the correct point, instead initially picking up on a more faint ultrasonic echo point mirrored symmetrically across the center line of the board. This was found to lead to errors up to 20 cm and is unavoidable when using the ultrasonic board.

In the study a 95% confidence interval was calculated to be $5.08\,cm \pm 0.67\,cm$. There were a total of 4 (21.5%) participants whose mean offsets were less than 4 cm, 10 (52.6%) less than 5 cm, 14 (73.7%) less than 5.5 cm and 17 (89.5%) less than 6.5 cm.

As earlier discussed the purpose of the user study was to validate the systems potential in positioning tactile feedback in a large volume in VR in the absence of alternative suitable methods. The results of the study have shown that the system is capable of delivering the feedback to within 5.08 cm on average. The visual accuracy of 1.50 cm provides a contrast enabling the significance of the accumulating errors at fault, discussed above, to be measured. The accuracy of the system is still determined as suitable for providing multi-sensorial feedback on low fidelity models across the large $1.5\,m^2$ volume in VR.

5   Conclusion

We set out to develop a proof of concept system capable of providing tactile feedback for VR users over a large volume to users in VR. We accomplished this by integrating an HTC Vive system, a Leap Motion, a Baxter robot and an Ultrahaptics Evaluation Kit. With the Baxter robot we are able to mobilize the Ultrahaptics board, enabling haptic feedback in an area greater than $1.5\,m^3$. This was done without compromising popular features in VR such as the user being able to see their hands.

To address the challenge of validating the precision of mid-air haptics in VR over the large volume a user study was conducted. Users were given the task of identifying the position of haptic feedback in space with their hands. The study across 19 participants produced a mean offset of 5.08 cm between where haptic points were placed and where they were participants determined them to be. We believe the accuracy of the system can be improved by changing from the calibration run at start time to a real time tracking system for the Ultrahaptics board. However, for low fidelity haptic shapes across large volumes in VR the system can be utilised effectively.

A demonstration has successfully been created where users can feel a range of 5 different haptic sensations matching different space-themed objects' meshes. The sensations are synced with objects in both size and movement. The demonstration provides tactile feedback for these objects in the absence of any wearables.

There are still opportunities for improvements regarding accuracy of our system. However, we have found, based on observation of users, that the system is viable in providing enjoyable multi-sensorial experiences for users.

6 Future Work

Using data collected on reaction times of people in our system an investigation has begun into developing the system into a reactionary one. This being such that the user does not need to wait for the robot to position itself, instead being able to dictate the position of the haptic feedback themselves through natural movements. It is anticipated this will entail advanced robot arm control and predictive analysis on human movement.

A second area of improvement for the system is to take the developed knowledge and apply it to a mobile robot. Though this would bring a lot of additional complexity to the system, it would permit the user to receive haptic feedback in a 360° field.

References

1. Blake, J., Gurocak, H.B.: Haptic glove with mr brakes for virtual reality. IEEE/ASME Trans. Mechatron. **14**(5), 606–615 (2009). https://doi.org/10.1109/TMECH.2008.2010934
2. Brice, D., Devine, S., Rafferty, K.: A novel force feedback haptics system with applications in phobia treatment (2017)
3. Carter, T., Seah, S., Long, B., Drinkwater, B., Subramanian, S.: Ultrahaptics: multi-point mid-air haptic feedback for touch surfaces. In: UIST 2013 Proceedings of the 26th Annual ACM Symposium on User Interface Software and Technology, UIST 2013. Association for Computing Machinery (ACM), October 2013. https://doi.org/10.1145/2501988.2502018
4. Corbett-Davies, S., Dünser, A., Clark, A.: Interactive AR exposure therapy. In: Proceedings of the 13th International Conference of the NZ Chapter of the ACM's Special Interest Group on Human-Computer Interaction, CHINZ 2012, p. 98. ACM, New York (2012). https://doi.org/10.1145/2379256.2379282
5. Covarrubias, M., Bordegoni, M.: Immersive VR for natural interaction with a haptic interface for shape rendering. In: 2015 IEEE 1st International Forum on Research and Technologies for Society and Industry Leveraging a Better Tomorrow (RTSI), pp. 82–89, September 2015. https://doi.org/10.1109/RTSI.2015.7325075
6. Cremer, S., Mastromoro, L., Popa, D.O.: On the performance of the Baxter research robot. In: 2016 IEEE International Symposium on Assembly and Manufacturing (ISAM), pp. 106–111, August 2016. https://doi.org/10.1109/ISAM.2016.7750722
7. Dexmo: Dexmo (2008). http://www.dextarobotics.com/. Accessed 14 Feb 2018
8. Dunbar, B., et al.: Augmenting human spatial navigation via sensory substitution. In: 2017 IEEE MIT Undergraduate Research Technology Conference (URTC), pp. 1–4, November 2017. https://doi.org/10.1109/URTC.2017.8284172
9. Garage, W.: MoveIt (2008). http://moveit.ros.org/. Accessed 14 Feb 2018
10. Geomagic: Phantom Omni (2011). http://www.geomagic.com/en/products/phantom-omni/overview. Accessed 14 Feb 2018
11. Guo, J., Guo, S.: A haptic interface design for a VR-based unskilled doctor training system in vascular interventional surgery. In: 2014 IEEE International Conference on Mechatronics and Automation, pp. 1259–1263, August 2014. https://doi.org/10.1109/ICMA.2014.6885880

12. Han, P.H., et al.: OoEs: playing in the immersive game with augmented haptics. In: ACM SIGGRAPH 2016 VR Village, SIGGRAPH 2016, p. 15:1. ACM, New York (2016). https://doi.org/10.1145/2929490.2929895

13. Hoshi, T., Takahashi, M., Iwamoto, T., Shinoda, H.: Noncontact tactile display based on radiation pressure of airborne ultrasound. IEEE Trans. Haptics **3**(3), 155–165 (2010). https://doi.org/10.1109/TOH.2010.4

14. HTC: HTC Vive (2011). https://www.vive.com/uk/. Accessed 14 Feb 2018

15. Iwamoto, T., Tatezono, M., Shinoda, H.: Non-contact method for producing tactile sensation using airborne ultrasound. In: Ferre, M. (ed.) EuroHaptics 2008. LNCS, vol. 5024, pp. 504–513. Springer, Heidelberg (2008). https://doi.org/10.1007/978-3-540-69057-3_64

16. de Jesus Oliveira, V.A., Brayda, L., Nedel, L., Maciel, A.: Experiencing guidance in 3D spaces with a vibrotactile head-mounted display. In: 2017 IEEE Virtual Reality (VR), pp. 453–454, March 2017. https://doi.org/10.1109/VR.2017.7892375

17. Knopp, S., Lorenz, M., Pelliccia, L., Klimant, P.: Using industrial robots as haptic devices for VR-training. In: 2018 IEEE Conference on Virtual Reality and 3D User Interfaces (VR), pp. 607–608, March 2018. https://doi.org/10.1109/VR.2018.8446614

18. Long, B., Seah, S.A., Carter, T., Subramanian, S.: Rendering volumetric haptic shapes in mid-air using ultrasound. ACM Trans. Graph. **33**(6), 181:1–181:10 (2014). https://doi.org/10.1145/2661229.2661257

19. Ma, Z., Ben-Tzvi, P.: RML glove–an exoskeleton glove mechanism with haptics feedback. IEEE/ASME Trans. Mechatron. **20**(2), 641–652 (2015). https://doi.org/10.1109/TMECH.2014.2305842

20. Martínez, J., García, A., Oliver, M., Molina, J.P., González, P.: Identifying virtual 3D geometric shapes with a vibrotactile glove. IEEE Comput. Graph. Appl. **36**(1), 42–51 (2016). https://doi.org/10.1109/MCG.2014.81

21. Motion, L.: Leap Motion (2008). https://www.leapmotion.com/. Accessed 12 June 2018

22. Neupert, C., Matich, S., Scherping, N., Kupnik, M., Werthschützky, R., Hatzfeld, C.: Pseudo-haptic feedback in teleoperation. IEEE Trans. Haptics **9**(3), 397–408 (2016). https://doi.org/10.1109/TOH.2016.2557331

23. Peng, W., Xiao, N., Guo, S., Wang, Y.: A novel force feedback interventional surgery robotic system. In: 2015 IEEE International Conference on Mechatronics and Automation (ICMA), pp. 709–714, August 2015. https://doi.org/10.1109/ICMA.2015.7237572

24. de Ribaupierre, S., Eagleson, R.: Editorial: challenges for the usability of AR and VR for clinical neurosurgical procedures. Healthc. Technol. Lett. **4**(5), 151–151 (2017). https://doi.org/10.1049/htl.2017.0077

25. Rethink Robotics: Baxter specifications (2008). http://www.rethinkrobotics.com/baxter/tech-specs/. Accessed 14 Feb 2018

26. Sodhi, R., Poupyrev, I., Glisson, M., Israr, A.: AIREAL: interactive tactile experiences in free air. ACM Trans. Graph. **32**(4), 134:1–134:10 (2013). https://doi.org/10.1145/2461912.2462007

27. Strickland, D., Hodges, L., North, M., Weghorst, S.: Overcoming phobias by virtual exposure. Commun. ACM **40**(8), 34–39 (1997). https://doi.org/10.1145/257874.257881

28. Suzuki, Y., Kobayashi, M.: Air jet driven force feedback in virtual reality. IEEE Comput. Graph. Appl. **25**(1), 44–47 (2005). https://doi.org/10.1109/MCG.2005.1

29. Thanh, V.D.H., Pui, O., Constable, M.: Room VR: A VR therapy game for children who fear the dark. In: SIGGRAPH Asia 2017 Posters, SA 2017, pp. 52:1–52:2. ACM, New York (2017). https://doi.org/10.1145/3145690.3145734
30. Tsai, W.L.: Personal basketball coach: tactic training through wireless virtual reality. In: Proceedings of the 2018 ACM on International Conference on Multimedia Retrieval, ICMR 2018, pp. 481–484. ACM, New York (2018). https://doi.org/10.1145/3206025.3206084
31. Ultrahaptics: Ultrahaptics technical white paper. Technical report, Bristo, l United Kingdom (2018)
32. van Wyk, E., de Villiers, R.: Virtual reality training applications for the mining industry. In: Proceedings of the 6th International Conference on Computer Graphics, Virtual Reality, Visualisation and Interaction in Africa, AFRIGRAPH 2009, pp. 53–63. ACM, New York (2009). https://doi.org/10.1145/1503454.1503465

Virtual Simulator for the Taking and Evaluation of Psychometric Tests to Obtain a Driver's License

Jorge S. Sánchez[1(✉)], Jessica S. Ortiz[1(✉)], Oscar A. Mayorga[1(✉)],
Carlos R. Sánchez[1(✉)], Gabrilea M. Andaluz[2(✉)],
Edison L. Bonilla[1(✉)], and Víctor H. Andaluz[1(✉)]

[1] Universidad de las Fuerzas Armadas ESPE, Sangolquí, Ecuador
{jssanchez, jsortiz4, oamayorga, crsanchez9, elbonilla,
vhandaluz1}@espe.edu.ec
[2] Universidad Internacional del Ecuador, Quito, Ecuador
gaandaluzor@uide.edu.ec

Abstract. This article describes the development of a virtual reality application that allows the application of psychometric tests to drivers, both professional and non-professional, who obtain their driver's license for the first time or are in the process of renewing or changing the category of their license; these implemented evaluations are aimed at measuring the reactions of drivers, as various audio-visual stimuli are applied in a more user-friendly manner. These implemented tests were developed in function of the traditional tests, that is to say complying with the objectives to be measured and evaluated; these evaluations have been developed with a Unity 3D graphic engine able to simulate the practical driving tests, to obtain results a numerical calculation software is used that presents graphs and statistics of the results obtained. The application is very attractive and allows users to perform their evaluations in a way that is close to reality, the tasks to be developed are performed with equipment that are very common for the user such as the steering wheel and pedals.

Keywords: Virtual reality · Psychometric tests · Drive test

1 Introduction

Traffic accidents can be understood as: "an event, general involuntary, generated at least by a moving vehicle, which causes damage to persons and property involved in it" [1]; human factors are the cause of the highest percentage of accidents having 77% according to the ANT (National Traffic Agency), due to circumstances how: reckless maneuvers, omission by the driver, speeding, physical health, loss of motor skills and reaction [2], these failures are not detected at the time of acquiring a license since drivers choose to pay third parties to perform psychosensometric tests to obtain their driver's license.

In several countries tests are carried out to evaluate the abilities of the driver at the time of controlling a vehicle, these tests are called psychosensometric, the characteristics that are evaluated of the person are his sensory abilities, his motor capacity, and his cognitive and psychological dexterity. Dividing the previous ones in activities such

© Springer Nature Switzerland AG 2019
L. T. De Paolis and P. Bourdot (Eds.): AVR 2019, LNCS 11613, pp. 138–149, 2019.
https://doi.org/10.1007/978-3-030-25965-5_11

as: *(i) multiple reactions under alert and non-alert conditions*, which measures the average reaction time and precision of response to different stimuli when the driver is in a condition of alert or permanent attention; *(ii) concentrated attention and vigilant resistance to monotony*, which measures concentration and the deterioration or decrease of aptitude conditions due to monotony; *(iii) eye-hand coordination*, which measures visio-perceptive-motor coordination, using both hands simultaneously; *(iv) speed perception and movement estimation*, which measures impulsivity in the face of unnecessary risks that generate accidents; and finally *(v) braking instruction and decision making*, which measures the capacity of the driver to brake in the face of an alert and non-alert condition [3].

In order to manoeuvre a motor vehicle, it is essential that the driver has technical training, as well as possessing certain qualities that are worthwhile and that guarantee the responsible driving of the motor vehicle.

The document enabling the driving of a car is the driver's license (different categories), to obtain or renew this document must submit certain tests, which can be performed efficiently in applications based on (VR) that thanks to the advancement of technology, and the combination of different areas such as electronics, mechanics and computing, have great potential [4] and aim to simulate the real world with virtual environments (VR) that have been applied in areas such as: medicine; for anatomy and surgery training [5]; aerospace engineering, for maintenance and repairing activities [6]; graphic design, for product design and manufacturing [7], in the automotive industry where the development of virtual environments has been found, mainly oriented to the following fields: *(i) Design where VR* can be used for the diagramming and evaluation of concepts during an early stage of the development process [8]; *(ii) Virtual Prototyping (VP)*, in VR it is possible to replicate physical models that allow the reduction of cost and time derived from omitting the construction of physical models; *(iii) Virtual Manufacturing (VM)* encompasses the processes of modeling, simulation and optimization of critical operations in a process related to automotive engineering; *(iv) Training in maintenance* tasks and automotive service [9] and skills improvement in immersive 3D environments [10]; *(v) Virtual Assembly (VA)* facilitates the assembly and disassembly of virtual objects, supplementing the training process.

In this context, with the development of several virtual environments in which the user performs the tests by means of the use of new devices that allow the stimuli and reactions [11] of the users to be better measured when interacting in immersive environments and thus simulate the driving of a motor vehicle, with easy-to-use devices that resemble reality, the idea is to replace the conventional way of performing the psychosensometric tests with scenarios in which real events that take place when driving are represented, the simulator is designed to assess the motor skills, dexterity and ability of a driver by using a haptic vehicle control device that performs the same function as the control controls on a real vehicle, so that by taking signals from the device the performance of the driver can be evaluated when performing a and determine whether or not a driver is fit to continue driving a motor vehicle.

This article is divided into 5 parts, the Introduction, in which the background and the importance of evaluating drivers prior to obtaining a driver's license are presented; in the second part, Formulation of the Problem, the problematic and general structure of the implemented system is described; in the third part is the Structure of the psychometric test simulator, a description of the hardware and software that make up the simulator is made, as well as the stages through which the tests will be carried out; the fourth part presents the experimental tests and results obtained, these tests were carried out in 4 scenarios in which the capacities of the driver in different scenarios and his performance during driving will be evaluated; and finally the fifth part, the Analysis of results in which the experience at the time of performing the driving tests is valued.

2   Problem Formulation

At the present time tests are carried out to obtain and renew the different types of driving license, these tests are: *(i) knowledge of traffic laws* and *(ii) psychometric tests aimed at measuring* the reactions of drivers to the application of various stimuli that are usually audio-visual, the aim of these tests is to see if the driver has physical, mental and coordination skills necessary to drive a car, depending on the type of license you want to obtain or renew, the tests are supplemented with a driving test with the respective vehicle of the category, specifically when the license are Type D, E and E1.

There are people who are not familiar with the equipment and instruments used in psychometric tests, due to the fact that there is no standardization of equipment in the different places enabled to obtain the license, as well as the management of the controls of the equipment is difficult for them. For this reason, one of the objectives of this article is the implementation of a simulator that allows to evaluate the reactions.

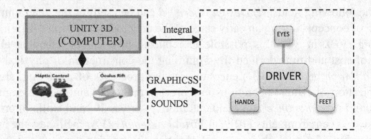

Fig. 1. Functioning of the system.

Figure 1 shows the block diagram of the overall system operation. The implemented application consists of two modules, oriented to measure: *(i) Mental capacity*, has as fundamental objective to examine if the individual has the capacity to respond to stimuli, to respond adequately with the environment, to maintain the sense of reality, reaction speed, orientation time - space, and decision making [12] *(ii) Integral motor coordination*, measures the capacity of the person to coordinate his movements and to control his own body to carry out specific actions. This series of tests includes: the aspirant's ability to perform precise and rapid actions using vision, hearing, upper and/or lower limbs [13], coordination of both hands, and coordination between acceleration and braking of a vehicle.

Figure 1 describes the interaction of the user (driver) with the proposed system, establishing as the main element of communication and feedback covering the two main actions within a virtual environment, observe and act.

The interaction between the user and the proposed system is established by means of a bilateral communication, that is to say, first through a graphical interface the implemented tests are shown that have as objective to measure the answers and reactions of the user; the user through the haptic control devices must comply with each one of the presented tests, and through a numerical calculation software the response curves and the weightings of the committed errors are obtained. The visual environment developed for this type of tests provides a comfortable environment for the user due to the immersion obtained when using virtual reality equipment and tools [14].

3   Structure of the Psychometric Test Simulator

This section describes the structure of the hardware and software implemented to simulate the practical driving tests interacting with the haptic devices and obtain the current level found by the user, the system consists of six stages as indicated in Fig. 2.

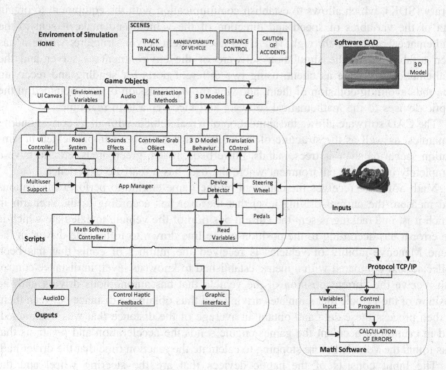

Fig. 2. Hardware structure and system software

The environment of simulation, contains the 3D environments of the graphical interface that allow to simulate a real driving environment, the characteristics of the environment are previously assigned as the climate, gravity, season, driving route, and other physical properties that simulate the real environment, where it incorporates a 3D land vehicle incorporated from the Assets of Unity with the main driving simulation task, there are four scenarios in which you can perform the simulation: *(i) Tracking*, contains a simulation path in which a guide line has been drawn by which the user must follow, avoiding moving away from it until completing the path previously drawn, *(ii) Maneuverability of vehicle*, in this scene the user must avoid the obstacles located in the simulation road and collect the largest amount of coins that are in the provided circuit, *(iii) Distance control*, contains a vehicle that starts automatically and the user in his simulated vehicle must always maintain the same distance that has been assigned to him before starting the circuit, *(iv) Caution of accidents*, in this last scenario the user demonstrates his ability to avoid accidents with pedestrians who unexpectedly cross the tracks by pressing the brake pedal of the simulated vehicle found in haptic devices; The interface allows us to observe the physical variables that the terrestrial vehicle possesses: speed, direction, position and skid.

The set of Scripts contains the code that allows the bidirectional communication between the haptic devices and the graphic engine, between the mathematical software and the graphic engine in real time. To interact with the input device is used proprietary library (SDK), which allows to establish communication with the equipment to obtain data of the variables of speed and direction of the vehicle and make it sent to the mathematical software through TCP/IP protocol client-server structure, where it has been established to the simulation software of the environment as server and the mathematical software as client, using two different ports for sending and receiving data thus avoiding collision of them, allowing the transmission of data obtained in the haptic devices to the mathematical software and vice versa is in real time.

The CAD software allows the simulation of a real vehicle with all its corresponding dynamics, as well as the structure of the objects that have been implemented in simulation scenarios such as trees, stands, flowerbeds, roads, green areas, etc., to have a completely immersive environment where the user has a realistic experience.

Math Software receives the data sent by the game engine to perform calculations and measure the current driving level that the user has according to the scenario in which it is, in Tracking is sent the current position of the vehicle to calculate which is the error it has according to the position of the line drawn to follow the driver, in the scene Maneuve-rability of vehicles is received the number of coins that has been collected and compared with a metric established to know its level, in distance control you receive the current position of the vehicle that has autonomous driving and the position of the vehicle that simulates driving and thus obtain the distance between them to then process these data and obtain an average of the distance that was maintained, and in caution of accident the games engine sends the acceleration and positions that was found the vehicle before stopping to calculate the reaction time that the driver had.

The Input consists of the haptic devices that are the steering wheel and the LOGITECH G29 simulation pedals, which allow sending speed and direction data to the video game engine according to the position they are in when running the simulation environment.

In the Exit it is obtained the real sound that emits a vehicle when being in movement, when it is in high speeds, low speeds and at the moment of crushing the brake, in addition a friendly environment of driving simulation is observed together with the feedback of the haptic device that allows a realistic immersion to the user in the simulation environment.

4 Experimental Tests and Results

The tests that were carried out in the 4 stages evaluate the capacity that a person has to obtain a driving license, for the experimental tests the virtual environments were carried out in Unity 3D 2018.2 and a kit of haptic devices was used such as: LOGITECH G29 steering wheel, pedalboard and gear lever, in addition a desktop computer with an AMD Ryzen 7 2700 Eight-Core processor, Ram memory of 32 GB, Óculus Rift.

Track Tracking
This test consists of a circuit to evaluate the capacity that the driver has for the hand-eye coordination by supplying the traditional Bimanual Coordination Test indicated in Fig. 3.

Fig. 3. Track tracking

First, the creation of a desired trajectory was carried out by means of points that must follow located in the center of the track, and then the trajectory described by the user will be obtained, thus obtaining the way to carry out an error calculation and the qualification at the end of the circuit indicated in Fig. 4.

In the virtual environment, you can see that the environment is familiar to a driver, with different traffic signals and a correct visualization of the road.

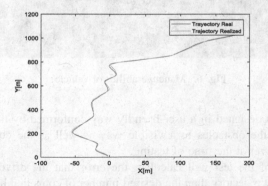

Fig. 4. Ideal route and the route described by the driver.

Analysis: The blue line describes the ideal trajectory that a driver should follow, while the red line represents the trajectory that was performed by a driver close to obtaining a driver's license.

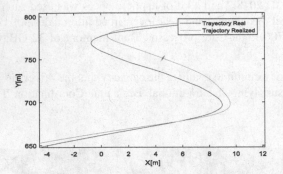

Fig. 5. Trajectory comparison. (Color figure online)

According to Fig. 5 there is no major difference between the ideal trajectory and the trajectory performed by the driver, demonstrating that the user has skills in this test.

Maneuverability of Vehicle

This test is designed to measure the ability to discriminate stimuli and the concentration capacity that a driver has on the road changing the usual way of Multiple Reactions test shown in Fig. 6, the circuit consists of several objects at different distances that the vehicle must dodge, but at the same time is trying to get the most coins that would come to become the desired trajectory.

Fig. 6. Maneuverability of vehicle.

The circuit was designed in a user-friendly way conformed by the different traffic signs and placing the obstacles in a visible way as well as the coins that must be acquired by the driver at the time of testing.

The last part of the test will calculate the error that the driver obtained when choosing a different trajectory than the desired number of coins that he obtained before finishing the circuit. The maximum number of coins that will be obtained in this circuit is 50 demonstrating 100% mastery in this test; the results obtained by the user indicate Fig. 7.

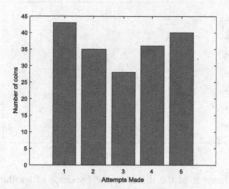

Fig. 7. Statistical data collected.

Analysis: Clearly the majority of drivers obtained a considerable amount of coins during the journey thus demonstrating their ability to drive a vehicle.

Distance Control

This circuit evaluates the time/space perception which is the user's ability to perceive speed and trajectories of a dynamic anticipation exercise and the ability to self-control replacing the Anticipation Test as shown in Fig. 8.

Fig. 8. Distance control.

This test coast of a vehicle that starts automatically at a standard distance that one vehicle must have from another, is followed by the user's vehicle which must maintain the distance that was initially assigned to it and its rating will depend on how close or far the vehicles are. This environment is composed of a circuit that has several curves that allow the driver's dexterity and prudence to be visualized, in the same way that the front of the vehicle can be clearly visualized.

Figure 9 indicates the distance obtained by the user according to the speed applied in the simulation.

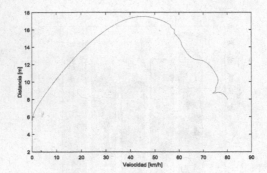

Fig. 9. Behaviour of the driver in the presence of another vehicle.

Analysis: A driver has control of his vehicle at lower speed, maintaining caution and distance from the front vehicle, when raising the speed the control decreases and can cause a collision.

Caution of Accidents
Figure 10 indicates the last test that evaluates the individual's capacity to react to external factors of the road, replacing the Braking Reaction Test. The test consists of presenting a mobile element with uniform displacement, in this case a pedestrian, an animal, a thing, etc. that will unexpectedly leave the environment, causing the driver to make the decision to brake quickly, softly or slowly, according to the distance that this external factor is found in the vehicle.

Fig. 10. Caution of accidents

This environment has a clear visualization of the imprudence that a pedestrian can eat when crossing the street, causing the driver to perform an unexpected maneuver depending on the distance at which he is.

Figure 11 shows the test performed by the user, who performed it at different speeds to see the reaction time of the vehicle.

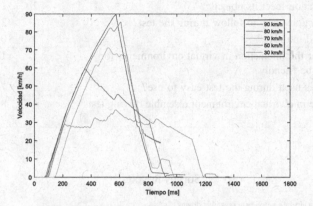

Fig. 11. Testing at different speeds.

Analysis: According to Fig. 11, the control of the vehicle before an external factor can be done at low speeds, avoiding a collision. At high speed, braking is sudden and there is no reliability to avoid collision.

5 Usability Testing

Usability is generally measured using a series of observable and quantifiable indicators from which tangible results can be obtained beyond intuition. One of these tests consists of quantifying the level of satisfaction by means of a questionnaire in the form of one or several questions that collect the impressions that the user has perceived as to the ease or difficulty of the general use of the application. The people chosen to carry out the study are chosen at random, the objective is to analyse the way in which users interact with the application to be evaluated. For this test, different questions have been selected, which have as weighting a scale that goes from "1" to "5", in which it is considered that 1 does not comply at all and 5 complies very satisfactorily (Table 1).

When tabulating the data it is observed that there is great acceptance in the use of the virtual tool, this tool is very intuitive and very easy to use. The evaluation of a virtual environment by users is important because it allows them to identify possible deficiencies and, at the same time, increase the use of headsets such as HTC VIVE, so that immersion in a virtual environment becomes an experience that facilitates learning and training in new tasks (Fig. 12).

Table 1. Representation of the results obtained

Questions	Weightings				
	1	2	3	4	5
Does the application meet its objective?		2	1	5	22
Were the instructions easy to follow during the test?			3	18	9
Is it efficient and intuitive?	2		10	12	6
Do you consider the interaction in virtual environments for driving tests to be friendly?		1	1	13	15
Were the devices used during the test easy to use?			7	12	11
Does experience in a virtual environment resemble driving tests in real life?		4	8	14	4

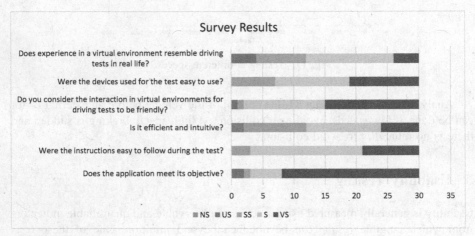

Nothing Satisfactory
Unsatisfactory
Somewhat Satisfactory
Satisfactory
Very Satisfactory

Fig. 12. Survey results

6 Conclusions

Virtual Reality tools are very flexible because they allow to build multiple scenarios according to the applications you want to give, by implementing virtual environments that resemble reality users can perform different psychometric tests in a more friendly way, by combining virtual applications with mathematical calculation software, different results are obtained that can be analyzed in order to observe the reactions of the users to the different audio-visual stimuli, as well as by means of the presentation of the results by means of curves, it is possible to analyze the driving behaviors that the drivers have in different circuits, and in this way it is possible to conclude whether or not the driver is able to obtain a driving license. These virtual tools combined with real or daily equipment are generally intuitive; however, it is convenient to validate the use of them either by direct observation or through surveys.

Acknowledgment. The authors would like to thank the Corporación Ecuatoriana para el Desarrollo de la Investigación y Academia – CEDIA for the financing given to research, development, and innovation, through the Grupos de Trabajo, GT, especially to the GT-eTURISMO; also to Universidad de las Fuerzas Armadas ESPE, Univer-sidad Técnica de Ambato, Escuela Superior Politécnica de Chimborazo, Universidad Nacional de Chimborazo, and Grupo de Investigación ARSI, for the support to develop this work.

References

1. Nutibara, C. (julio de 2018). Laborum IPS. Recuperado el 17 de enero de 2019, de http://www.laborum.com.co/prueba_psicosensometrica.php
2. Ruiz, J., López, E., Norza, E.: Social y Jurídica I: Percepción de seguridaden Jóvenes Colombianos. Bogotá (2012)
3. SEGOB. Accidentes y sus factores. México (2013)
4. Coburn, J.Q., Freeman, I., Salmon, J.L.: A review of the capabilities of current low-cost virtual reality technology and its potential to enhance the design process. J. Comput. Inf. Sci. Eng. **17**(3), 031013 (2017)
5. Jang, S., Vitale, J.M., Jyung, R.W., Black, J.B.: Direct manipulation is better than passive viewing for learning anatomy in a three-dimensional virtual reality environment. Comput. Educ. **106**, 150–165 (2017)
6. Dini, G., Mura, M.D.: Application of augmented reality techniques in through-life engineering services. Proc. CIRP **38**, 14–23 (2015)
7. Berg, L.P., Vance, J.M.: Industry use of virtual reality in product design and manufacturing: a survey. Virtual Real **21**(1), 1–17 (2017)
8. Lawson, G., Salanitri, D., Waterfield, B.: Future directions for the development of virtual reality within an automotive manufacturer. Appl. Ergon. **53**, 323–330 (2016)
9. Borsci, S., Lawson, G., Broome, S.: Empirical evidence, evaluation criteria and challenges for the effectiveness of virtual and mixed reality tools for training operators of car service maintenance. Comput. Ind. **67**, 17–26 (2015)
10. Ortiz, J.S., et al.: Teaching-Learning Process Through VR Applied to Automotive Engineering
11. Diemer, J., Alpers, G.W., Peperkorn, H.M., Shiban, Y., Mühlberger, A.: The impact of perception and presence on emotional reactions: a review of research in virtual reality. Front. Psychol. **6**, 26 (2015)
12. Jayes, M., Palmer, R., Enderby, P., Sutton, A.: How do health and social care professionals in England and Wales assess mental capacity? A literature review. Disability and rehabilitation, pp. 1–12 (2019)
13. Sun, Q., Xia, J., Nadarajah, N., Falkmer, T., Foster, J., Lee, H.: Assessing drivers' visual-motor coordination using eye tracking, GNSS and GIS: a spatial turn in driving psychology. J. Spat. Sci. **61**(2), 299–316 (2016)
14. Wu, C.M., Hsu, C.W., Lee, T.K., Smith, S.: A virtual reality keyboard with realistic haptic feedback in a fully immersive virtual environment. Virtual Reality **21**(1), 19–29 (2017)

Development of the Multimedia Virtual Reality-Based Application for Physics Study Using the Leap Motion Controller

Yevgeniya Daineko, Madina Ipalakova, and Dana Tsoy[✉]

International IT University, 050040 Almaty, Kazakhstan
yevgeniyadaineko@gmail.com, m.ipalakova@gmail.com,
danatsoy@gmail.com

Abstract. In this paper we suggest to use new gestures in controlling of the objects in virtual physical laboratory. These custom gestures are based on Leap Motion controller SDK and allows creating any of desirable hand motions. Such an approach allows creating the most intuitive and friendly interaction between human and software. In educational applications as in our case, this way of interaction provides the highest level of knowledge adoption because of native way of information purchasing.

Keywords: Physics · Gestures · Leap motion controller · Virtual reality

1 Introduction

In recent years, we have witnessed how new technologies have been introduced into our lives. And here, first of all, one would like to highlight the virtual reality technology, which is a virtual world created with the help of technical and software, transmitted to a person through touch, hearing, and also vision and, in some cases, smell. It is the combination of all these influences on the feelings of a person in the sum that is called the interactive world. Virtual reality relies on technologies such as: wide-angle stereoscopic displays; 3D computer graphics in real time; tracking browser (especially the head); binaural sound; tactile feedback; tracking hand gestures; sound. The use of virtual reality is multifaceted: from the training of soldiers and the imitation of battles to the reconstruction of future buildings and their elements, interior design, from the training of surgeons' skills to entertainment in 3D games and simulators.

A particularly important application of virtual reality technology is the educational environment. Today, education is no longer a pen, a pencil and a book, but a system of using interactive technologies that help to transfer knowledge and understanding. An unofficial worldwide study by Burdea [1] showed that in 2003 only 148 universities offered courses using virtual reality technology. Since then, the number of such courses has increased, and most of them use VR in lectures and in the form of laboratory assignments. For example, in [2] it was described the practical application of virtual reality both at lectures and as an experimental component. Miyata et al. developed an educational environment designed for undergraduates that allows you to create VR applications in a team [3]. Project presented in [4] was developed for students of both

L. T. De Paolis and P. Bourdot (Eds.): AVR 2019, LNCS 11613, pp. 150–157, 2019.
https://doi.org/10.1007/978-3-030-25965-5_12

technical and humanitarian areas. It consists of lectures, demonstrations, research presentations, group laboratory work, which include the development of a virtual environment using HMD and WorldUp software. Another interdisciplinary VR project, presented in [5], is designed for students of mechanical engineering, electrical engineering, computer science, physics and engineering management. In [6], the effectiveness of using immersive virtual reality for learning the effects of climate change, in particular for studying the acidity of sea water, was studied. Studies have shown that after the experience of immersion in virtual reality, people showed a good level of knowledge, curiosity in the field of climate science, and in some cases showed a more positive attitude towards the environment. A specially developed and tested virtual reality learning environment [7] offers medical students access to educational materials, as well as the possibility of their re-learning, thus improving education. In addition, the use of such training tools will make medical education more accessible. In [8], an analysis of the virtual environment is considered, which is carried out to support students studying the development of tourism and related impacts. Virtual reality is used to fully immerse students in the environment. This type of experience is able to provide participants with a holistic experience of the real environment, which in reality is expensive, especially for groups with a large number of people.

The International IT University (Almaty, Kazakhstan) has experience in developing e-learning systems with the use of new technologies. Thus, the authors developed a virtual physical laboratory, which is a software package for studying the behavior of object models [11, 12]. The authors also developed applications for studying physics using the technology of augmented reality [13, 14]. This article is devoted to the development of multimedia applications using virtual reality technology for the study of physics, which is implemented as an application in three languages: Kazakh, Russian and English.

Thus, the use of virtual reality opens up many new opportunities in education, allowing upgrading the learning process.

2 System Implementation

The implemented system is based on the integration between the Leap Motion Control and Unity Game Engine [9]. Unity Game Engine is a cross-platform computer game development environment from Unity Technologies. Unity allows creating applications running under more than 20 different operating systems, including personal computers running under Windows, for which this application was developed.

Unity Editor has a simple Drag & Drop interface that is easy to configure, consisting of various windows, so it is possible to debug the game directly in the editor. The engine supports two scripting languages: C #, JavaScript. Physical calculations are made by the PhysX physical engine from NVIDIA.

As a virtual reality device, the Leap Motion controller was chosen, designed to track the movements of hands and fingers in space and can be used for human-computer interaction. Leap Motion is a device about 12.7 × 12.7 × 5.1 cm in size, which is connected via USB to a computer (Fig. 1). The device includes three infrared LEDs and two cameras, and its principle of tracking is stereoscopy (the reflected light from the LEDs is visible from two different points of view, and the distance from the sensor is calculated respectively). Thanks to the SDK libraries, one can get the information about tracking both hands in the space above the device at a height of 15–60 cm. Library subprograms can recognize both hands and transmit information about the location of each bone segment.

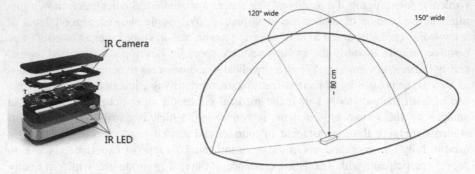

Fig. 1. Disassembled view of the device leap motion movement [10].

3   Results

Developing of technologies and different human-computer interaction tools led the world to the new level of software. Education system is also under this technological revolution. It would be a huge lack not to apply these new tools in many kinds of educational applications. Using of usual computer mouse and keyboard is becoming outdated. A lot of various manipulators, joysticks and other similar things is a new trend in interaction between users and programs. The most novel and fresh approach is using of your own hands as an input device.

In case of this project the proposed way of communication of a person and software is presented as a custom set of gestures based on Orion – Leap Motion controller SDK. Leap Motion controller has a set of basic gestures as circle, swipe, key tap and screen tap.

They are presented in Fig. 2. Moreover, Leap Motion controller provides user with simple and usual grasp. However, interaction with object through default grasp looks strange and non-stable (Fig. 3).

This set of new gestures allows user to work with the laboratory equipment in more convenient and natural way. Exclusion of mouse during the work over laboratory gives students more rich experience. Such approach simulates not only a real laboratory equipment and data that can be extracted from it, but also strong knowledge.

Other advantage of interaction with gestures is its tuning. Gestures themselves can be set up in any of necessary variants. As example, gesture can be created according to the shape of object.

Leap Motion provides programmers with Orion SDK. It allows customizing and developing of new gestures for applications. Developer gets access to different details

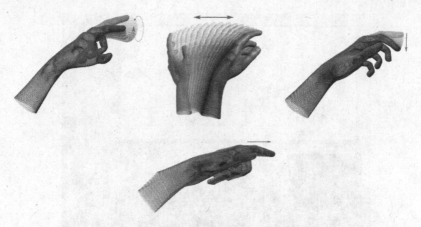

Fig. 2. Leap motion embedded gestures

Fig. 3. Example of built-in grasping

such as palm size, position, velocity, fingers id, position, hand id, wrist position, rotation etc. All of these values help to create new grasps and gestures.

The developed application offers students the opportunity to explore the virtual laboratory work "Studying of Malus's law", the aim of which is receiving and graphic research of plane-polarized and elliptically polarized light. In this laboratory work several custom gestures were set. For example in the lab, there is a need in rotation of polarizer and analyzer (Figs. 4 and 5). In previous version of the lab, these manipulations were done using both mouse and keyboard.

Nonetheless, this time with Leap Motion it was decided to control equipment only using controller and hands of the user.

So the most appropriate approach to the implementation of such control was to divide roles of the hands. This method was proposed in the [15]. The right hand

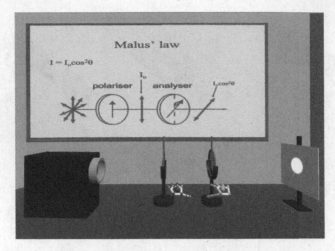

Fig. 4. Common view on the laboratory scene

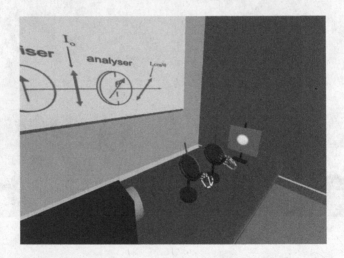

Fig. 5. Common view on the lab from the second camera

became dominant and is used for picking up of the object that need to be controlled. The left hand became functional and is used for the action itself.

In Fig. 6 the rotation of the polarizer by hand rotation is presented. When user picks up the necessary object of the lab by tap of the right hand he/she can rotate the chosen object. Direction of hand rotation and rotation of polarizer/analyzer are the same. It means that to turn the object to the right user needs to rotate hand to the right.

The laboratory work belongs to optic and it is a necessity to see what happens by changing of analyzer/polarizer position. This became a reason to add second camera on the scene. However, to use both cameras the user needed some certain action that would change a camera view. To control switching between cameras another

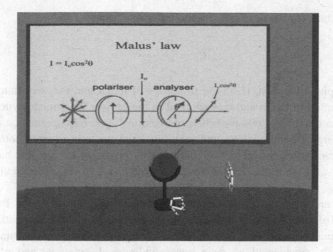

Fig. 6. Demonstration of polarizer rotation using hand

customized gesture was added. The pinch was chosen because it has a high accuracy of detection by controller and ease of the gesture itself. So the user can switch view by simple pinch. If pinch was done even number of times then camera 1 will be turned on, odd number of pinches will turn on the camera 2.

For now, we are continuing to add some gestures. They are aimed to be more precise and accurate to work with different by size and shape laboratory pieces. They will be used not only in this certain lab but also in all of our works.

4 Methods

In this paper we presented use of pinch and palm gestures Fig. 7.

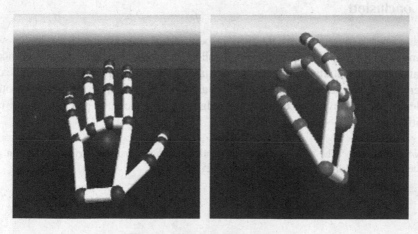

Fig. 7. Palm and pinch gestures

Pinch gesture can be described by the equation below:

$$C_p = \frac{t1 + t5}{2} \tag{1}$$

where C_p is pinch location, t1, t5 are fingertips of index and index of thumb.

It is presented in Orion and was adopted in the project for camera switching. Palm gesture was defined for rotation of the polarizer/analyzer. Implementation of this was done by getting of palm position and rotation. Palm position is represented by Hand class and LeapQuaternion class. By defining the position of the left hand and its rotation by x axis we rotate analyzer/polarizer by y axis to simulate a real laboratory process.

Now we are working on developing of gestures that are based on the shape of objects. In [16] authors propose methodology aided for manipulation and interactive assemble virtual objects. This methodology allows interacting with virtual shapes in natural way. It considers three main poses: cylindrical, spherical and pinch. These poses also can be represented in mathematical equations and adopted for Leap Motion controller data. We are going to develop a new pose for manipulating with cubical objects that cannot be controlled with grasp.

5 Testing

In the project was used manual testing. The source code of the application consists of two modules: logic of the physical process and logic of human-computer interaction including Leap Motion controller. The test was conducted gradually and errors in each module were fixed separately. This method allows testing features of the project in the fastest and convenient way. In the end, the whole functionality was tested and approved.

6 Conclusion

In this paper a new way of interaction between a human and a computer was demonstrated. Using of Leap Motion controller was chosen because of its high accuracy and ability to intuitively control the virtual laboratory equipment. This complex of cooperation between Leap Motion and virtual lab software allows students and other people to get rich experience and knowledge that are close to practice in the real conditions.

Acknowledgments. The work was done under the funding of the Ministry of Education and Science of the Republic of Kazakhstan (No. AP05135692).

References

1. Burdea, G.: Teaching Virtual Reality: Why and How? Presence Teleoperators Virtual Environ. **13**(4), 463–483 (2004)
2. Stansfield, S.: An introductory VR course for undergraduates incorporating foundation, experience and capstone. ACM SIGCSE Bull. **37**(1), 197–200 (2005)
3. Miyata, K., Unemoto, K., Higuchi, T.: An educational framework for creating VR application through groupwork. Comput. Graph. **34**(6), 811–819 (2010)
4. Zimmerman, G., Eber, D.: When worlds collide! ACM SIGCSE Bull. **33**(1), 75–79 (2001)
5. Fallast, M., Obershmif, H., Winkler, R.: The implementation of an interdisciplinary product innovation project at Graz University of Technology. INTED 2007. 1 (ed). IATED, Valencia (2007)
6. Markowitz, D.M., Laha, R., Perone, B.P., Pea, R.D., Bailenson, J.N.: Immersive virtual reality field trips facilitate learning about climate change. Front. Psychol. **9**, 2364 (2018). https://doi.org/10.3389/fpsyg.2018.02364
7. King, D., Tee, S., Falconer, L., Angell, C., Holley, D., Mills, A.: Virtual health education: scaling practice to transform student learning. Nurse Educat. Today **71**, 7–9 (2018)
8. Schott, C., Marshall, S.: Virtual reality and situated experiential education: a conceptualization and exploratory trial. J. Comput. Assist. Learn. **6**(34), 843–852 (2018)
9. Unity Company (n.d.). https://unity3d.com/company. Accessed 1 Mar 2019
10. Colgan, A.: How does the leap motion controller work? Leap Motion Blog, 9 August 2014, http://blog.leapmotion.com/hardware-to-software-how-does-the-leap-motion-controller-work/ (27 July 2016)
11. Daineko, Y., Dmitriyev, V., Ipalakova, M.: Using virtual laboratories in teaching natural sciences: an example of physics. Comput. Appl. Eng. Educ. **25**(1), 39–47 (2017)
12. Daineko, Y.A., Ipalakova, M.T., Bolatov, Z.Z.: Employing information technologies based on.NET XNA framework for developing a virtual physical laboratory with elements of 3D computer modeling. Program. Comput. Softw. **43**(3), 161–171 (2017)
13. Daineko, Y., Ipalakova, M., Tsoy, D., Shaipiten, A., Bolatov, Z., Chinibayeva, T.: Development of practical tasks in physics with elements of augmented reality for secondary educational institutions. In: De Paolis, L.T., Bourdot, P. (eds.) AVR 2018. LNCS, vol. 10850, pp. 404–412. Springer, Cham (2018). https://doi.org/10.1007/978-3-319-95270-3_34
14. Duzbayev, N.T., Zakirova, G.D., Daineko, Y.A., Ipalakova, M.T., Bolatov, Z.Z.: The use of information and communication technologies in education. In: Dortmund International Research Conference, 29–30 June 2018, pp. 37–41 (2018)
15. Cui, J., Fellner, Dieter W., Kuijper, A., Sourin, A.: Mid-air gestures for virtual modeling with leap motion. In: Streitz, N., Markopoulos, P. (eds.) DAPI 2016. LNCS, vol. 9749, pp. 221–230. Springer, Cham (2016). https://doi.org/10.1007/978-3-319-39862-4_21
16. Valentini, P.P.: Interactive virtual assembling in augmented reality. Int. J. Interact. Des. Manuf. (IJIDeM) **3**, 109–119 (2009). https://doi.org/10.1007/s12008-009-0064-x

Virtual Reality and Logic Programming as Assistance in Architectural Design

Dominik Strugała🆔 and Krzysztof Walczak(✉)🆔

Poznań University of Economics and Business,
Niepodległości 10, 61-875 Poznań, Poland
{strugala,walczak}@kti.ue.poznan.pl,
http://www.kti.ue.poznan.pl/

Abstract. In this paper, we describe a new approach to designing architectural spaces, called SADE, in which virtual reality and logic programming techniques are used to support and simplify the architectural design process. In the presented system, users have the possibility to design, configure and visualize architectural spaces within an immersive virtual reality environment, which helps them get better understanding of the created spaces. The system simplifies the design process by taking into account formal design rules, which describe domain knowledge, such as the construction law, technical conditions, design patterns, as well as preferences of a designer. The use of additional knowledge represented in the design rules can significantly shorten and improve the design process.

Keywords: Virtual reality · Logic programming · Architecture ·
3D content · Unity

1 Introduction

The increasing use of interactive 3D technologies, such as virtual reality (VR) and augmented reality (AR), for building immersive multimodal human-computer interfaces is possible mainly due to the significant progress in the efficiency of computing and graphics hardware, continually increasing bandwidth of computer networks, as well as increasingly sophisticated forms of presentation and interaction with 3D content available on end-user's equipment.

Technical progress is largely driven by the quickly growing market of computer games based on 3D user interfaces. However, the use of these techniques is not only limited to applications in the entertainment industry. They can be – and in fact often are – successfully used also in other areas, such as rapid prototyping of products, e-commerce, medicine, tourism, education, and training. The use of 3D/VR/AR techniques may not only lead to the implementation of improved versions of existing services, but it makes it also possible to implement a new class of applications and services, which otherwise would not be possible.

In the architectural industry, digital techniques have been widely used for a long time. 3D presentation and interaction methods largely improve the process

© Springer Nature Switzerland AG 2019
L. T. De Paolis and P. Bourdot (Eds.): AVR 2019, LNCS 11613, pp. 158–174, 2019.
https://doi.org/10.1007/978-3-030-25965-5_13

of creating architectural designs and their subsequent visualization, and enable better parameterization of individual components (BIM) [4,10,11]. The next step is virtual reality, which enables photorealistic presentation of the created designs in real time. Currently, VR techniques are used in architecture mostly for visualization of designs already created with some other design tools. The real challenge, however, is to enable the use of VR directly in the design process. A designer would be able to modify an architectural design while being immersed in its photorealistic VR visualization. However, there is a mismatch between the complexity of the design process and the availability and the precision of interaction techniques used in VR. The method of designing must be simplified to enable it to be performed directly in VR.

In this paper, a new approach to designing 3D architectural spaces is presented. In this approach, called SADE (*Smart Architectural Design Environment*), virtual reality and logic programming techniques are used to support and simplify the design process. In the presented system, users can design, configure and visualize architectural spaces within an immersive virtual reality environment, which helps them get better understanding of the created spaces. The system simplifies the design process by taking into account formal design rules, which contain domain knowledge, such as the construction law, technical conditions, design patterns, as well as preferences of a designer. The use of additional knowledge represented in the design rules has the potential to significantly shorten and improve the design process. The application architecture along with a prototype of the architectural design tool, implemented as an extension to the Unity IDE, are also presented.

The rest of the paper is organized as follows. Section 2 presents an overview of the state of the art in architectural design. In Sect. 3, the concept, the requirements and the implementation of the Smart Architectural Design Environment are discussed. Section 4 presents a step by step design example. Finally, Sect. 5 concludes the paper and discusses possible further development and research directions.

2   State of the Art

The advancement of technology has allowed the development of various approaches to the design of architectural spaces – both 2D and 3D. Many of these approaches aim at providing comprehensive design environments, while reducing the amount of information that must be supplied at the design stage and decreasing the duration and complexity of the creation process. This, however, is often associated with the increase of the complexity of the software used in the design stage. Commonly used tools use several approaches described below.

2.1   2D Drawings

2D design applications rely on the use of 2D drawings for modeling of architectural spaces. These applications enable creating 3D spaces with the use of 2D elements of various types and complexity, starting with simple lines in Autodesk AutoCAD and ending with complex multi-layer elements such as walls, e.g., in Graphisoft ArchiCAD or Autodesk Revit. The complexity of creating a space in this way depends on the degree of integration between the elements. For example, comparing two software packages from the same company – AutoCAD and Revit, one can conclude that creating complex architectural designs with AutoCAD is more time consuming than with Revit [3,18]. This is due to different ways these two applications handle the designed content. First of all, AutoCAD is a purely planar application that can be compared to a sheet of paper. Changes made in one place must be marked separately in another. In turn, Revit automatically transfers each change to the other planes (cross-section, elevation, projection). This can be a key aspect when the time for creating a design is critical. Nevertheless, each of these tools in its own environment has advantages over the other. The possibility of editing each element separately gives a developer full control over the drawings. On the other hand, automatic transfer of a single change to all related drawings facilitates and accelerates work on a project.

There is a possibility to build three-dimensional spaces with the use of 2D tools, such as AutoCAD, in a similar way as it is possible with Revit or Archi-CAD, which are specialized for creating 3D models, but the process is complicated, and – therefore – 2D tools are most often used in conjunction with specialized 3D modeling applications. Such an approach gives more freedom and introduces a hierarchy to the architectural space design. By focusing on individual design stages, it is possible to have greater control over the entire process, however, this is also often associated with the increase in the duration of the development due to the need for incorporating changes in both the 2D drawings and the 3D model.

2.2   Visual Content Modeling

The continuous technical progress has caused a significant increase in the availability of 3D modeling environments. There are many examples of 3D modeling tools, but the most commonly used in both the entertainment and architectural industries are: 3ds Max, 3D-Coat, Blender, Maya, Modo, Rhino, and Zbrush. These sophisticated environments intended for professional users provide a wide variety of tools and operations, which enable any object to be modeled. However, their comprehensive capabilities require that users have experience in 3D modeling. Nonetheless, creation of 3D content does not always have to be difficult. Proper orientation on specific domains and constraining available functionality permits creation of design environments that are user-friendly. Examples of such environments are: AutoCAD Civil 3D, Ghost Productions, SketchUp, and Sweet Home 3D. Designed with specific type of application domain in mind, they enable efficient creation of 3D content without the need of extensive experience in 3D modeling. However, it also results in a significant reduction of the generality of the content creation process.

2.3   Parametric Modeling

In the field of architecture, very popular is the parametric modeling approach, i.e., creating content with the help of parameterized complex content templates, instead of direct free shape modeling used in typical 3D content creation applications [22,23]. In this way, content is generated on demand using a specific set of parameter values. With the help of ready-made templates, the designer creating a given piece of content needs only to ascribe values to specific parameters, which does not require comprehensive knowledge. This limits to a large extent the flexibility of content creation, but also reduces the amount of information a designer needs to provide. These types of templates are often inspired by natural world phenomena such as the Voronoy diagram, which resembles arrangement of cells on a leaf. One of the applications that enable generation of this type of content is the Grasshopper extension to the Rhino modeling package [8].

2.4   Augmented Reality

Augmented reality is a promising technique in the field of architectural design [1,2,24]. However, currently the use of AR in this domain is low. This is due to several factors. Firstly, the available AR applications do not permit professional design of architectural spaces. Secondly, high cost and low availability of AR devices such as Magic Leap and Microsoft HoloLens, which facilitate interaction within the AR environments, discourage potential users and result in low market penetration. Devices which are widely available and popular, such as tablets and smartphones, do not provide the quality and the efficiency that would allow for effective creation of 3D architectural designs. Some companies provide simple AR design applications through application markets, such as the Google Play or the AppStore, but they are mainly limited to browsing and matching specific products (e.g., furniture), without the possibility of performing full architectural design.

2.5   Virtual Reality

Although virtual reality offers great benefits in architectural visualization [14,16], currently there is very limited access to applications permitting design of architectural spaces in VR [20]. There are software packages and extensions to modeling programs that enable generating a VR environment from an existing architectural model. In such an environment, a user can freely navigate to examine the designed space [5,7,9], but without the possibility to create, modify or even annotate the visualized space. Applications that enable creation of objects in VR are very limited in their functionality.

The currently used mainstream design process is the preparation of a ready-made 3D model in a program purely aimed at creating 3D architectural designs and exporting the model to a generally available game engine, such as Unity 3D [19] or Unreal Engine [6]. Consequently, to achieve the expected result, there is a need to posses knowledge both in the field of developing 3D designs and in operation of game engines. Often, there are issues related to compatibility between the modeling programs and particular game engines. By importing an architectural design into a game engine, a higher-quality visualization can be achieved, but still there are no means of modifying the design.

2.6 Design Patterns in Architecture

Design patterns are recognized and proven ways to manage the designed space. Patterns are based mainly on previously developed projects and current design trends [17]. Their usage is not strictly required, but non-usage of a relevant pattern is usually treated as a mistake. In cases when it is not possible to obtain an acceptable solution based on the applicable patterns, it is allowed to omit them. The use of design patterns can be seen as an attempt to introduce standardization into architectural design. One of the most well-known sets of design patterns in architecture is Feng shui [13].

2.7 Summary

High complexity of 3D content creation software and the large number of aspects to be considered during the design process make creation of architectural designs a sophisticated task on many levels. The above described approaches are suitable for solving different problems in various contexts. In a practical approach, it is necessary to combine several of these methods to ensure the expected final effect. Even this, however, does not guarantee achieving a satisfactory result due to limitations in the presentation of the final designs. There is an evident lack of an approach that would enable to combine professional architectural design process with the benefit of doing this in a virtual reality environment with the support of rules describing the architectural design process.

3 Smart Architectural Design Environment

In Sect. 2, several different approaches to creating 3D architectural designs have been presented. Each of them offers some specific advantages, but there is clearly a lack of an integrated solution that could facilitate architectural design, by enabling real-time visualization and modification of the design in an immersive VR environment, while simplifying the design process to mitigate the deficiencies of VR as a design user interface.

In this section, we describe the concept of the *Smart Architectural Design Environment* – the main contribution of the paper. First, motivation for the work is provided, followed by requirements, and a description of the approach itself.

3.1   Motivation

During the architectural design, highly complex 3D models are being created. There is a lot of freedom in this process, but there are also constrains. Properly modeled constraints help to simplify the process by lowering the variability of possible results. An example of well-used constraints is parametric modeling, described in Sect. 2.3, which hugely simplifies the design task. However, parametric modeling takes into account only physical properties of objects.

Further simplification of the design process may be possible by taking into account other rules, such as construction law, technical conditions, design patterns, and preferences of a designer. These rules, however, require a specific approach. Firstly, they do not describe only individual objects, but relationships between objects and groups of objects. Secondly, they operate on a different (higher) level of abstraction. Finally, the constraints are often changing, e.g., when new construction materials and elements are introduced, new law comes into force, or a new style of design is used by an architect.

These new constraints must be translated into information processable by a computer program. Construction law and technical conditions are legal concepts, which rise the problem of their interpretation. It is necessary to describe them in the form of unambiguous rules. The design patterns, which are principles developed based on experience, are more challenging because of their subjective nature. They can suggest particular combinations of choices (e.g., colors) or relationships between objects (e.g., relative location).

The applicable laws and principles must be assigned to individual objects, but they must reflect relationships with other objects or types of objects. Therefore, a taxonomy of objects is required and rules must be programmed in a way enabling dynamic composition and linking of rules for all objects present in a given space. For this purpose, formal specification of an ontology and logic programming languages (e.g., Prolog) may be used. Logic rules may be associated with classes of architectural spaces, classes of objects or even specific objects in the ontology, thus enabling expression of constraints at different levels.

The use of a content library is needed to simplify the management of objects and rules – both at the creation stage and during their later use. To enable easy access and sharing of both particular objects and designs by design teams, a remotely accessible shared repository may be used. There is a large variety of modeling techniques and programs currently used by the designers (e.g., 3ds Max, Blender or SketchUp), and therefore it is important to enable importing models in various data formats.

An important decision is the choice of the user interface style. On the one hand, classical user interfaces (screen, mouse, keyboard) provide a convenient and familiar working environment for architects. On the other hand, VR interfaces – even if they are constrained in their interaction capabilities and precision – clearly provide an added value in permitting to immediately see the changes in a realistic visualization. It is important, therefore, to enable the use of both interface styles. A specific UI must be provided within the VR space for presentation of messages and warnings. This may be accomplished with pop-up windows, annotations, or changing the color of individual objects.

3.2   Requirements

The motivation presented in the previous section can be converted into a list of requirements, as presented below.

1. The system should enable immersive presentation of an architectural space and natural navigation of users and designers within the space.
2. The system should enable modification of the architectural space, while a designer is immersed in the space. In general, the space being designed can be a part of a more comprehensive 3D virtual environment (e.g., a kitchen being a part of a flat, a flat being a part of a building).
3. The system should allow for using a blueprint or an empty model of an architectural space, but also for using elements taken from existing content libraries. It is important that these elements could be imported in various formats, not only limited to currently popular ones such as .3DS, .FBX or .OBJ, but also formats used in other applications.
4. The system should enable parameterization of objects. Parameters should describe individual objects as well as connections between objects.
5. The system should permit categorization of objects to enable associating rules with specific categories of objects and creating dependences between interdependent elements of the designs (e.g., washing machine and bathtub).
6. The system should enable the use of semantics as a formal way of describing the role, the structure, and the interaction of particular design components to permit automatic reasoning as well as efficient search and parameterization of components.
7. The system should provide support for expressing design rules, which represent architectural design patterns, common practices, and other constraints.
8. The rules should be independently associated with spaces, objects and their categories, and then combined ad-hoc during the design process, thus forming a set of rules applicable to a particular part of the design.
9. The system should enable labeling of selected regions in order to assign them to specific semantic categories, which may be affected by different sets of design rules.
10. The design environment should enable access to a repository of already designed and described components, at the same time providing mechanisms to modify them.
11. Finally, the system should enable definition of actions to be executed as a result of evaluating rules (e.g., suggesting an element, limiting a value, informing the designer about errors).

Requirement 1 is met by existing game engines and VR environments, but the use VR is not currently within the mainstream of architectural design practices. Requirement 2, in a general case, is difficult to meet due to the complexity of the 3D design process. Therefore, in the current practice, permitted modifications are limited to predefined surfaces (e.g., painting colors on a view of a wall) and basic changes to particular elements, without the reference to other objects inside the environment [12, 15]. Requirement 4 (parameterization) is partially supported by the existing Building Information Model (BIM) environments. The use of parameters, however, is limited to simple values (e.g., weight of a bathtub) and it does not support complex relationships (e.g., the distance between an electric device and a bathtub). Requirements 3, 5 and 10 are available within VR/AR design environments, such as the Unreal Engine and Unity. Requirements 7, 8, 9 and 11 are at an early conceptual stage or are not considered in current systems. Finally, requirement 6 has no reference in current VR/AR or architectural modeling environments.

3.3   The SADE Approach

The use of VR for designing architectural spaces requires simplifying the process as much as possible. The SADE approach is an attempt to partially handle this problem by implementing the domain knowledge into a set of well-defined rules. The application of rules does not only facilitate the use of VR interfaces, but most importantly it supports design experts in their daily work through the simplification and safeguarding of the design process. To some extent, it also enables design or configuration of proper architectural spaces by non-experts.

In the SADE approach, an immersive VR system is used both for the design and for the visualization (requirements 1 and 2). The design is performed using semantically described components, with well-defined meaning, parameters, and interactions (requirements 3, 4, 5, 6). The components are stored in shared semantic on-line repositories (requirement 10). The design process is supported and simplified by verification of formally defined rules, which describe possible actions and therefore limit the complexity of the design process (requirements 7, 8, 11). While designing a space, a designer associates it with a specific class of spaces (e.g., kitchen, office) thus activating appropriate sets of rules to use (requirement 9). Rules may be also associated with classes or instances of architectural objects used within the spaces. All rules are merged in real time enabling proper interaction between all components in the space and the space itself.

The SADE approach uses a common ontology of concepts. A fragment of the ontology with classification of concepts is presented in Fig. 1. The ontology describes both architectural spaces and particular classes of objects, which may be present in the spaces. The starting point is an abstract class *Architectural Entity*. It can be either an *Architectural Space* or an *Architectural Object*. Subclasses of the Architectural Space represent different categories of areas with different sets of design rules. The first example subcategory is *Room*. It includes categories such as *Living Room*, *Kitchen* and *Bathroom*. Some categories can have specific variants, e.g., *Large Living Room*. In some cases, rules for several categories can be combined in a subcategory, e.g., *Living Room with Kitchen*. Another type of architectural spaces is represented by the *Open Space* class. These are fragments of larger spaces, which cannot be clearly assigned to any specific type of room.

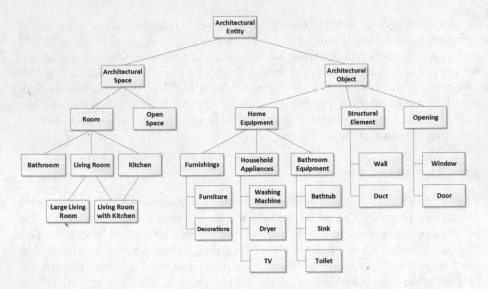

Fig. 1. Classification of concepts in the SADE Ontology.

The second type of Architectural Entity is represented by the *Architectural Object* class, which includes *Structural Elements*, *Openings* and *Home Equipment*. Structural elements are static parts of the construction, e.g., walls or ducts. Opening, as the name suggests, contains all structural openings, i.e., windows and doors. The third type of architectural objects are elements of the *Home Equipment*, encompassing such classes as *Furnishings*, *Household Appliances*, and *Bathroom Equipment*.

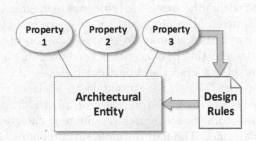

Fig. 2. Architectural entity with properties and design rules.

Each of the entities may be associated with a specific set of *Design Rules* (Fig. 2) governing the use of the entity in a design. The design rules are expressed as clauses in the Prolog language. The clauses may be parameterized with values of properties of the entities. During the design process, a property value may be set by the designer explicitly (e.g., type or weight of the object) or implicitly (e.g., position or size of the object). Properties and rules may be associated with any kind of architectural entity, including both spaces and objects.

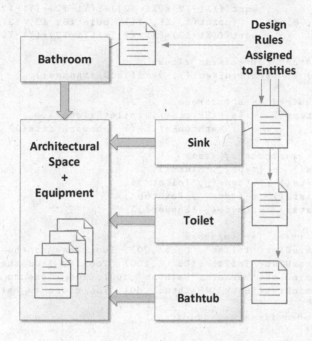

Fig. 3. Merging design rules in SADE.

When an architectural space is being designed, all rules (implemented as Prolog clauses) associated with the space and particular objects that are used in the space are merged. Since all clauses use a common ontology, they can be combined automatically into a consistent set of rules, describing the space being designed. The process is illustrated in Fig. 3. An architectural space of class *Bathroom* is being designed. The class has its own set of rules, which are automatically associated with the space. The designer adds tree items of equipment – a *Sink*, a *Toilet*, and a *Bathtub*. All these elements have their own sets of rules, which are combined with the set of rules specific for bathroom. The resulting joint set of rules is used during the design process. Whenever the designer adds another element, the set of rules is updated. An example of a merged set of rules is presented in Listing 1.1

Listing 1.1. Example of merged design rules in Prolog

```
1   %Generic utility clauses
2   minDist(A, B, X) :- (point(A, X1, Y1), point(B, X2, Y2),
3                        sqrt(((X1-X2)*(X1-X2))+((Y1-Y2)*(Y1-Y2)))>X).
4   maxDist(A, B, X) :- (point(A, X1, Y1), point(B, X2, Y2),
5                        sqrt(((X1-X2)*(X1-X2))+((Y1-Y2)*(Y1-Y2)))<X).
6
7   %Main bathroom correctness clause
8   bathroom() :- verifiedItems(), verifiedDistances().
9
10  %Items required in a bathroom
11  verifiedItems() :- (sinkExists(), toiletExists(),
12                      ( bathtubExists(); showerExists()) ).
13
14  %Checking existence of items
15  sinkExists() :- (type(_,'Sink')).
16  toiletExists() :- (type(_,'Toilet')).
17  bathtubExists() :- (type(_,'Bathtub')).
18  showerExists() :- (type(_,'Shower')).
19
20  %Distance rules for bathroom
21  toVerify(minDist('Toilet','Duct',20),'Too close to the duct!').
22  toVerify(maxDist('Toilet','Duct',100),'Too far from the duct!').
23  toVerify(minDist('Washer','Bathtub',100),'Too close to bathtub!').
24  toVerify(minDist('Sink','Bathtub',50),'Too close to bathtub!').
25
26  %Created when items are added
27  type('ID1', 'Duct').
28  type('ID2', 'Sink').
29  type('ID3', 'Bathtub').
30  type('ID5', 'Toilet').
31  type('ID4', 'Washer').
32
33  %Set when items are placed
34  point('Duct', 0, 0).
35  point('Sink', 0, 50).
36  point('Bathtub', 0, 120).
37  point('Toilet', 60, 0).
38  point('Washer', 100, 0).
39
40  %Attemt to falsify any of the rules and write a message
41  failedVerify() :- toVerify(Rule,_), \+call(Rule), !, true.
42  :- (toVerify(Rule, Msg), \+call(Rule) -> writeln(Msg); true).
43
44  %Verified if not possible to falsify any of the distance rules
45  verifiedDistances() :- \+ failedVerify().
46
47  %Check the main rule and write a message
48  :- (bathroom() -> write('All correct'); write('Not correct')).
```

In the example, the first two clauses (lines 2–5) are used to verify the minimum and the maximum distances (X) between two items (A and B). The main bathroom correctness clause (line 8) indicates that a bathroom design is correct when there are all required items and there are proper distances between the items. In the example, items required in a bathroom are a sink, a toilet, and either a bathtub or a shower (lines 11–12). Next four clauses (lines 15–18) verify whether there are objects of these classes in the design. The names of classes are taken from the SADE Ontology (Fig. 1). Lines 21–24 contain distance rules that must hold in a correctly designed bathroom. Clauses 2–24 are taken from the design rules assigned to the *Bathroom* class (Fig. 3). Clauses 27–38 indicate that items of particular classes are assigned to the bathroom design and set their positions. The clauses are taken from rules assigned to the particular classes of items (e.g., *Sink, Toilet, Bathtub*) with specific values taken from properties. In the listing, they are grouped for readability. The clause 41 is an attempt to falsify any of the distance rules (lines 21–24). If any of the distance rules fails, appropriate message is displayed (line 42). If none of the clauses fails, it means that distances are correct (line 45). The last clause (line 48) is used for testing if the main bathroom rule holds.

3.4 SADE Implementation

The design process is performed in an open distributed environment that enables the use and sharing of content. Design is understood here as building a space or a part of a space from scratch (which may involve importing libraries of objects) or re-using elements previously created – also by other users – and finishing with ready-made design templates.

The method of categorization and parameterization of components in SADE builds upon the previous work on SemFlex [21]. The main distinguishing novelty is the use of formally defined *design rules*, as described in Sect. 3.3. The design rules are expressed using constraint logic programming in Prolog. Prolog is a declarative programming language – the program is expressed in terms of clauses, representing facts and rules. It is the most popular logic programming language, with several free and commercial implementations available. Most importantly, Prolog enables to freely combine sets of rules given that the common ontology of concepts is provided. This feature enables implementation of separate sets of design rules for specific entities in the SADE Ontology – both spaces and objects (cf. Fig. 1), as well as combining different sources of rules, e.g., regulations, guidelines, design patterns, preferences, etc. Programming in Prolog is declarative and does not require advanced IT expertise or effort.

The SADE environment has been implemented as an extension to the Unity IDE (Fig. 4) communicating with shared repositories (SCRs) implemented as REST services operating on web servers. The Unity IDE was chosen as the starting point because of its popularity and the multitude of features it has as a cross-platform game engine and IDE. Unity enables to freely expand the design environment with new menus, custom component editors and editor windows using C# and JavaScript.

SADE can be operated in two ways. First, it can be used as a typical IDE with classic user interface providing all features of the Unity editor together with SADE extensions. Second, it can be used as a fully immersive VR environment, in which a user can directly interact with the designed space using an HMD

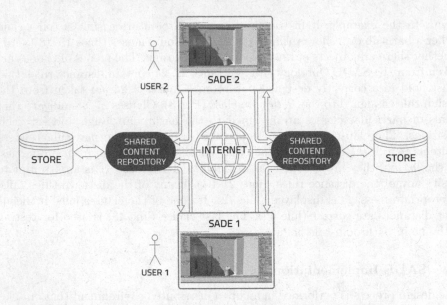

Fig. 4. Architecture of the Smart Architectural Design Environment.

and controllers. With the help of the implemented design rules, all changes are monitored, which enables the system to limit possible actions and to inform the user about potential design errors or non-conformances to applicable regulations.

4   Design Example

Creating an architectural space in SADE starts with importing a 2D projection or an empty model. On the left side of the application window there is a hierarchical list of elements within a given scene. At the beginning, it contains only basic objects, such as: Camera, Source of Light, Room Projection. The right side of the screen contains a preview of the 3D scene being designed. In the bottom left corner of the screen located is the SADE tool window. A button available in the window enables selection of the class of the architectural space being designed. When the button is pressed, a pop-up menu appears with a list of classes taken from a shared repository (Fig. 5). After selecting the class, a class window is displayed, which enables either to edit or to apply the class to the architectural space (Fig. 6a). When the class is applied, a scalable surface is added to the scene, enabling the designer to set the boundaries of the architectural space of the selected class (Fig. 6b).

After a designer defines the boundaries of the space, the actual design process starts (Fig. 7). The process consists of several phases. First, the designer needs to choose objects that are to be located in the designed space. The objects are downloaded from a repository in the form of packages (.unitypackage) along with the design rules assigned to them. After importing packages with objects, it is

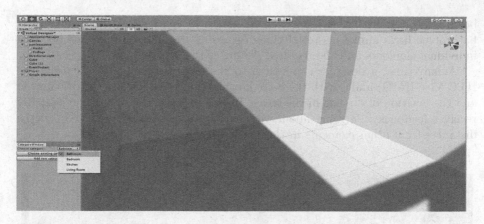

Fig. 5. SADE IDE – selecting the class of the architectural space.

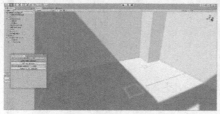

(a) Applying the selected class   (b) Indicating the area of applicability

Fig. 6. Indicating the area of applicability of a class of architectural spaces.

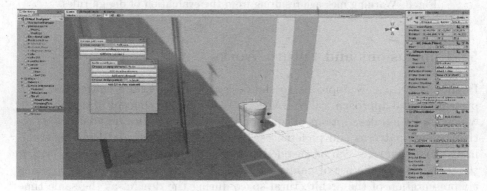

Fig. 7. Editing an item of equipment in the SADE IDE.

possible to add individual elements to the room (washing machine, toilet, etc.) by using a drop-down menu in the application.

During the design process, the SADE system verifies design rules in a given space. When the system identifies that the design is not consistent with the rules, the designer receives a warning. In the presented example, the room has been associated with the *Bathroom* architectural space class. In architecture, bathrooms have numerous restrictions regarding the mutual relations between individual elements and the required distances (Fig. 8).

At any time, a designer can switch to the immersive mode (e.g., using the HTC Vive HMD) and – with the help of controllers – can adjust the orientation and the position of objects in the space. In this process, the designer repeatedly receives feedback informing about verification of the design rules (e.g., if the distances from other elements in the space are correct).

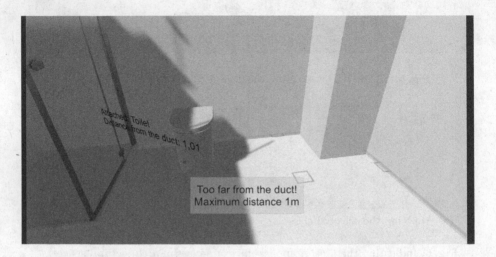

Fig. 8. Warning message resulting from the evaluation of design rules.

5   Conclusions and Future Works

In this paper, a new approach to the design of architectural spaces has been presented. The approach, called SADE, is based on the concept of design rules, which are formal descriptions of commonly used architectural regulations, practices, design patterns and preferences. By employing constraint logic programming the design rules can be encoded into precise clauses that can be verified at the design time. The use of immersive mode helps the designers to see a realistic presentation of the architectural space during the design, at the same time verifying correctness of the space.

The use of the presented approach enhances the process of designing architectural spaces by helping to avoid design errors and reducing the designer's effort through automatic verification of rules. Thanks to the use of shared repositories, it is possible to use a wide range of products available on the market as library components, and also to share objects and designs with other designers.

Currently, the SADE IDE prototype – implemented as an extension to the Unity IDE – enables importing content and rules from repositories, assigning classes of architectural spaces to selected areas of designs, and creating instances of architectural objects. Programming of design rules is performed manually by editing Prolog scripts associated with the classes or objects in the repository.

As the next development step, we plan to extend the prototype by enabling definition of rules at different levels of abstraction, thus enabling implementation of the construction law, design patterns, design styles, designers preferences and technical properties of objects. The second direction of development is to expand the system of rules to automatically propose layouts of elements within a given space, using typical solutions and specific constraints.

In the future, we plan to extend the approach to support machine learning techniques, through the analysis of designers' choices and existing designs. This may further simplify the design process.

References

1. Abboud, R.: Architecture in an age of augmented reality: opportunities and obstacles for mobile AR in design, construction, and post-completion. NAWIC International Women's Day Scholarship Recipient (2013)
2. Anders, P., Lonsing, W.: AmbiViewer: a tool for creating architectural mixed reality. Association for Computer Aided Design in Architecture, pp. 104–113 (2005)
3. Dib, C.: 5 key differences between AutoCAD and Revit! (2017). LinkedIn https://www.linkedin.com/pulse/5-key-differences-between-autocad-revit-carole-dib
4. Engineering.com: BIM 101: what is building information modeling? (2016). https://www.engineering.com/BIM/ArticleID/11436/BIM-101-What-is-Building-Information-Modeling.aspx
5. Enscape: homepage (2018). https://enscape3d.com/architectural-virtual-reality/
6. Epic Games: unreal engine (2019). https://www.unrealengine.com/en-US/what-is-unreal-engine-4
7. Eyecad VR: homepage (2018). https://eyecadvr.com/pl/
8. Grasshopper homepage. https://www.grasshopper3d.com/
9. IrisVR: homepage (2018). https://irisvr.com/
10. ISO: BIM - Part 1: concepts and principles, ISO 19650–1:2018 (2018). https://www.iso.org/standard/68078.html
11. ISO: BIM - Part 2: delivery phase of the assets, ISO 19650–2:2018 (2018). https://www.iso.org/standard/68080.html
12. Kaleja, P., Kozlovská, M.: Virtual reality as innovative approach to the interior designing. Sel. Sci. Papers - J. Civil Eng. **12**(1), 109–116 (2017)
13. Mak, M.Y., Ng, S.T.: The art and science of Feng Shui - a study on architects' perception. Build. Environ. **40**(3), 427–434 (2005)
14. Paes, D., Arantes, E., Irizarry, J.: Immersive environment for improving the understanding of architectural 3D models: comparing user spatial perception between immersive and traditional virtual reality systems. Autom. Constr. **84**, 292–303 (2017)
15. Pixel Legend: virtualist (2019). https://www.pixellegend.com/virtualist/
16. Portman, M.E., Natapov, A., Fisher-Gewirtzman, D.: To go where no man has gone before: virtual reality in architecture, landscape architecture and environmental planning. Comput. Environ. Urban Syst. **54**, 376–384 (2015)

17. Salingaros, N.A.: Design Patterns and Living Architecture. Sustasis Press, Portland (2017)
18. Tobias, M.: How do AutoCad and Revit compare? (2017). https://www.ny-engineers.com/blog/how-do-autocad-and-revit-compare
19. Unity Technologies: unity game engine v. 2018.3.3f1 (2018). https://unity3d.com/unity
20. VRender: top 5 virtual reality applications for architects (2018). https://vrender.com/top-5-virtual-reality-applications-for-architects/
21. Walczak, K.: Semantics-supported collaborative creation of interactive 3D content. In: De Paolis, L.T., Bourdot, P., Mongelli, A. (eds.) AVR 2017. LNCS, vol. 10325, pp. 385–401. Springer, Cham (2017). https://doi.org/10.1007/978-3-319-60928-7_33
22. Walczak, K., Cellary, W.: X-VRML for advanced virtual reality applications. Computer 36(3), 89–92 (2003)
23. Walczak, K., Wojciechowski, R., Wójtowicz, A.: Interactive production of dynamic 3D sceneries for virtual television studio. In: The 7th Virtual Reality IC VRIC - Laval Virtual, pp. 167–177, Laval (2005)
24. Wang, X.: Augmented reality in architecture and design: potentials and challenges for application. Int. J. Architectural Comput. 02(7), 309–326 (2009)

Effects of Immersive Virtual Reality
on the Heart Rate of Athlete's Warm-Up

José Varela-Aldás[1,2]($\boxtimes$) ⓘ, Guillermo Palacios-Navarro[2] ⓘ,
Iván García-Magariño[3] ⓘ, and Esteban M. Fuentes[1] ⓘ

[1] Grupo de Investigación en Sistemas Industriales, Software y Automatización
SISAu, Facultad de Ingeniería y Tecnologías de la Información y la
Comunicación, Universidad Tecnológica Indoamérica, Ambato, Ecuador
josevarela@uti.edu.ec, tebanfuentes@gmail.com
[2] Department of Electronic Engineering and Communications,
University of Zaragoza, Zaragoza, Spain
guillermo.palacios@unizar.es
[3] Department of Software Engineering and Artificial Intelligence,
Complutense University of Madrid, Madrid, Spain
igarciam@ucm.es

Abstract. An adequate warm-up prior to intensive exercise can bring benefits
to athletes, these requirements may vary depending on the physical activity and
the training needs. The immersive virtual reality could have benefits in the
warm-up and it can be determined by physiological data of the athlete. This
work presents a mobile virtual reality application to stimulate the warm-up of an
athlete using a standard treadmill, where the developed application is composed
of a pleasant and stimulating environment. A Smartphone and the Gear VR are
used as an HMD device, and wireless headphones are placed in the users, in
addition, the heart rate of the athletes is monitored using a Polar H7 sensor.
Experimental results are obtained in athletes with similar characteristics and
conditions, identified a direct relation of the virtual environment with the pulsations per minute (ppm), denoting pulsations greater than usual in case of the
stimulating environment and lower pulsations for the pleasant environment.
Finally, a usability test is performed that shows the level of sociability of the
system.

Keywords: Virtual reality · Immersion · Heart rate · Warm-up · Usability

1 Introduction

Immersive virtual reality offers the possibility of comparing the human response in
activities and processes within virtual environments with respect to situations presented
in the real world, in order to analyze the perception, preference, and behavior of the
user. The concurrent areas in the use of this technology are medicine, psychotherapy,
and education, advantageously these immersion systems are available to almost all
researchers [1]. In medicine, the user can improve the learning processes of the human
anatomy, reviewing the internal composition of human parts, including complex

© Springer Nature Switzerland AG 2019
L. T. De Paolis and P. Bourdot (Eds.): AVR 2019, LNCS 11613, pp. 175–185, 2019.
https://doi.org/10.1007/978-3-030-25965-5_14

elements of the brain [2]. It is also possible to study decision-making processes by applying subjective and objective evaluation techniques [3]. On the other hand, virtual neurological rehabilitation allows the patient to improve their cognitive functions even when the lesion is in the subacute period [4] and in motor rehabilitation when the patient has lost mobility, the positive effects of these applications have been demonstrated using different methods [5]. In addition, virtual reality has allowed diagnosing physical and psychological states; studies determine the consumption of methamphetamines through social virtual environments and analyzing neurophysiological signals [6]; stress levels are qualified by the variability of the heart rate in situations of tension within immersive virtual environments [7]. The development of these applications requires a correct insertion of the immersion components, as well as an adequate coherence to the responses of the virtual environment, studies show that discomfort occurs in the user due to lack of coordination [8].

To determine the benefits and usability of a virtual reality system, it is necessary to rely on surveys and physiological data that validate this information. Current technology has facilitated acquiring sensory information from the user, reducing the size of the sensors, new communication devices, and connectivity, allowing the use of multiple contact devices which are widely used in medicine [9]. The data commonly recorded are heart rate, systolic blood pressure, diastolic blood pressure, electroencephalography, core temperature, eye movement, and so forth. [10], These data allow to measure states of relaxation and stress, especially in situations of great pain and discomfort [11]. Several proposals manage to quantify the concentration of the user through real-time information and some works manage to design telemetric monitoring systems for patients [12, 13].

Sport is a human activity with a high degree of influence on the quality of life, for this reason there are multiple virtual reality games to encourage exercise and study the user's physical responses, thus, Exergame includes the virtual tracking of the body movements of the user, combining physical activity and visual feedback, achieving favorable responses according to studies [14]. Soccer experts have evaluated first-person immersive virtual training systems [15], recommending its use. To study the daily exercise or any sport activity, several data of the user-application system are required, from knowing the movements made in real time to measuring the physiological conditions affected [16], where it had seen that the more immersive the application of virtual reality, the greater stimuli are generated in the human body and reactions similar to those produced by real exercises are achieved [17]. An element of incidence are the landscapes which appear in the virtual environment, because physical activities are usually carried out in confined places, virtual training allows to improve the comfort of the user and highlight the benefits of exercising in spaces of reality virtual [18]. The intensity of exercise in active virtual reality games has been studied by oxygen consumption and heart rate, obtaining important data to determine the metabolic rate of the activity and making decisions by specialists according to the requirements [19]. Finally, the users of a football training software with immersive virtual reality presented performance improvements of up to 30% in 3 days of evaluation [20].

This work performs the analysis of warm-up on a treadmill using visual stimulation using immersive virtual environments. The purpose is to stimulate the walk by presenting landscapes and measuring the heart rate in each case, the application is developed using the Gear VR and the raw data is acquired using the Polar H7 sensor. The document has been organized in the following manner: *(ii) formulation of the problem*, the problem to be solved is analyzed and a proposed structure is presented; *(iii) Proposed system*, the components of the system are detailed; *(iv) Virtual environment development*, the implementation of virtual environments is described; *(v) Results*, the results obtained through experimentation and applying a usability test are presented; *(vi) Conclusions*, the final proposals are presented.

2 Formulation of the Problem

In physical training, the first step recommended before running exercises is warm-up, where the purpose is to prepare the muscular system for more violent or forceful movements. Technically, most athletes do not have control of the level of warm-up required for their training and the components of gyms and sports areas do not motivate the proper development of this stage of training. In professional sports, a careless warm-up can cause serious injuries and affect the health of the athlete. On the other hand, an inadequate warm-up could generate discomfort in the later phase, affecting performance. For these cases, improving the environment surrounding of the athlete could generate changes in their performance, and it is possible to influence him/her. The level of involvement of the immersive virtual reality in the warm-up phase of an athlete depends on the recreated virtual components and the devices used, this influence could be positive or negative in the user, so it is important to design properly the virtual environments to be used.

The effect produced by the virtual environment in an athlete while performing physical warm-up on a treadmill can be analyzed through a survey or using physiological data. This information can be used by specialists who make the decisions that best suit the athlete. In this context, it is proposed to develop a virtual reality application to stimulate warm-up using a treadmill while acquiring heart rate data in the user, as shown in Fig. 1. The HMD device works by means of a smartphone, using additionally a waterproof belt type sensor called Polar H7.

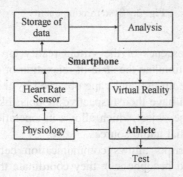

Fig. 1. Proposed structure

The implementation of virtual environments aims to stimulate the athlete through a landscape of relaxation and excitement, influencing the heart rate and motivating the mood of the user. The application can be used on a treadmill or controlled trips in fully structured areas. Finally, the information collected is compared to normal warm-up to determine the variations.

3   Proposed System

The structure of the proposed system is split into three groups: the game objects inserted in Unity, the control scripts for the virtual elements, and the external input and output devices. Figure 1 shows the structure of the system, where each component is presented (Fig. 2).

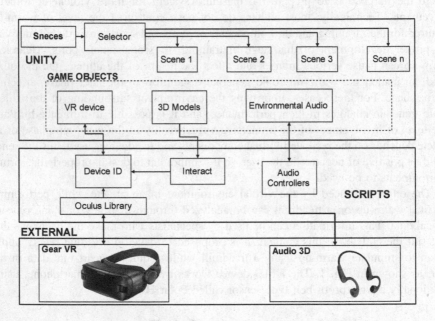

Fig. 2. System components

In Unity there are the scenes of the virtual reality environment, depending on whether it is necessary to stimulate the state of relaxation or excitement. In this stage, the 3D models which make up the landscape observed by the athlete are inserted, the scenes are independent and have their respective models; likewise, the virtual reality device is defined by a game object which allows its operation, and finally, there are game objects linked to spatial audio sources.

On the other hand, scripts allows communication between Unity scenes and external devices (HMD and headphones); they coordinate the functions of the game objects and the actions of the game; and the oculus libraries manage the Gear VR recourses. The input devices are the internal sensors of the Smartphone and the output devices are the mobile virtual reality glasses and the wireless headphones.

For the acquisition of heart rate data, an additional mobile application which receives the sensor information is used. The device used is the polar H7, this sensor is characterized by its robustness and resistance to conditions of almost any sports activity, including swimming. Figure 3 illustrates the heart rate sensor used.

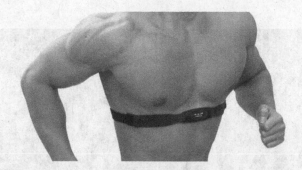

Fig. 3. Heart rate sensor

4   Virtual Environment Development

Two types of environments are developed to implement the application and the functions of the virtual elements. The scene of relaxation consists of sunny days and dense vegetation, and the stimulating environment with dark scenes.

Fig. 4. Scene of Relaxation

Figure 4 shows the images of the relaxation place, a trail-like environment for the athlete's walk surrounded by relaxing attractions. The components have been entered using prefabs from multiple assets and have been placed in the scene by position parameters and rotation parameters of the scene inspector.

Figure 5 shows the images of the stimulating environment, an abandoned place which is in flames. The components have been entered in a similar way to the relaxation area and fire effects have been inserted using the Unity particle system.

Fig. 5. Excitement environment

Regarding the movement of the user, a control script is created which allows the movement of the virtual camera in the walking area, and through *player.transform.- position*, the displacement has been designed in a cyclical manner, that is, when the player reaches the end of the route, he/she is located again at the beginning.

5   Results

Next, the results obtained using immersive virtual reality in the physical warm-up phase of athletes using a treadmill are presented. The results are evaluated using the data acquired by the Polar H7 sensor and applying a usability test of the application. Given the random and constant movement due the sport activities, for the visual output is used a Galaxy S8+ Smartphone with Gear VR, while for the output are used the Z8 wireless headphones. Figure 6 shows an experienced user the virtual reality system.

The experiments are carried out in 4 athletes who regularly attend the gymnasium of the Universidad Tecnológica Indoamérica, all male between 22 and 24 years old. The tests are performed on a standard treadmill at a constant speed of 2.5 km/h and with a duration of 10 min. In the first test, the users perform the exercise without the

Fig. 6. Athlete testing the system

Gear VR, and their cardiac responses are monitored, obtaining the data of Fig. 7, where it is observed that the athletes reach maximum peaks of 143 ppm in this period, maintaining a range of pulsations between 128 and 143 ppm in the stationary stage.

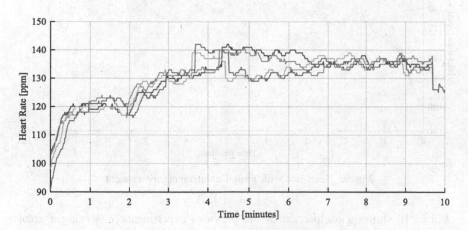

Fig. 7. Heart rate without virtual environment

Evaluating the scenes of relaxation, the individuals generated the cardiac pulses of Fig. 8, observing a maximum heart rate of 136 ppm, and maintained a range of pulsations between 122 and 136 ppm in the stationary phase, which indicates a reduction of heart rate respect to normal exercise conditions, due to the effect of relaxing place on the walk.

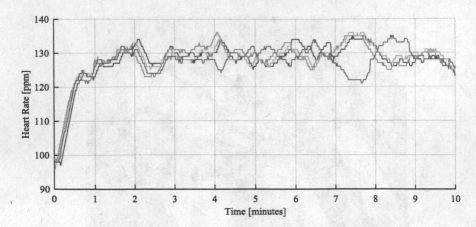

Fig. 8. Heart rate with virtual scenes of relaxation

From the experiment with the stimulating virtual environment, the heart rate signals of Fig. 9 were obtained, showing a maximum heart rate of 154 ppm, and a range of heart beats between 131 and 154 ppm in a permanent period, which indicates an increase in average heart rate with respect to normal walking conditions.

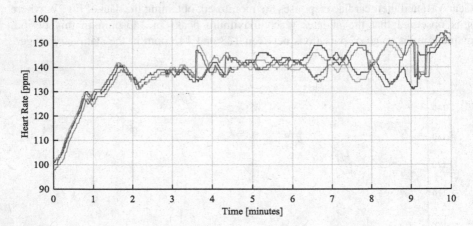

Fig. 9. Heart rate with virtual excitement environment

Figure 10 shows a comparison of the previous experiments, observing a greater heart rate in the case of experiencing a virtual environment of excitement, and lower pulsations when individuals visualize the virtual environment of relaxing. Demonstrating a clear influence of the immersive virtual reality in the physical warm-up with the conditions exposed in this work.

To evaluate the usability of this application, a SUS test is applied, widely used in mobile applications, for which the procedures detailed in [21] are used. Table 1 shows the results of the test applied in the 4 test subjects, obtaining a total of 27.25 and

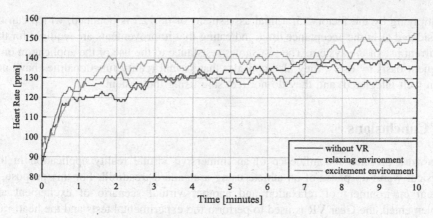

Fig. 10. Comparison of heart rate responses

Table 1. Test SUS results

Question	Score 1	Score 2	Score 3	Score 4	Mean	Operation
I think I would like to use this system frequently	2	3	2	2	2.25	1.25
I find this system unnecessarily complex	1	1	2	1	**1.25**	**3.75**
I think the system is easy to use	4	4	4	4	4	3
I think you would need technical support to make use of the system	1	2	2	3	2	3
I find the various functions of the system quite well integrated	3	4	2	3	3	2
I have found too much inconsistency in this system	2	1	1	2	1.5	3.5
I think most people would learn to make use of the system quickly	4	3	4	4	3.75	2.75
I found the system quite uncomfortable to use	2	2	2	1	1.75	3.25
I have felt very safe using the system	2	2	2	1	1.75	0.75
I would need to learn a lot of things before I can manage the system	1	1	1	1	1	4
Total						27.25

multiplying by the factor 2.5, a final assessment of 68.125 is obtained, which can be considered as in the acceptance limit, indicating that improvements are required for the recurrent use of the system. The test shows resistance to the use of the application on a frequent basis by users, they also consider that virtual reality functionalities have not been well integrated, and they do not feel safe using the system.

6 Conclusions

The work presents the influence of an immersive virtual reality application in the physical warm-up phase of an athlete using a standard treadmill. For this purpose, a virtual environment of relaxation and another virtual scenario of excitement are implemented, the Gear VR is used to perform the experimental tests and the heart rate is monitored by a specialized sensor. The virtual environments are developed using the basic characteristics of Unity, considering that the user experiment in first person, where wireless sports headphones are used to increase the level of immersion.

The results obtained show notable differences in the heart rate of the athletes, starting from the analysis under normal training conditions which allow to compare the effects of the system on the user. The virtual environment of relaxation maintains a heart rate lower than the normal and the exciting virtual environment produces a higher heart rate, conditions which can be used by a specialist to obtain the desired effects in the subsequent intensive training. In addition, the usability of the applications is evaluated, demonstrating that some shortcomings in the system can be improved, mainly related to comfort and user satisfaction.

References

1. Smith, J.W.: Immersive virtual environment technology to supplement environmental perception, preference and behavior research: a review with applications. Int. J. Environ. Res. Public Health. **12**, 11486–11505 (2015)
2. Izard, S.G., Juanes Méndez, J.A.: Virtual reality medical training system. In: ACM International Conference Proceeding Series. pp. 479–485. Association for Computing Machinery, Salamanca (2016)
3. De-Juan-Ripoll, C., Soler-domínguez, J.L., Guixeres, J., Contero, M., Álvarez Gutiérrez, N., Alcañiz, M.: Virtual reality as a new approach for risk taking assessment. Front. Psychol. **9**, 1–8 (2018)
4. Dahdah, M.N., Bennett, M., Prajapati, P., Parsons, T.D., Sullivan, E., Driver, S.: Application of virtual environments in a multi-disciplinary day neurorehabilitation program to improve executive functioning using the Stroop task. NeuroRehabilitation **41**, 721–734 (2017)
5. San Luis, M.A. V., Atienza, R.O., San Luis, A.M.: Immersive virtual reality as a supplement in the rehabilitation program of post-stroke patients. In: International Conference on Next Generation Mobile Applications, Services, and Technologies, pp. 47–52. IEEE Computer Society, Cardiff, Wales (2016)
6. Wang, M., Reid, D.: Using the virtual reality-cognitive rehabilitation approach to improve contextual processing in children with autism. Sci. World J. **2013**, 9 (2013)

7. Ham, J., Cho, D., Oh, J., Lee, B.: Discrimination of multiple stress levels in virtual reality environments using heart rate variability. In: Proceedings of the Annual International Conference of the IEEE Engineering in Medicine and Biology Society, EMBS, pp. 3989–3992. Institute of Electrical and Electronics Engineers Inc., Island (2017)
8. Skarbez, R., Brooks, F.P., Whitton, M.C.: Immersion and coherence in a stressful virtual environment. In: Proceedings of the ACM Symposium on Virtual Reality Software and Technology, VRST, pp. 1–11. Association for Computing Machinery, Tokyo (2018)
9. Yetisen, A.K., Martinez-hurtado, J.L., Khademhosseini, A., Butt, H.: Wearables in medicine. Adv. Mater. **30**, 1706910 (2018)
10. Da SiLva Neves, L.E., et al.: Cardiovascular effects of Zumba performed in a virtual environment using XBOX kinect. J. Phys. Ther. Sci. **27**, 2863–2865 (2015)
11. Gerber, S.M., et al.: Visuo-acoustic stimulation that helps you to relax: A virtual reality setup for patients in the intensive care unit. Sci. Rep. **7**, 1–10 (2017)
12. Bombeke, K., et al.: Do not disturb: psychophysiological correlates of boredom, flow and frustration during VR gaming. In: Schmorrow, D.D., Fidopiastis, C.M. (eds.) AC 2018. LNCS (LNAI), vol. 10915, pp. 101–119. Springer, Cham (2018). https://doi.org/10.1007/978-3-319-91470-1_10
13. Kohani, M., Berman, J., Catacora, D., Kim, B., Vaughn-cooke, M.: Evaluating operator performance for patient telemetry monitoring stations using virtual reality. In: Proceedings of the Human Factors and Ergonomics Society 58th Annual Meeting, pp. 2388–2392. Human Factors an Ergonomics Society Inc., Chicago (2014)
14. Schmidt, S., Ehrenbrink, P., Weiss, B., Kojic, T., Johnston, A., Sebastian, M.: Impact of virtual environments on motivation and engagement during exergames. In: International Conference on Quality of Multimedia Experience, pp. 1–6. Institute of Electrical and Electronics Engineers Inc., Sardinia (2018)
15. Gulec, U., Yilmaz, M., Isler, V., Connor, R.V.O., Clarke, P.M.: Computer Standards & Interfaces a 3D virtual environment for training soccer referees. Comput. Stand. Interfaces. **64**, 1–10 (2018)
16. Yoo, S., Gough, P., Kay, J.: VRFit: an interactive dashboard for visualising of virtual reality exercise and daily step data. In: ACM International Conference Proceeding Series, pp. 229–233. Association for Computing Machinery, Melbourne (2018)
17. Vogt, T., Herpers, R., Askew, C.D., Scherfgen, D., Strüder, H.K., Schneider, S.: Effects of exercise in immersive virtual environments on cortical neural oscillations and mental state. Neural Plast. 2015 (2015)
18. Solignac, A.: EVE: exercise in virtual environments. In: IEEE Virtual Reality Conference, pp. 287–288. Institute of Electrical and Electronics Engineers Inc., Arles (2015)
19. Gomez, D.H., Bagley, J.R., Bolter, N., Kern, M., Lee, C.M.: Metabolic cost and exercise intensity during active virtual reality gaming. GAMES Heal. J. Res. Dev. Clin. Appl. **7**, 310–316 (2018)
20. Huang, Y., Churches, L., Reilly, B.: A case study on virtual reality american football training. In: ACM International Conference Proceeding Series. pp. 1–5. Association for Computing Machinery, Laval (2015)
21. Quevedo, W.X., et al.: Market study of durable consumer products in multi-user virtual environments. In: De Paolis, L.T., Bourdot, P. (eds.) AVR 2018. LNCS, vol. 10850, pp. 86–100. Springer, Cham (2018). https://doi.org/10.1007/978-3-319-95270-3_6

The Transdisciplinary Nature of Virtual Space

Josef Wideström[(✉)]

Chalmers University of Technology, Göteborg, Sweden
Josef.widestrom@chalmers.se

Abstract. This paper presents a transdisciplinary view on virtual space, through a description of how different domains of knowledge inform the concepts of virtuality and space. The aim is to show how these different perspectives come together in the virtual space that facilitates combining science and technology with cultural aspects coming from arts and other domains of knowledge. The argument leads to two models of the understanding of virtual space. The first model is an explanation of virtual space as a hybrid that has emerged from both nature (represented by sciences) and culture (represented by arts). The second model puts the observer in the center, exploring the physical-virtual space through an embodied interaction. The contribution of this paper is twofold. First, it presents virtual space as a platform for transdisciplinary work, exposing its underlying processes from both theoretical and practical point of view. Second, it introduces a model for the way transdisciplinarity can inform the understanding of virtuality that is taking increasing part of our everyday lives as well as variety of knowledge production in form of advanced visualizations, simulations and virtual reality approaches.

Keywords: Virtual space · Transdisciplinary · Natural philosophy

1 Introduction

Virtual Reality research is a multidisciplinary domain of knowledge that includes computer science, applied information technology, cognition, aesthetics and design. This paper aims to contribute to bridging the gap between the two separate academic cultures, the sciences and the humanities, using Virtual Reality with its spaces as a transdisciplinary platform. The two academic cultures are on one hand physics, chemistry, biology, astronomy and computer science, and on the other literature, history, philosophy, art and art practice. It is hard to find a more resistant division in the academic world, than the mutual alienation of natural scientists and humanists. Nicholas Maxwell explains in his In Praise of Natural Philosophy how transdisciplinary research has been lost, and how vital it is for the scientific communities to find their way back to the ideals of natural philosophy [1]. Transdisciplinary research strategies cross disciplinary boundaries to create a holistic approach, and apply to research efforts focused on problems that combine two or more disciplines. Here transdisciplinary research, especially where philosophy meets science, can fulfil the "need to recreate natural philosophy – a synthesis of science and philosophy" [1]. The development of contemporary natural philosophy is encouraged by the possibilities of quick and efficient communication among different scientific, humanistic and cultural

L. T. De Paolis and P. Bourdot (Eds.): AVR 2019, LNCS 11613, pp. 186–202, 2019.
https://doi.org/10.1007/978-3-030-25965-5_15

disciplines which made visible the pressing need for better understanding and common semantics. Nowadays there is an alarming division and separation of knowledge fields that live in their isolated worlds, without awareness of each other. This fragmented understanding of the world presents a problem for the further development of sciences, humanities and arts. Transdisciplinarity occurs when two or more disciplines transcend each other to form a new holistic approach. The outcome will be completely different from what one would expect from the addition of the parts. Transdisciplinarity results in a type of heterogenesis where output is created as a result of disciplines integrating to become something completely new.

Concurrently with the tremendous developments of research and technologies, the emerging phenomenon of virtual space is winning new grounds. The understanding of virtual space requires new insights in the relations between human and space, from philosophical, cultural and artistic, as well as technological and scientific (cognitive, biological, neuroinformatic, etc.) point of view. This paper investigates an understanding of virtuality through a transdisciplinary approach, where sciences meet philosophy. The methodology originates in semiotics and hermeneutics, connecting the production of signs with a human-centered view on interpretation and creation. The contemporary phenomenon of virtual reality with its virtual spaces provides a platform for conceptualizing natural philosophy through multidisciplinarity. In order to support this idea, the paper investigates an understanding of virtuality through a combination of natural sciences, cognitive science, philosophy, and art. The aim is to show how these different perspectives come together in a holistic view of virtual space. The focus on 'space' is motivated by its wide range of connotations and applications in different research disciplines and practices. Space is here used as a concept for the structural properties of an entity where objects and events are related. Space is the conceptual framework that gives the conditions for these relations, while at the same time being constituted by these relations.

2 Physical Space

Through the history of humankind space has been experienced and investigated in different ways, in the traditions of a variety of specialist fields. Space has been measured in distance, connected to real-time and related to motion, through explorations in natural sciences. Space has been the subject of extensive research and literature, where philosophy is deeply connected to physics [2]. The fundamental concepts for space created in the 17th and 18th centuries (by Leibniz, Newton, Kant and others) have formed our knowledge of space to this day. From these theories 'Space' can be seen as an abstract and discrete set of objects and voids formed by relations, a continuous and measurable entity formed by forces, or a synthetic framework for organizing experiences. It is the combinations and relations between these different explanations, theories and models that have formed our understanding and knowledge of space.

However, the scientific and philosophical theories do not cover all our understanding of space. Space has also been explored culturally and understood through literature, architecture and art. Spatial representations in visual art, fiction movies and books have contributed to the understanding of space as physical phenomenon. Our perception of space and the relations between what we see and what there is has also been formed by ground-breaking artistic work, such as Picasso's cubism or Magritte's surrealism. Architects have developed theories and skills how to analyze and create physical spaces for human life. The dimension, organization and shape of physical spaces are created in relation to certain needs and certain contexts. Different human needs and human activities require different spatial structures, and these structures are shaped by technological, topological, economic, environmental, social and other conditions.

3 Virtual Space

Today we have a completely new world of virtual reality with its spaces. New rules and conditions apply for the development of virtual spaces. There are different conditions for dimension, structure and gravity in virtual spaces compared to physical ones. Contemporary knowledge about virtual spaces is based on natural sciences and developed through computer graphics, architecture, art, interaction design, cognitive science, semiotics, hermeneutics, and social science.

The term 'virtual' is loosely defined. With all its widespread use in both popular culture and academic discourse, what does this term actually mean? As a starting point in the framing of 'virtual' in this text, there are some definitions that have to be made. The word 'virtual' could stand for anything that is seemingly unreal or intangible, yet maintains some kind of existence on some other level of reality, in other words, something that exists on a metaphysical level. This can be related to the idea of alternative realities parallel to ours (as found in physics), which remain 'virtual', until they 'actualize'. The view (as found in fiction and poetry) is that the world in which we live might be nothing more than a result of our imagination, and thus what we call 'reality' is virtual as well. The meaning of 'virtual' as something unreal or metaphysical is however not used in this paper. Other uses of the word 'virtual' may have a somewhat metaphysical sense as well. For example, virtual can be used to describe how objects and spaces are imagined through books, music, or other media. This notion of 'virtual space' as a space that is envisioned or visualized internally is not used here, but rather referred to as cognitive space. In quite a similar way, virtual is commonly used to describe our experiences when we browse the Internet or as the third space between two people communicating over distances. In all these cases there seems to be some other dimension in which the contents of these experiences exist – beyond the vibrations in the air, the printed letters on the paper, or the electric signals running through computers – and we need a name for it. So, the word 'virtual', with its inherent ambiguity, often satisfies us as a replacement for a wide range of different things going on. But virtuality is not primarily about computers and definitely not about metaphysics. Here 'virtual space' is not a matter of imagination but rather perception, interpretation and experience. A written text is therefore not a virtual space in this sense, because the reader would have to create a mental picture fundamentally different from the pattern of black and white on the book page.

For the framing of the term 'virtual space' I highlight the concepts 'image space' and 'digital space'. Image space is a space that is not physical, and yet not imaginary. Image space is the abstract space that is accessed through and in images, the overall space of all pictorial media. The notion of 'image space' suggests that what we see in a pictorial image is located in another space that is neither physical nor imaginary. Image space is not just the space of a particular picture, but rather the overall space of all pictures, and of all pictorial media. This is stated in analogy with physical space that is the overall space of all physical places. Image space is the abstract space that is accessed through and in images. Structurally, images work as interfaces to image space where the semiotic code forms the language of creating and reading images. Meaning is produced in a communicative process that involves context, space, representation, and interpretation, Consequently, 'digital space' is the overall space of all spaces created by digital media.

In this text I make a distinction between (mutually overlapping, but still distinct) spaces: physical space, virtual space, image space, and digital space. The focus on digital space is motivated by a media perspective, where the means of production and modes of interaction are forming both the experience and the structure of the space. Present day communication of knowledge is done via media – journals, online, etc. and thus this digital space of knowledge communication plays a role in creating, communicating and adopting knowledge. The notion of 'image space' is related to human knowledge such as found in art, art history, visual culture, while 'digital space' relates to natural sciences such as computer science, computer graphics, systems and simulations. The intersection of image space and digital space is used to explain the use of 'virtual space' (Fig. 1). This model is an explanation of virtual space as a *latourian hybrid* that has emerged from both culture (art) and nature (science) [3].

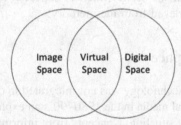

Fig. 1. Virtual space as the intersection of image space and digital space

Seeing the image as an interface to image space emphasizes the importance of understanding the functions of the image: its dimensions, properties, and its function as a sign. An important function of the image is visualization, in the sense of representing something externally in visual form. That 'something' might be concrete information or abstract ideas, real or imaginary, but its representation in visual terms is physical and concrete. Visualization is the act of communicating this something using the image as medium. This communication does not follow any given rules or simple recipe, but is rather open to a continuous negotiation and development of new concepts. In the evolution of images, from the first cave paintings to the high-resolution, interactive visualizations of today, the development has not only been technological but of course also conceptual. Contemporary image media have not only created new ways of accessing image space but they have also transformed the space. New ways of creating and interacting with images have made new spaces possible. The image is not just an interface that is disconnected from image space but rather closely interconnected with the 'something' it is communicating.

Now, what does 'virtual' mean in relation to 'real'? Is not Virtual Reality just another medium or technology and as such a subset of the real world? Is the virtual a simulation of the real or a representation of an imaginary world? Or the both at the same time? Is it the real that stands for something else or is perceived as something else? The distinction of real and virtual, reality versus virtuality, is an idea that has been investigated and represented over thousands of years, one of the most fundamental being Plato's cave. Pierce defines virtual as "something that is 'as if' it were real" [4]. Virtuality as a philosophical concept in Deleuze builds on Proust's idea of a memory as "real but not actual, ideal but not abstract", developed in the following formulation:

> "The virtual is opposed not to the real but to the actual. The virtual is fully real in so far as it is virtual. Exactly what Proust said of states of resonance must be said of the virtual: 'Real without being actual, ideal without being abstract'; and symbolic without being fictional." [5]

The virtual is a potentiality that becomes fulfilled in the actual. Hence, it is not material but still real. If we search the answer to the question of virtual vs. real in cognitive science, "the difference between real (actual) and virtual is not as sharp as one might believe" [6]. Already Minsky in his *Society of Mind* reminds us that even our everyday experiences are not direct and they are not even happening in "real time" [7]. There is always a time delay between the event in the world and our perception of that event, that relies on memory. When observing a scene in "real time" we actually observe only a small part of the scene which is expected to be changing, while the majority of the scene is retrieved from the memory.

4 Physical-Virtual Space

In the beginning, computer technology was not integrated in our physical environment. With the emergence of digital media in the 1980–90's an explosion of development has led to a completely different situation. Concepts from information technology, such as the Internet and computer graphics, are closely related to television, film, and radio. In entertainment areas, like computer games, the two worlds are completely unified. Today there are no important distinctions between digital media and computer technology. More and more physical objects and spaces become digital, computers are becoming ubiquitous, embedded in our everyday objects and environments and embodied in the way we experience them in our everyday life. In human-computer interaction the concept of *embodied interaction* is a way to resolve this physical-digital divide [8, 9]. The concepts of 'physical space' and 'digital space' have been developed further into the *Four Space Model*, including also 'interaction space' and 'social space' [10].

In our everyday life, in our homes and work places, we are not always present only in a physical environment. We also experience virtual environments, mediated through different devices. In certain situations, both professional and otherwise, the relations between physical and virtual spaces become essential for the experience and understanding of the spaces.

Using Deleuze's terminology, the virtual is a surface effect produced by actual causal interactions at the material (physical) level. When one uses a computer, the screen displays an image that depends on physical interactions happening between the actor (user) and the computer (at the level of hardware). The virtual space is nowhere in actuality of the outside world, but is nonetheless real and can be interacted with as it is present in our cognition. Simultaneously, the actor is present in a physical space, where the screen works as a window into the virtual world. An actor who interacts with both a physical and a virtual space simultaneously, can be said to be present in a physical-virtual space.

In order to investigate the relations between physical and virtual space I focus on the *experience of space* in the phenomenological sense and the *structure of space* in the architectural sense. In the holistic approach presented here, virtual space is the inter-section between 'image space' and 'digital space'. Virtual space is seen as separate from physical space in an architectural (structural) sense, but the two worlds co-exist in an interdependent relation. An actor/user/observer can experience presence in both physical and virtual space simultaneously, through an interaction space that involves both physical and virtual space, meaning that this actor interacts in physical-virtual space through an embodied interaction (Fig. 2).

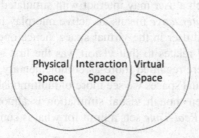

Fig. 2. Physical-virtual space as the interaction space

Physical space and virtual space are entities that exist in reality as subsets of the wider entity of space. The co-existence of physical and virtual space makes it possible to experience both physical and virtual space, even simultaneously, creating a unified physical-virtual space in a phenomenological sense [11]. From a phenomenological perspective, the level in the hierarchy of physical space and virtual space is equal, so one is not a subset of the other.

The emergence of augmented reality and mixed reality spaces has led to new experiences and possibilities. From a practice-based perspective, we can see that this co-existence of physical and virtual space also creates a challenge for designers, architects, and artists that work with spaces for human interaction and experience. In the domain of interaction design, the physical-digital divide has been resolved with the notion of embodied interaction, and connected to space through an increased interest in presence [12].

5 Seeing Virtual Space Through Perception: "Being There"

Virtual space can best be studied when created by Virtual Reality (VR), an invention, engineered by advanced computer technology. VR technology presents both a "tool" and a "world". VR is a computer medium used as a tool to convey a message to a user, just like any other medium. At the same time, as a medium, VR can be so perceptually persuasive and interactive that the user/actor can experience *presence in* the virtual environment which thus plays a role of a world. Using Virtual Reality as a way of exploring what a virtual space is, and what it can be, goes back to the pioneers of VR [13]. The technical definitions were stipulated in the 1980's by researchers in the field of computer science and neuroscience [14]. An important conclusion of these views was made in *The Metaphysics of Virtual Reality* [15] that analyzed virtual reality into seven different concepts: simulation, interaction, artificiality, immersion, telepresence, full-body immersion, and network communication.

A typical technical description of Virtual Reality reads *Interactive Visual Real-Time Computer Simulation*. Hence, in order to be able to claim an environment "virtual" we need to fulfill these five conditions (interactive, visual, real-time, computer-based, simulation). There are numerous variations on this definition, such as in *The American Heritage Science Dictionary* "A computer simulation of a real or imaginary world or scenario, in which a user may interact with simulated objects or living things in real time" [16]. Here *Interactive* means an active interplay between user and virtual space or between user and user in the virtual space, hence open for intervention from the user. The term *Visual* relates to that vision was the first sense to be used in VR, while the other senses were regarded more as complementary modes of virtualization [14]. In contemporary virtual spaces we see more of multimodal interaction, using aural and haptic interfaces, even though visual simulation is almost without exception in focus. The technical term *Real-Time* sets a limit for what is considered to be immediate response.

Virtual space is changing the way we live our daily lives, both as a society and as individuals. We can be present in virtual worlds and have access to virtual institutions and work places. Through the technology of Virtual Reality, Augmented Reality and Mixed Reality we get new experiences and gain new knowledge. New hybrid physical-virtual spaces emerge with new possibilities for interaction. The majority of research and development in Virtual Reality has been to use it as a way to simulate physical reality [17]. Yet VR is a medium that has the potential to go far beyond anything that has been experienced before in terms of transcending the bounds of physical reality, through transforming your sense of place, and through non-invasive alterations of the sense of our own body. In other words, virtual reality has rarely been seen as a medium in its own right, as something that can create new forms of experience, but rather as a means of simulating existing experience [18, 19]. VR needs to be handled as something with its own unique conventions and possibilities that provide a medium where people respond with their whole bodies, treating what they perceive as real.

Virtual spaces give new experiences. In a virtual space, we could for example see temperature and air flow in a room, listen to molecules, walk around in buildings that are about to be built, alter the chain of events in a historical scene, or fly through galaxies experiencing the birth of stars. Through Virtual Reality all these things can be communicated perceptually and not by suggestion, dreams or hypnosis. Concerns about Virtual Reality and other digital spaces are raised, that these offer a "low-resolution" life [20], which refers to the low granularity or low media richness of multimodal sensory input in comparison with "real life". Virtual spaces are here seen as "almost real". These concerns are valid for situations where certain aspects of realism or face-to-face communication are lost, but on the other hand we must realize that Virtual Reality also makes new experiences possible. We can be tele-present with others over long distances and augment our senses with new representations and layers. Virtual spaces are hence also "more than real".

Understanding virtual space through cognition, we need to focus on the user's experience of immersion and the concept of *presence*, the sense of "being there". Studies have shown that the degree of immersion in a virtual space has a positive relation to the degrees of user performance, communication and collaboration in VR applications [19, 21], meaning there is an objective to take the technology further, hence "more virtual". In this area of experience-oriented definitions, I see five factors connected to presence that are important to present here. These factors are *Perception, Transparency, Transportation, Attention* and *Social factors*. Here *Perception* means the sum of all sensory input that together give the user a sense of being in a space, other than the physical space that the user is physically present in. A higher quality of sensory input is regarded to lead to a higher degree of presence [17]. Image, sound and touch can today be virtualized to an almost life-like level, so that the user will have trouble telling the physical from the virtual, in a mere sense of perception. However, the *Transparency* of the medium is not always as high as one strives for in order to keep a high level of immersion. Computer screens can have poor resolution, there might be cables that users get tangled in, there can be delays in the communication or low frame rate. Apart from technical problems there can be disturbing real-world noise or light or the user can get nauseous. These are all examples of presence-breaking factors due to low transparency. *Transportation* is a factor that in my view actually reaches the core of Virtual Reality. It has to do with the sense of being in another place, to move away from or beyond physical space and "travel" to a virtual space. The comparison with Cyberspace is not far away here, in a very everyday meaning. When we use the Internet, we use metaphors such as "visiting" a website and "surfing" the Internet, even though we just download data from a server to our own computer. It is the same sort of agreement that a user can make with a virtual space, if the environment uses those sets of metaphors that encourage traveling. However, these three presence factors mentioned above could all be over-ruled by the *Attention* factor. This issue has to do with how interesting and meaningful the environment is for the user. It does not matter if the VR application runs on a giant screen in real-time, completely wireless and immediate in response, if the user is not interested or if it does not make sense. And the other way around, if the user is completely focused on or is fascinated by the content in virtual space, a lot of perception and transparency failures will be forgiven. The coherence of agreements and experiences creates the plausibility of the virtual space [17]. We all know how we start noticing what the chair feels like in a movie theatre if the film is boring, or how we can forget thirst or hunger when we get lost in an exciting book. I have on many occasions seen people so excited about a virtual world that they laugh out loud, cry, jump back or even fall over, from just a crude set of polygons shown in the right way at the right time. *Social factors* are also very important for the degree of presence, due to the obvious reason that we are social beings and as such we are affected by other peoples' interactions. If there are other virtual subjects (avatars) users can meet and interact with, the user will feel more present in the virtual space, in the sense of "being there together" [22, 23].

Virtual Reality is often seen as a medium where the human body is detached, that an actor in a virtual space is disembodied. One reason for this is that VR has a background in the ideas about cyberspace, which is explored by the mind rather than the body. Another reason is the conceptual and technical background of VR in its early military and scientific use, where the actor in a virtual space is regarded primarily as a camera with a point-of-view and secondly as a hand with some type of interaction device. The actor is actually somewhere else, outside the virtual space. It can of course be argued that, no matter how transparent the interface is, the user is always in front of a screen or looking into the virtual space, he or she is not actually "there". But what is actually the difference? Isn't it true that we see, hear and interact with a virtual world using our bodies and senses just as we do in the physical world? We are as humans trapped in our own bodies; we can never really be disembodied [24]. We can always in our dreams and fantasies leave our physical reality, but when it comes to perception of an outside world, there is no fundamental difference between reality and virtuality. What Husserl says about our *life-world* applies well to how we experience the virtual as real and vice versa. Husserl's idea of *lebenswelt (life-world)* shows how everyone lives primarily in a subjective world of cognitive space, rather than in directly in a shared physical one [25].

A key aspect of presence in virtual space is the difference between watching and acting. One of the fundamental concepts of VR as realistic simulation of physical environments, is that the user is understood as a viewer that gets access to the virtual world through a camera (point-of-view) in the virtual model. Here presence is measured through the degree of immersion in the virtual environment by realism, in framerate and screen resolution. This immersion creates a "place illusion" that gives the user presence, in the sense of "being there" [17]. The presence can be broken by inconsistencies in behaviors and actions. Therefore, a high degree of presence also requires that the "plausibility illusion" is fulfilled. This does not mean that the virtual space has to be realistic, but rather coherent in relation to the agreements that are made between actor and space. Virtual space becomes a place for human life through the cognitive processes of navigation and identification.

6   The Role of the Observer

In the current renaissance of natural philosophy triggered by the rise of multidisciplinarity, we want to understand the relation between the observer and the world. In this context, the works on phenomenology by Husserl and Heidegger have found new interest. Husserl's phenomenological reduction (the suspension of judgment about the natural world and focus on subjective experience) and Heideggers concept of Dasein or "being-there" are now used to create new frameworks and extended theories for the relations between the observer and the world, philosophy and natural science, between culture and nature. The human is not only an observer, but importantly also an actor in relation to the world.

Brier's transdisciplinary theory of *Cybersemiotics* [26], presents an attempt to meet this challenge of connecting the observer with the outside world. By combining the 19th century Piercean semiotics, with contemporary theories of phenomenology and cognition, Brier constructs a non-reductionist framework for the integration of natural sciences with first-person experiences (cognition) and social interactions (culture). Cybersemiotics sets semiotic cognition in the centre for the understanding of reality, connecting to the four aspects; surrounding physical nature, biological corporality, subjective experience, and our social world. Through Brier's distinct analysis it becomes clear that humanities and sciences enrich each other and that this mutual dependence create not only a wider perspective but also a deeper understanding.

Rössler's Endo-physics presents philosophical extension and interpretation of the natural sciences [27] that is important to the understanding of virtual space. Much as Rössler proposed in endo-physics, reality as the interface between outside (exo) and inside (endo) worlds, Virtual Spaces are understood from within the spaces, through their interfaces. Observer and interface are therefore just as central issues in endo-physics as in conceptualization of virtual space. The observer is represented in the virtual space as a camera or a viewpoint that changes the space. From the exo-perspective, virtual spaces can be measured in bytes, polygons or pixels, while it is only from the endo-perspective that the space can be subjectively experienced. The human being is therefore part of this virtual universe, and the world is the interface between the observer and the rest of the world, using Rössler's terminology. The difference between virtual space and the actual world that Rössler discusses is that we as creators of virtual space have access to the interface, and design the interface, meaning that we can actually step outside the virtual world into the remaining (actual/physical) world. Still, the observed reality from within relies on subjectivity as the observer inevitably distorts the world or *actively* perceives and constructs the world locally. Virtual space has the potential to work as model worlds that simulate exo-models of endosystems.

All of these post-modern theories strive to find meaningful analysis of complex systems, without reducing these systems to mere physics and/or information. They show that one can include first-person experience and thought as well as social communication in natural science without making it arbitrary or random. Virtual space is a true *hybrid* in Latour's meaning, with its emergence from nature and culture: "Nature and culture shape each other, producing hybrids" [3].

This first-person perspective in natural philosophy connects to the human-centered understanding of virtual space. These connections are different in different domains, which contribute to a diverse understanding of virtual space. From a semiotic perspective, new connections can be made between the codes of the overlapping domains that inform the knowledge of virtual space. The relations between natural philosophy and virtual space extend the semiosis (sign process) in these domains. When natural science meets philosophy and arts in this context, new knowledge is created. This production of new knowledge does not only happen by random connections, but also from intentional, designed efforts by the communities (both theorists and practitioners) from the different domains.

When the *observer* discussed by Rössler in terms of interface with endo- and exo-reality, is the observer in virtual space, its interactions are often focused on visual aspect (optics). The observer in a three-dimensional virtual space is represented as a camera that changes the projected view from the observer's perspective by the rules of optics. In analogy, sound and haptic feedback can adapt to the observer's location in the virtual space. In interaction design (the practice and theory of designing interactive digital products, environments and systems) the human is understood as the *user*. From this user-centered perspective, the focus is on human-computer interaction and behaviors. This means that interaction design synthesizes digital space, physical space, interaction space and social space through an embodied interaction between human and space [10]. In performance arts the relations between human and space is articulated by the triangular relations between actor, stage and spectator [28]. The *actor* makes use of the stage in relation to a narrative with the spectator as audience. In visual arts the human is seen as a *viewer* and/or creator that relates to the work of art in different ways. This view of the creating and observing human puts virtual space in an artistic discourse, leading to the understanding of virtual space as image.

7   Science Related to Arts and Aesthetics in Virtual Space

From the ancient time of Aristotle, via Newton, Leibniz and Kelvin, natural philosophy or philosophy of nature was the philosophical study of nature and the physical universe, where universe was synonymous with "reality". With the development of new technologies as well as new scientific methods including simulations and advanced visualisations, the immediate connection between the observer and the world "out there" has become increasingly complex. More and more of information we perceive about the world is heavily "pre-processed", it is also entering complex associations with other knowledge. Thus, increasing part of information in our knowledge about the world is not directly perceived but is mediated through steps involving instruments and equipment including computers that transform original data observed "in the world". With the computer technology of today we have the possibility to both dig deep into the microscopic world and see the big macroscopic picture, and also connect different areas of knowledge belonging to various domains and levels of abstraction. This means that a deeper understanding of images and virtual spaces as an interface to the world becomes important for natural science.

Natural science has always been subject to aesthetic concepts such as symmetry, harmony, simplicity and complexity, and aesthetic values such as finding beauty in nature and the beauty of truth [29]. These aesthetic concepts and values create subjective relations between the scientist and the science. Today the connections between science and aesthetics go even further and are more concrete, with the increased use of advanced visualizations in science. These visual representations have developed from simple graphs to images, animated images, simulations, and virtual environments. Advanced visualizations are increasingly becoming the interface to the world that is observed in science. As such they are developed in a design process that includes decisions about color, framing, grid, perspective, etc., using the aesthetic concepts of symmetry, harmony, parsimony and similar. This also means that these images and spaces are related to other images made and thus part of our visual culture. What is observed by the scientist is perceived logically and rationally but also intuitively through various filters of senses and aesthetic judgement. The visualizations are not only representations of data and relations but also carriers of aesthetic, artistic qualities. They have become a medium for representing and exploration of data generated by observations/measurements and theoretical models.

Also, the multimodality of these interfaces has become richer in the use of not only visual, but also auditory and haptic interfaces, turning them into interactive virtual spaces. This development of virtual spaces is a result of transdisciplinary progress in technology, sciences and arts. Virtual spaces are as such not only technological constructions with functional purposes but also designed artifacts, and subject to values and aesthetics.

Artworks, as well as virtual spaces, are meant to generate perception. Virtual spaces take form as representations of intentions and are interacting with and being experienced by users (actors, viewers) as a gestalt. These aspects imply that virtual spaces have an object-subject relation with the user, that is similar to the relation of an artwork with its observer. In addition to this, virtual spaces cannot be "looked at" in the same way as a painting or a photo. It is the experience of the virtual space from *within* that gives meaning. This means that we can talk about the *endo-aesthetics* of virtual space, leading to research in media art.

One of the most important issues of media art research is the relation between the viewer and the work. The pioneer works by Heilig (Sensorama) in the 1960's for the head-mounted display, by Weibel ("Inverse Space", "Tangible Image") in the 1970's for the works on observer-dependent worlds, and by Davis (Osmosis) in the 1990's for the immersive VR-cave have shown how virtual spaces can work as alternative artificial worlds and their interfaces as windows to another world [30, 31]. These metaphors of windows (and doors) to another world are important for the semantics as well as aesthetics of virtual space, since the observer acts within two parts of reality; the perception of virtual space, and the consciousness of acting in a simulation. The interactions of the observer result in spatial and temporal experiences that then lead to new interactions in this endo-system. However, for the discussion of endo-aesthetics of virtual space it is important to see that the observers' presence in the virtual space is only part of their cognitive processes. Another part is still controlling the presence in the physical (actual) world outside. An aesthetic experience of virtual space is dependent not only on the endo-system but also on the exo-system where the world outside constitutes the context. Therefore, we can talk about a degree of presence on a continuum from virtual to actual, keeping in mind that presence is dependent on physical interaction, whether in virtual or actual space.

This makes the semantics of virtual space quite complex. From a semiotic perspective, the semiosis (production of signs) takes place in an interplay between the experienced virtual space, the observer's physical space and body, the observer's cognitive processes, and social/cultural context. Here Cybersemiotics [30] can be used to analyse these relations. Although the internal observer (inter-actor) is physically located in the real world, he/she contributes to the creation of an artificial model world in which the observer (actor) participates. The observer is in fact "in the picture" while his/her body remains in the actual physical space. This means that a coherent and understandable space for an observer (actor) is dependent on the semiotic code created in an interplay between endo- and exo-system. The experience depends on a double-duality; on one hand between world-observation and self-observation, and on the other hand between the immateriality of the virtual space and the materiality of the physical body. These various levels of reality (endo and exo) show the double game played by endo-semiotics (endo-aesthetics). The observer-dependent reality, that is the reality as the interface between the observer and the other world, in combination with the distinction between internal and external observers' phenomena [43] create conditions for the development of an endo-aesthetics; the aesthetics of self-reference, of virtuality (the virtual space), of interactivity (the actions and the role of the observer within the system), and the interface (the conception of the world as the interface). As such, endo-aesthetics enable an analysis of virtual space from a media art perspective, where the observer (viewer/audience) is located in the system where it interacts.

This understanding of virtual spaces from an endo-aesthetic perspective evolve from Welsh's concept of an "aesthetics beyond aesthetics" [32] and from the transition from art to space to system. These spaces can be described from various perspectives as complex, flexible, context-conditioned, hypermedia, and multidisciplinary systems. From the endo-aesthetic perspective these virtual spaces "exist" (make sense and appear) as such only through an active relationship between actors and the (actual or virtual) system. The virtual space as system is always potential, and does not exist autonomously. It is constructed based on semantic/semiotic/aesthetic conventions where user has possibility of changing or choosing the "rules of the game" that govern the space. Understanding virtual spaces from an endo-aesthetic perspective enables creation of virtual spaces and realities as systems or model worlds. It supports flexibility of observer-dependent systems, and the integration of internal observers into a virtual system that can be observed from the external perspective.

For example, a condition for the endo-aesthetics of virtual space concerns mixed reality, where both internal and external participants are inside a virtual space in which they exchange messages in order to generate new communication structures that become constitutive elements of the simulated world. An endo-aesthetics of virtual space is reliant on the relativity of an observer-dependent world and the possibilities resulting in reference to internal observers, to the world as interface, and to the relationship between physical and virtual spaces. The phenomena of telepresence and co-presence, where the interactors physically located at different places come together as telepresent inhabitants of the same virtual space, create semantic and aesthetic conditions unique for virtual space. Also, alternative biological-virtual interfaces open up the interaction to natural processes of the body such as eye movement and breathing. This embodied interaction unfolds the observer's self-perception via the self-controlled activity of the body, giving the interactor the impression of taking part in a natural fashion in the virtual space. This integration of body and space provides new conditions for the semantics and aesthetics of virtual space [28].

Virtual space can be seen as a system where art meets science. In this space the actor becomes part of what (s)he observes. Distortion triggered by the observer in the reality of the environment is provoked likewise by an actor participating in the artificial, interactive system. In a simulated artificial world, the internal observers have access to certain actions and interventions of which the effects allow them to draw conclusions for their own environment. When these actions, interventions and effects are different in virtual space compared to physical space, a different semiotic code is established. This code forms the fundament for the space and for the agreements and experiences that are made. The complexity of the semantics of virtual space shows how aesthetics, as subordinate to semantics, becomes complex and different from aesthetics in the actual world. Hence the "aesthetics beyond aesthetics" is fundamentally different in virtual space, and form conditions for an aesthetic experience of a setting in virtual space different from the aesthetic experience of the corresponding physical setting.

8   Discussion

In the current renaissance of natural philosophy and transdisciplinarity, we need to interconnect the best theoretical knowledge and best practice from different domains. Contemporary phenomena, such as virtual spaces, create common platforms where computer science and technology can meet human sciences and arts. We can see that the concepts of virtual space relate to a wide range of research disciplines; natural sciences, philosophy, psychology, cognitive science, social science, and fine arts. The relations between virtual space and these disciplines go two ways: (a) virtual space can connect transdisciplinary research in different combinations of these disciplines providing visualizations and simulations, and (b) all of these disciplines inform the understanding of virtual space as a phenomenon and technology.

We can see how science and arts come together, not only side by side, but also transcending in new forms in the wide range of expressions and applications of virtual space. For natural science, a deeper understanding of images and virtual spaces as an interface to the world is vital. With the development of new technologies as well as new scientific methods including simulations and advanced visualisations, the connection between the scientist and the world has become increasingly complex. Information in our knowledge about the world is not directly perceived but is mediated through interfaces that transform original data observed "in the world". This transformation is dependent on the semantics of virtual space and its semiotics.

With the emergence of a variety of virtual environments and hybrid physical-virtual spaces in our everyday life, a deeper understanding of virtuality is needed. This calls for transdisciplinary approach on virtual space, in the line of Maxwell's ideas. This perspective is important for the understanding of virtual space, since it embraces both natural sciences, philosophy, and art. I highlight the potentials of the emerging area of virtual space to become a platform for transdisciplinary research. The transdisciplinary nature of virtual space is demonstrated by its richness in aspects, connecting the inherent concepts of virtual space with a wide range of knowledge and practice domains. Here the contemporary phenomenon of virtual space provides a platform for conceptualizing natural philosophy through multidisciplinarity.

An aesthetic perspective on virtual space is also vital, since aesthetics (beyond aesthetics) shows the relations between actors and objects and the specific conditions of virtual space. From the endo-aesthetic perspective these virtual spaces exist as such only through the interactions between actors and the system.

On the other hand, understanding virtual space requires a broad philosophical perspective, with insights where physical nature, digital information, corporal embodiment, first-person experience, and social aspects come together. Holistic and transdisciplinary approaches like Brier's *Cybersemiotics* provide important framework for understanding virtual space. It becomes clear how meaning in virtual space is created in a semiosis of nature, embodiment, language and subjective experience. The discussion shows how these different perspectives come together in understanding of virtual space. In this synthetic view, virtual space is seen as a framework for making experiences and agreements. The human is put in the center of virtual space, and therefore in the center of the understanding of natural science and philosophy in this context.

Virtual Space as a *hybrid* combines aspects that would be considered to belong to the traditionally separate natural and social realms. For Latour, the distinctive characteristic of modern societies is that they differentiate between nature and society, whereas the premodern civilizations did not make this difference. Latour opposes this duality, and argues that our culture needs to reconnect the natural and social aspects. The hybrid of 'virtual space' successfully accomplishes this synthesis for us. The notion of 'image space' is related to human knowledge fields such as art, art history, visual culture, while 'digital space' relates to technical and natural sciences such as computer science, computer graphics, systems science and simulations. This virtual space hybrid bridging traditionally separate fields facilitates transdisciplinary research in new fields such as interaction design and cognitive science and offers possibilities for studies of variety of other transdisciplinary, cross-disciplinary and multidisciplinary fields.

9 Conclusions

This paper is based on arguments, examples and discussions resulting in two models of the understanding of virtual space. For the first model, the paper elaborates the notion of virtual space offering the suggestion that virtual space can be understood as the intersection of 'image space' and 'digital space'. This view has the potential to give new insights in virtuality, as a contemporary example of a Latourian hybrid. The second model shows how the physical-digital divide can be resolved for spaces, from a focus on embodied interaction and the role of the observer. The contribution of this paper is twofold; first it presents how virtual space creates a common platform for transdisciplinary collaborations crossing the boundaries of sciences, philosophy, humanities, and arts, and secondly in the way transdisciplinarity informs the understanding of virtuality.

Acknowledgments. I would like to thank Professor Gordana Dodig-Crnkovic for her sustained encouragement and constructive criticism.

References

1. Maxwell, N.: In praise of natural philosophy a revolution for thought and life. Philosophia **40**, 705–715 (2012)
2. Jammer, M.: Concepts of Space: The History of Theories of Space in Physics. Courier Corporation, North Chelmsford (1954)
3. Latour, B.: We Have Never Been Modern. Harvard University Press, Cambridge (1993)
4. Pierce, C.: Dictionary of Philosophy and Psychology, pp. 763–764. Macmillan, New York (1902)
5. Deleuze, G.: Difference and Repetition. Columbia University Press, New York (1994)
6. Dodig-Crnkovic, G.: Cognitive revolution, virtuality and good life. AI Soc. **28**(3), 319–327 (2013)
7. Minsky, M.: Society of Mind, Simon & Schuster (1988)
8. Dourish, P.: Where the Action Is-the Foundations of Embodied Interaction. The MIT Press, London (2001)
9. Ehn, P., Linde, P.: Embodied interaction – designing beyond the physical-digital divide (2006)
10. Eriksson, E.: Spatial explorations in interaction design. In: Interaction Design (2011)
11. Meleau-Ponty, M.: Phenomenology of Perception. Routledge and Kegan Paul Ltd., London (1962)
12. Hallnäs, L., Redström, J.: From use to presence: on the expressions and aesthetics of everyday computational things. ACM Trans. Comput. Hum.-Interact. **9**(2), 106–124 (2002)
13. Sutherland: The ultimate display. In: Proceedings of the IFIP Congress (1965)
14. Ellis, S.R.: Nature and origins of virtual environments. Comput. Syst. Eng. **2**, 321–347 (1991)
15. Heim, M.R.: The Metaphysics of Virtual Reality. Oxford University Press, Oxford (1993)
16. The American Heritage Science Dictionary. Houghton Mifflin Harcourt (2005)
17. Slater, M.: Place illusion and plausibility can lead to realistic behaviour in immersive virtual environments. Philos. Trans. R. Soc. B Biol. Sci. **364**, 3549–3557 (2009)
18. Brooks Jr., F.P.: What's real about virtual reality? Comput. Graph. Appl. IEEE **19**, 16–27 (1999)
19. Axelsson, A.-S., Wideström, J.: Cubes in the cube: a comparison of a puzzle-solving task in a virtual and a real environment. CyberPsychol. Behav. **4**(2), 279–286 (2001)
20. Taylor, J.: Psychology today. (2011). https://www.psychologytoday.com/us/blog/the-power-prime/201105/technology-virtual-vs-real-life-you-choose. Accessed 14 Oct 2018
21. Mel Stater, R.S.: Small group behaviour in a virtual and real environment: a comparative study. Presence: J. Teleoperators Virtual Environ. **9**, 37–51 (2000)
22. Wideström, J., Schroeder, R.: The collaborative cube puzzle: a comparison of virtual and real environments. In: The 3rd Collaborative Virtual Environments, San Francisco (2000)
23. Schroeder, R.: The varieties of experiences of being there together. In: Being there together (2010)
24. Wideström, J., Muchin, P.: The pelvis as physical centre in virtual environments. In: ACM Conference on Presence, Delft (2000)
25. Stanford Encyclopedia of Philosophy (2016). https://plato.stanford.edu/entries/husserl/. Accessed 14 Oct 2018
26. Brier, S.: Cybersemiotics: a new foundation for transdisciplinary theory of information, cognition, meaningful communication and the interaction between nature and culture. Integral Rev. **9**(2), (2013)
27. Rössler, O.E.: Endophysics: The World as an Interface. World Scientific, Singapore (1998)

28. Ljungar-Chapelon, M.: Actor-spectator in a virtual reality arts play. University of Gothenburg (2008)
29. Ivanova, M.: Aesthetic values in science. Philos. Compass (12) (2017)
30. Davies, C.: (1995). www.immersence.com/osmose/. Accessed 14 Oct 2018
31. Davies, C.: Changing space: virtual reality as an arena of embodied being. In: The Virtual Dimension: Architecture, Representation, and Crash Culture, pp. 144–155. Princeton Architectural Press, New York (1998)
32. Welsch, W.: Undoing Aesthetics. SAGE Publications, Thousand Oaks (1997)

Automatic Generation of Point Cloud Synthetic Dataset for Historical Building Representation

Roberto Pierdicca[1(✉)], Marco Mameli[2], Eva Savina Malinverni[1],
Marina Paolanti[2], and Emanuele Frontoni[2]

[1] Dipartimento di Ingegneria Civile, Edile e dell'Architettura,
Universitá Politecnica delle Marche, 60100 Ancona, Italy
{r.pierdicca,e.s.malinverni}@staff.univpm.it
[2] Department of Information Engineering, Università Politecnica delle Marche,
Via Brecce Bianche 12, 60131 Ancona, Italy
{m.paolanti,e.frontoni}@staff.univpm.it

Abstract. 3D point clouds represent a structured collection of elementary geometrical primitives. They can characterize size, shape, orientation and position of objects in space. In the field of building modelling and Cultural Heritage documentation and preservation, the classification and segmentation of point clouds result challenging because of the complexity and variety of point clouds due to irregular sampling, varying density, different types of objects. After moving into the era of multimedia big data, machine-learning approaches evolved into deep learning approaches, which are a more powerful and efficient way of dealing with the complexity of semantic object classification. Despite the great benefits that such approaches brought in automation, a great obstacle is to generate enough training data, which are nowadays manually labeled. This task results time-consuming for two reasons: the variety of point density and geometry, which are typical for the Cultural Heritage domain. In order to accelerate the development of powerful algorithms for CH point cloud classification, in this paper, it is presented a novel framework for automatic generation of synthetic dataset of point clouds. This task is performed using Blender, an open source software which permits to access to each point in an object creating one in a new mesh. The algorithms described allow to create a great number of point cloud synthetically, simulating a virtual laser scanner at a variable distance. Furthermore, these two algorithms not only work with a single object, but it is possible to create simultaneously many point clouds from a scene in Blender also with the use of an existing model of ancient architectures.

Keywords: Point cloud · Synthetic dataset · Cultural Heritage

1 Introduction

The advancement of 3D point cloud acquisition techniques have enabled the use of 3D models from real world in several domains such as building

L. T. De Paolis and P. Bourdot (Eds.): AVR 2019, LNCS 11613, pp. 203–219, 2019.
https://doi.org/10.1007/978-3-030-25965-5_16

industry, urban planning or Cultural Heritage (CH) management and conservation. Nowadays, laser scanners or photogrammetry tools produce accurate and precise point clouds, up to millions of points, structured or not of CH goods. A burdensome task is the hand-written labelling of a point cloud that represents a building, considering that the research in recent years was oriented toward the "HBIM" paradigm, which is the acronym of Historical (or sometimes Heritage) Building Information Modeling [21].

3D point clouds represent a robust collection of elementary geometrical primitives. They can characterize size, shape, orientation and position of objects in space. These features may be augmented with further information obtained from other sensors or sources, such as colours, multispectral or thermal content, etc. For a relevant exploitation of point clouds and for their better comprehension and interpretation, it is important to proceed with segmentation and classification procedures. In particular, the segmentation groups points in subsets (generally called segments) defined by one or more characteristics in common (geometric, radiometric, etc.). Instead, the classification task defines and assigns points to specific classes according to different criteria. The classification and segmentation of point clouds result challenging because of the complexity and variety of point clouds due to irregular sampling, varying density, different types of objects; this is true especially in field of building modelling and CH documentation and preservation.

However, since moving into the era of big data, machine-learning and feature-based approaches have evolved into deep learning approaches, which are a more powerful and efficient way of dealing with the massive amounts of data generated from modern approaches and coping with the complexities of point clouds [19, 20]. Deep learning has taken key features of the machine-learning model and has even taken it one step further by constantly teaching itself new abilities and adjusting existing ones [8,11,24].

Deep Convolutional Neural Networks (DCNNs) have promptly grown as the core technique for a whole range of learning-based image analysis tasks [6]. Their success for multimedia data analysis is mainly due to easily parallelisable network architectures that facilitate training from millions of images on a single GPU and the availability of huge public annotated dataset (e.g. ImageNet [10]). Segmentation and classification are compelling issues faced among several fields [13,18,22].

While DCNNs obtained considerable success for image classification and segmentation, it has been less so for 3D point cloud interpretation. In fact, the supervised learning is extremely arduous for 3D point clouds because of the sheer size of millions of points per data set, and the irregular, not grid-aligned, and in places very sparse structure, with strongly varying point density (in terms of both geometrical and physical features).

Nowadays, while the acquisition is straight-forward, the main obstacle is to generate enough manually labeled training data, needed to learn good models, which generalize well across new, unseen scenes. Furthermore, the number of classifier parameters is larger in 3D space than in 2D because the additional dimension, and specific 3D effects such as variations or occlusion in point density conduct to different patterns for identical output classes. This complicates training, and more training data are generally needed in 3D than in 2D. In contrast to images, which are quite easy to obtain and annotate, the automatic interpretation (through classification procedures) of 3D data is more challenging. In fact, defining common patterns among objects can be quite impervious, given that for the same object point clouds can differ from one another. CH suffers this aspect the most, given that shapes are more complex than nowadays architectures (just think to a capital of to decoration) and, even when repeatable (just think to the architectural style) objects are unique since produced by handcraft and not by a serialized production. Beside this, objects dates back hundreds years and can be affected by a decay and erosion process of surfaces.

In order to accelerate the development of powerful algorithms for CH point cloud processing, in this paper, it is presented a novel approach for automatic generation of synthetic dataset of point clouds, specifically designed for CH field. The automatic generation is performed in Blender, an open source software which permits to access to each point in an object creating one in a new mesh. This allows to generate an empty mesh where the points of the starting object is added and visualized to check the right production of the mesh. If the points in the object are few, the algorithm permits to increase the number of points by subdividing the edge between point. After the point cloud generation, two CSV files is extracted and these files contain the list of the point coordinates and the label of the single point in the first file.

The main contributions of this paper are: (i) the development of two algorithms for synthetic point cloud generation of historical buildings; (ii) an approach that easily handles the irregularity of point clouds, and encodes diverse local structures; (iii) the possibility to enables effective deep learning, which is useful for recognition tasks, especially in the low-data regime of CH field.

The paper is organized as follow: Sect. 2 gives a brief overview of the past similar research work; Sect. 3 is devoted to the description of synthetic data set generation which is the core of this work; Sect. 4 presents the results. Finally, in Sect. 5, we draw conclusions and discuss future directions for this field of research.

2 Related Work

Synthetic datasets provide detailed ground-truth annotations. The generation of these data are widely adopted in research community because of their affordability to label images manually [5]. The first mention for automatic data set generation appear in [15] where the author generate a set of data for the environment of testing code. This generation is based on a priori knowledge of the software structure and test case.

In [16], Myers et al. use the automatic data set generation for studying the genome and for developing *CELSIM*. It is a software that allows the description and the generation of a target DNA sequence, with a great variety of repeated structures. This work has the main goal of generating a shotgun data set that might be sampled from the target sequences.

An entirely synthetic dataset is MPI Sintel. It is derived from a short open source animated 3D movie. The Sintel dataset contains 1064 training frames and currently it is the largest dataset available. It comprehends adequately realistic scenes including natural image degradation such as fog and motion blur. However, for training a DCNN, the dataset is too small yet [1].

Pei et al. propose two approaches for synthetic data generation [17]. They have demonstrated that the multidimensional synthetic data generation has great importance for clustering application and outlier analysis.

In [9], the authors have developed Thermalnet, a DCNN for transformation of visible range images into infrared images. This network is based on the SqweezeNet CNN. The Thermalnet network architecture is inspired by colorisation DCNN and it is devoted to the augmentation of existing large visible image datasets with synthetic thermal images.

In [3], Dosovitskiy et al. have trained DCNN for optical flow estimation on a synthetic dataset of moving 2D chair images superimposed on natural background images.

The authors in [14] have extended the concept of optical flow estimation via convolutional networks to disparity and scene flow estimation. They have proposed three synthetic stereo video datasets to enable training and evaluation of scene flow methods.

Fan et al. address the problem of generating the 3D geometry of an object based on its single image. In particular, they have explored generative networks for 3D geometry based on a point cloud representation. Their final solution is a conditional shape sampler, able of predicting multiple plausible 3D point clouds from an input image [4].

In [12], the authors use an unsupervised approach that aggregate the open doors input point cloud of city from LiDAR into point clusters by geometric attributes from which the object segments was obtained and labeled. Their approach consists of two main phases: the automatic generation of training data and the point cloud classification and optimization.

Furthermore, synthetic data have been used to learn large CNN models. Wang et al. and Jaderberg et al. use synthetic text images to train word-image recognition networks [7,23].

In [3], the authors have generated a large synthetic Flying Chairs dataset. This dataset is used for comparing two architectures: a generic architecture and a network including a layer that correlates feature vectors at different image locations.

The learning of generative models are performed with synthetic data. In [2], Dosovitskiy et al. train inverted CNN models to render images of chairs. In [25], Yildirim et al. instead use DCNN features trained on synthetic face renderings to regress pose parameters from face images.

In respect of this context, this paper investigates the synthesis and processing of point clouds, presenting a novel system for 3D point cloud generation of historical building for CH domain.

3   Methods

In this section, we introduce the framework as well as the Blender platform used for the synthetic dataset generation. The phases of this process is schematically depicted in Fig. 1. In particular, starting from a OBJ document that contains the 3D model of an object, the algorithm extracts a label based on the name of the object. The first step consists of the research of every center of a face of the object and after this a point for every face is created. The second step extracts the coordinates of vertices of edge in the object and then the point in the coordinates of this object is created. After this extraction procedure, an empty mesh is created and to this mesh the set of point extracted in the previous step is added to it. This mesh permits to visualize the point cloud generated. The last step is the extraction of the two CSV file, where in the first one there is a coordinates list of every point and in the second one, there is the label of the coordinate in the same row in the first file. Further details are discussed in next subsections.

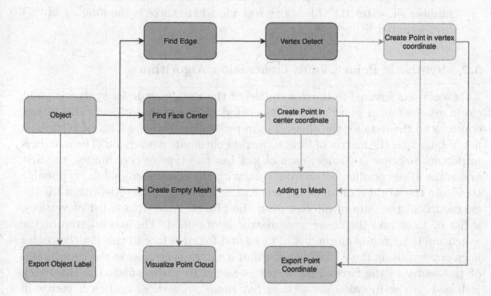

Fig. 1. Phases of the algorithm for synthetic point clouds generation.

3.1   Blender Platform

Blender[1] is an open source 3D modeling software, based in Python. It is widely used as not commercial software for 3D modeling and development with a great community and a wide range of plugin and scripting. It has a development

---

[1] https://www.blender.org, last access July 16, 2019.

configuration and allows the access to all the structure of an object from vertex to material. This leads to the write once and execute everywhere paradigm to permit to all readers to try this code. At the base of automatic synthetic point cloud data set generation, we have considered that all the 3D object modelled is composed by vertices. Blender allows the access to this vertex and allows to copy them to a new empty object that is made only of vertex without lines which joins the vertices to create a face.

In addition to the vertices of a face is used a vertex that describes the position of the face in the original object obtained by the position of the center of the face. Blender allows the creation of object piece by piece and this is useful for the generation of point cloud of the single part of an object. After the creation of the point clouds, the results is saved in a three separated "txt" file in point comma separated style:

- "_coordinates.txt" file contains all the coordinates of the point clouds generated.
- "_label.txt" file contains all the label of each vertex in first file.
- "_number_of_vertex.txt" file is one row file where there is the number of vertices generated.

3.2   Synthetic Point Clouds Generation Algorithm

This section is devoted to the description of the Python code for synthetic point cloud generation. A point cloud is generated from an input object. This object contains all the data for the algorithm, in particular there are a list of vertices, a list of faces and the matrix of local to world coordinate conversion. The matrix is important because in Blender each object has two type of coordinates: the first one is the local coordinate and from this, with the conversion matrix, is possible to obtain the world coordinates in scene. The first step of the Algorithm 1 is the extraction of the data of interest from the object. After that a list of vertices, a list of faces and the conversion matrix is obtained. The second step of the execution is to iterate on the list of faces and for each face in this list the center is calculated using the API of blender and a vertex is created at the coordinates of the center of the faces. This vertex is added to *verticesList*. The third step is based on the iteration on vertices list *mesh_obj.vertices* and each vertex in this list is added to *verticesList*. As described in Algorithm 1 before a vertex is added to the list an error is calculated based on the function error of the 3D laser scanner LEICA (See Sect. A). The first step for the generation of point cloud is done to obtain the position of the center of a faces and then by applying the error to it, at the end this vertex is added to a list of vertexes. The second step uses the list of vertices and from each vertex it is applied the error and after the matrix vector multiplication the world coordinates is added to the list of vertex. At the end of the algorithm, all the obtained vertices are added to the point cloud object. The Algorithm 2 performs as first step the extraction of the list of vertices, the list of faces and the conversion matrix from the object. The second step is the *subdivision step*, in which a subdivision of the edges is done for every edge in the list *slectedEdges* based on the *NumberOfCut* parameter. The third step iterate on the list of faces, for each face in the list *mesh_obj.faces* the center of the face is computed using the API of blender and a vertex is created at the center computed. This center is added to *verticesList*. The fourth step iterate on the augmented vertices list *mesh_obj.vertices* to add it in the *verticesList*. As explained the Algorithm 1 differs form the Algorithm 2 in the step of the subdivision application. This step is used to increase the number of point for each object in the scene. This is useful when the objects in the scene have few vertices. Few vertices objects are commonly in the scene creation but this type of object is not useful for a point cloud data set realization because a good data set require a huge number of point in it. The subdivision algorithm is applied to all the edges in the list of edges in the object input. A subdivision is done stating from a single edge with the division of it a number of edges equal to *Number of point*.

---

input : Object
input : Error parameter: center, variance
output: Point Cloud Object

$meshObj \longleftarrow object.data$;
$matrixWorld \longleftarrow meshObj.matrixWorld$;
$verticesList \longleftarrow emptyList$;
for $face \in meshObj.faces$ **do**
> $faceCoordinates \longleftarrow face.calc\_enter\_median()$;
> $error \longleftarrow normal(center, variance, 3)$;
> $faceCoordinatesNoised \longleftarrow faceCoordinates + error$;
> $faceWorldCoordinate \longleftarrow matrixWorld \times faceCoordinateNoised$;
> $verticesList.add(faceWorldCoordinate)$;

end
for $vertex \in meshObj.vertices$ **do**
> $error \longleftarrow normal(center, variance, 3)$;
> $vertexLocationNoised \longleftarrow vertex.coordinates + error$;
> $vertexWorldCoordinate \longleftarrow$
> $matrixWorld \times vertexLocationNoised$;
> $verticesList.add(vertexWorldCoordinate)$;

end
$PointCloudObject.add(verticesList)$;

Algorithm 1. Automatic point cloud algorithm without subdivision edges

---

For the sake of completeness, in Fig. 2 a comparison between the two algorithms is reported, described in the shape of flowcharts.

```
input  : Object
input  : Error parameter: center, variance
input  : Number of cut
output: Point Cloud Object

meshObj ⟵ object.data;
matrixWrld ⟵ meshObj.matrixWorld;
verticesList ⟵ emptyList;
selectedEdges ⟵ emptyList;
for edge ∈ meshObj.edges do
    selectedge;
    selectedEdges.add(edge)
end
subdivideEdges(meshObj, selectedEdges, NumberOfCut);
for face ∈ meshObj.faces do
    faceCoordinates ⟵ face.calc_enter_median();
    error ⟵ normal(center, variance, 3);
    faceCoordinates_noised ⟵ faceCoordinates + error;
    faceWorldCoordinate ⟵ matrixWorld × faceCoordinateNoised;
    verticesList.add(faceWorldCoordinate);
end
for vertex ∈ meshObj.vertices do
    error ⟵ normal(center, variance, 3);
    vertexLocation_noised ⟵ vertex.coordinates + error;
    vertexWorldCoordinate ⟵
    matrixWorld × vertexLocationNoised;
    verticesList.add(vertexWorldCoordinate);
end
PointCloudObject.add(verticesList);
```

Algorithm 2. Automatic point cloud algorithm with the use of the subdivision edges to increase the number of vertices

4   Results and Discussion

In this section, the results of the algorithm described in the previous section are reported. Figure 3 demonstrates that from a column composed by one single object (Fig. 3(a)) which is composed by the vertices shown in Fig. 3(b) creates a unique point cloud of the column showed in Fig. 3(c) and (d).

Figure 4 shows the starting point of a point cloud obtained from an object composed by two or more simple object. This subdivision of the object allows the result showed in Figs. 5 and 6 which demonstrate how the algorithm has preserved the composition of the complete object and has generated two distinct point cloud.

These Figures demonstrates the effectiveness and the suitability of the proposed approaches as well as the perfect generation of a good synthetic point clouds dataset from a single object or a scene done by multiple object.

The results of the Algorithm 2 is shown in Fig. 7(c) and (b). In particular, the following steps are performed:

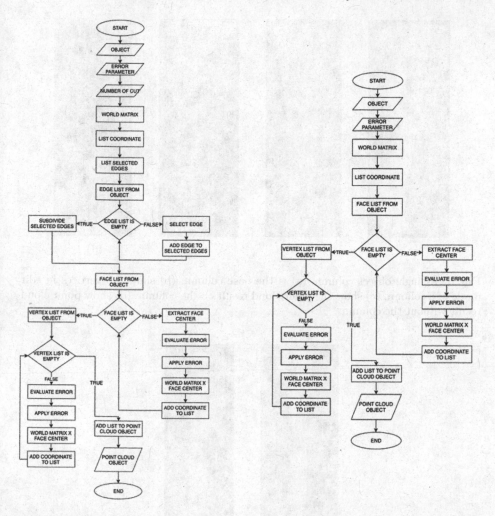

Fig. 2. The figure presents the flow chart of the algorithms. The flow chart on the left is relative to the Algorithm 2. The flow chart on the right is relative to the Algorithm 1.

- The generation of a subdivision of the edges that compose the object with the application of the point cloud generation algorithm, where the number of cut is set to 10.
- The application of a subdivision of the edges generate an increment of the number of vertices in the object (Fig. 7(c) and (b)).

The result is the generation of a dense point cloud with more vertices as shown in Fig. 7(d).

The generation of point cloud from a scene require the building of an example scene. The scene used is illustrated Fig. 8(a).

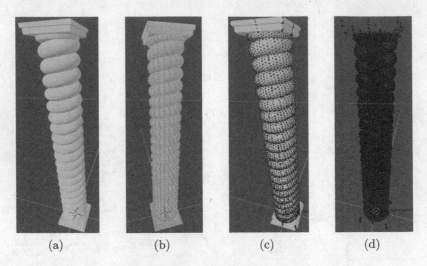

Fig. 3. A single object column. (a) is the base column. (b) show the vertices in edit mode of a column. (c) show the point cloud result on the column. (d) show point cloud result without the column

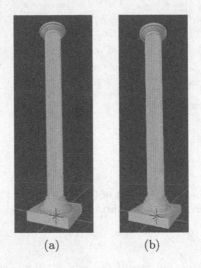

Fig. 4. A column composed of two simple object. (a) show the base of the column selected. (b) show the column selected.

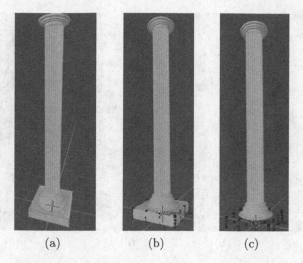

(a)   (b)   (c)

Fig. 5. A column composed of two simple object with the point cloud of the base object. (a) is the vertices of the base in edit mode. (b) is the point cloud of the base depicted on the base. (c) is the point cloud of the base showed without the object.

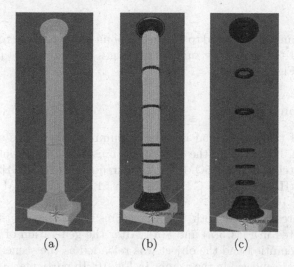

(a)   (b)   (c)

Fig. 6. A column composed of two simple object with the point cloud of the column object. (a) is the vertices of the column in editmode. (b) is the point cloud of the column depicted on the column. (c) is the only point cloud of the column.

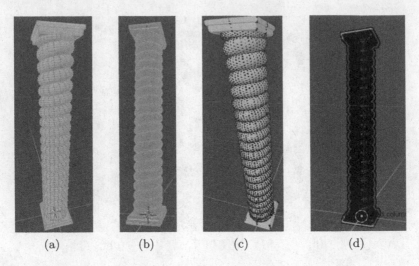

(a)   (b)   (c)   (d)

Fig. 7. A single object column. (a) is the column in editmode without subdivision of
the edges. (b) is the column in edit mode with subdivision of the edges. (c) is the point
cloud without subdivision of the edges. (d) is the point cloud with the subdivision of
the edges.

The algorithm was executed to create the point cloud and the result is showed
in Fig. 8(b) and the use of the subdivision edges create a dense point cloud as
represented in Fig. 9.

4.1   Execution Time

The creation of the point cloud is time consuming and in this section a time
analysis is done. In particular the configuration used is a MacBook Pro 15″ mid
2012 with an Intel Core i7-3615QM with base frequency of 2.30 GHz, a Dedicated
Video Card NVIDIA GeForce GT 650M with 512 MB of VRAM and 16 GB of
DDR3 RAM at 1600 MHz.

The tests are executed both for Algorithms 1 and 2 and using the single
object represented in Figs. 3(a) and 8(a). After the generation of a point cloud,
the memory is emptied and the object was reloaded as the scene.

The results are showed in the graphs in Fig. 10. In particular Fig. 10(a) is the
execution time of a one object point cloud generation without subdivision and
in Fig. 10(b) the execution time of one object point cloud generation with the
application of the subdivision applied. After this generation, the average time
is calculated and the execution without the subdivision applied has a 1.44 s and
the algorithm with the subdivision need 3.00 s of average time.

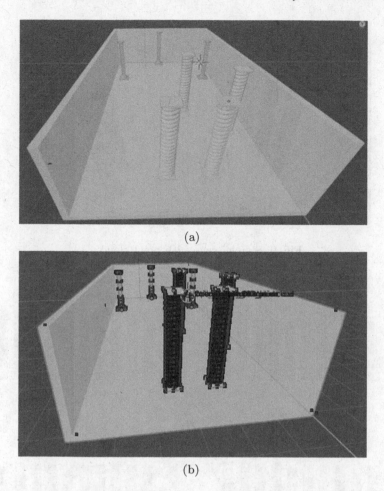

(a)

(b)

Fig. 8. An example scene. (a) show the starting scene. (b) show the result of point cloud generation.

Figure 10(c) shows the execution time of the generation of a point cloud from the scene in Fig. 8(a) with an average time of 3.25 s.

The generation of the scene point cloud with the subdivision applied has an average execution time of 48.59 s, as represented in the graphs.

It is possible to infer from the graphics and from the average time the generation of a dense point cloud with the subdivision applied of a scene need more time as expected because the Algorithm 2 subdivide any edge that composes an object in the scene and this is time consuming for the execution.

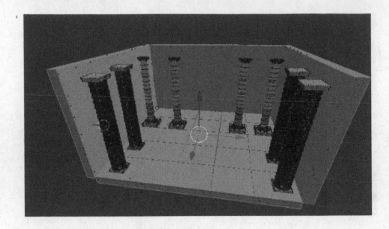

Fig. 9. Simple scene with point cloud.

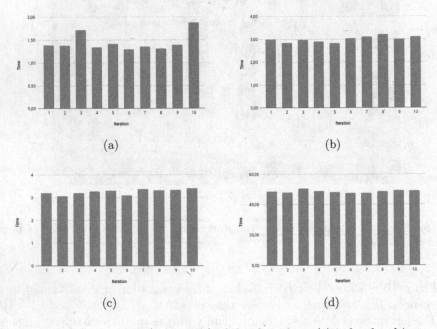

Fig. 10. The execution time (in seconds) of the algorithms. (a) is for the object point cloud generation without subdivision. (b) is for the object point cloud generation with subdivision. (c) is for the scene point cloud generation without subdivision. (d) is for the scene point cloud generation with subdivision.

5   Conclusion and Future Works

In this paper it is presented a novel approach to generate automatically synthetic point clouds dataset from a 3D object using an open source software. With the algorithms described is possible to create a great number of point cloud synthetically varying the distance of virtual laser scan.

The algorithm presented works not only with a single object but it is possible to create simultaneously many point cloud from a scene in Blender, also with the use of a quite old hardware architecture.

This approach will be particularly useful in the field of point cloud classification, where the main bottleneck is represented during the training phase. Since the main reason of this issue is that, nowadays, the labelling phase of each architectural class is performed manually, the automatic generation of such dataset might represent a turnkey.

Future works will be devoted to the creation of a plugin for Blender that executes this algorithm with options on laser scan distance and the possibility to chose if its execution has to be applied on a single object or on the entire scene and the possibility to choose if apply or not the subdivision to obtain a dense point cloud. Further investigations will involve the creation of a script that allows the automatic scene generation from the starting scene with reorganizing the object in the scene with respect to architectural constrains.

Finally, the synthetic dataset generated with our approach, will be tested in already existing DCNN in order to evaluate their benefits for the classification. By comparing synthetic datasets with those collected with real tools (e.g. laser scanner), will provide the research community with a pathway for speeding up the annotation phase, for the CH domain as well as in other research fields.

A   Appendix: Error

To create a more realistic data set and close to what you would get from a laser scanner an error function was added. This error function simulates the error of a laser scan based on the distance of scanning. The function definition came from the data sheet[2] and is based on the function in (1)

$$y = ax^2 + bx + c \tag{1}$$

where

- $a = -1e - 5$
- $b = 4e - 4$
- $c = 1e - 3$

which are obtained from the specifications sheet of LEICA BLK360 (See footnote 2). We have choose this tools since it is a medium level one and widespread among workers dealing with surveying.

---

[2] https://lasers.leica-geosystems.com/eu/sites/lasers.leica-geosystems.com.eu/files/leica_media/product_documents/blk/prod_docs_blk360/leica_blk360_spec_sheet.pdf.

Table 1. The values of the parameters of the system reported in Eq. 2

Error	Distance
4	10
7	20

The Eq. 1 is developed by a simple parametric system (Eq. 2). The values of the parameters are reported in Table 1.

$$\begin{cases} 4e - 3 = a * 10^2 + b * 10 + c \\ 7e - 3 = a * 20^2 + b * 20 + c \end{cases} \tag{2}$$

where c is chosen $c = 1e - 3$. In the code environment this function is used as the variance description of the normal distribution used to simulate the error generation.

References

1. Butler, D.J., Wulff, J., Stanley, G.B., Black, M.J.: A naturalistic open source movie for optical flow evaluation. In: Fitzgibbon, A., Lazebnik, S., Perona, P., Sato, Y., Schmid, C. (eds.) ECCV 2012. LNCS, vol. 7577, pp. 611–625. Springer, Heidelberg (2012). https://doi.org/10.1007/978-3-642-33783-3_44
2. Dosovitskiy, A., Brox, T.: Inverting visual representations with convolutional networks. In: Proceedings of the IEEE Conference on Computer Vision and Pattern Recognition, pp. 4829–4837 (2016)
3. Dosovitskiy, A., et al.: FlowNet: learning optical flow with convolutional networks. In: Proceedings of the IEEE International Conference on Computer Vision, pp. 2758–2766 (2015)
4. Fan, H., Su, H., Guibas, L.: A point set generation network for 3D object reconstruction from a single image. In: 2017 IEEE Conference on Computer Vision and Pattern Recognition (CVPR), pp. 2463–2471. IEEE (2017)
5. Gupta, A., Vedaldi, A., Zisserman, A.: Synthetic data for text localisation in natural images. In: Proceedings of the IEEE Conference on Computer Vision and Pattern Recognition, pp. 2315–2324 (2016)
6. Hackel, T., Savinov, N., Ladicky, L., Wegner, J.D., Schindler, K., Pollefeys, M.: Semantic3D.NET: a new large-scale point cloud classification benchmark. arXiv preprint arXiv:1704.03847 (2017)
7. Jaderberg, M., Simonyan, K., Vedaldi, A., Zisserman, A.: Synthetic data and artificial neural networks for natural scene text recognition. arXiv preprint arXiv:1406.2227 (2014)
8. Klokov, R., Lempitsky, V.: Escape from cells: deep kd-networks for the recognition of 3D point cloud models. In: 2017 IEEE International Conference on Computer Vision (ICCV), pp. 863–872. IEEE (2017)
9. Kniaz, V., Gorbatsevich, V., Mizginov, V.: Thermalnet: a deep convolutional network for synthetic thermal image generation. Int. Arch. Photogramm. Remote Sens. Spat. Inf. Sci. **42**, 41 (2017)

10. Krizhevsky, A., Sutskever, I., Hinton, G.E.: Imagenet classification with deep convolutional neural networks. In: Advances in Neural Information Processing Systems, pp. 1097–1105 (2012)
11. LeCun, Y., Bengio, Y., Hinton, G.: Deep learning. Nature **521**(7553), 436 (2015)
12. Li, Z., Zhang, L., Zhong, R., Fang, T., Zhang, L., Zhang, Z.: Classification of urban point clouds: a robust supervised approach with automatically generating training data. IEEE J. Sel. Top. Appl. Earth Obs. Remote Sens. **10**(3), 1207–1220 (2017)
13. Liciotti, D., Paolanti, M., Pietrini, R., Frontoni, E., Zingaretti, P.: Convolutional networks for semantic heads segmentation using top-view depth data in crowded environment. In: 2018 24th International Conference on Pattern Recognition (ICPR), pp. 1384–1389. IEEE (2018)
14. Mayer, N., et al.: A large dataset to train convolutional networks for disparity, optical flow, and scene flow estimation. In: Proceedings of the IEEE Computer Society Conference on Computer Vision and Pattern Recognition, vol. 2016-January, pp. 4040–4048 (2016)
15. Miller Jr, E., Melton, R.: Automated generation of testcase datasets. In: ACM SIGPLAN Notices, vol. 10, pp. 51–58. ACM (1975)
16. Myers, G.: A dataset generator for whole genome shotgun sequencing. In: ISMB, pp. 202–210 (1999)
17. Pei, Y., Zaïane, O.: A synthetic data generator for clustering and outlier analysis. Department of Computing Science, University of Alberta, Edmonton, AB, Canada (2006)
18. Pierdicca, R., Malinverni, E.S., Piccinini, F., Paolanti, M., Felicetti, A., Zingaretti, P.: Deep convolutional neural network for automatic detection of damaged photovoltaic cells. In: International Archives of the Photogrammetry, Remote Sensing and Spatial Information Sciences - ISPRS Archives, vol. 42, pp. 893–900 (2018)
19. Qi, C.R., Su, H., Mo, K., Guibas, L.J.: PointNet: deep learning on point sets for 3D classification and segmentation. In: Proceedings of the IEEE Conference on Computer Vision and Pattern Recognition, pp. 652–660 (2017)
20. Qi, C.R., Yi, L., Su, H., Guibas, L.J.: PointNet++: deep hierarchical feature learning on point sets in a metric space. In: Advances in Neural Information Processing Systems, pp. 5099–5108 (2017)
21. Quattrini, R., Pierdicca, R., Morbidoni, C.: Knowledge-based data enrichment for HBIM: exploring high-quality models using the semantic-web. J. Cult. Herit. **28**, 129–139 (2017)
22. Sturari, M., Paolanti, M., Frontoni, E., Mancini, A., Zingaretti, P.: Robotic platform for deep change detection for rail safety and security. In: 2017 European Conference on Mobile Robots (ECMR), pp. 1–6. IEEE (2017)
23. Wang, T., Wu, D.J., Coates, A., Ng, A.Y.: End-to-end text recognition with convolutional neural networks. In: 2012 21st International Conference on Pattern Recognition (ICPR), pp. 3304–3308. IEEE (2012)
24. Xie, S., Liu, S., Chen, Z., Tu, Z.: Attentional shapecontextnet for point cloud recognition. In: Proceedings of the IEEE Conference on Computer Vision and Pattern Recognition, pp. 4606–4615 (2018)
25. Yildirim, I., Kulkarni, T.D., Freiwald, W.A., Tenenbaum, J.B.: Efficient and robust analysis-by-synthesis in vision: a computational framework, behavioral tests, and modeling neuronal representations. In: Annual Conference of the Cognitive Science Society, vol. 1 (2015)

Semantic Contextual Personalization
of Virtual Stores

Krzysztof Walczak[✉][iD], Jakub Flotyński[iD], and Dominik Strugała[iD]

Poznań University of Economics and Business,
Niepodległości 10, 61-875 Poznań, Poland
{walczak,flotynski,strugala}@kti.ue.poznan.pl
http://www.kti.ue.poznan.pl/

Abstract. Virtual stores and showrooms gain increasing attention in
e-commerce, marketing and merchandising to investigate customers'
behavior, preferences and the usefulness of shopping and exhibition
spaces. Although virtual stores may be designed using numerous available 3D modeling tools and game engines, efficient methods and tools
enabling development and personalization of virtual stores are still lack-
ing. In this paper, we propose a novel approach to the development of
personalizable contextual virtual stores that can be generated and configured on-demand, using interfaces based on semantic web technologies.
A virtual store model is created as a combination of three elements:
an exposition model, a collection of product models, and a virtual store
configuration. The first element visually reflects an existing or imaginary
3D store layout. The second element contains 3D models of all products
that can be presented in the exposition. The third element is an ontology,
which connects the two previous elements using domain-specific knowledge and reasoning. Based on a virtual store model, a personalized virtual
store is generated in response to a specific user's request.

Keywords: Virtual reality · Stores · Showrooms ·
Immersive visualization · User interfaces · E-commerce · Marketing ·
Merchandising

1 Introduction

Significant progress in the performance and the presentation capabilities of contemporary IT equipment offers the possibility to transfer various physical spaces
into virtual reality. One of prominent and interesting examples are virtual reality models of shopping spaces. Realistic three-dimensional visualization of stores
and showrooms in VR has two main purposes.

The first purpose are merchandising tests to verify how different product
arrangements in a modeled physical exposition influence perception of the products by customers. The goal is either to increase the total store turnover (by

© Springer Nature Switzerland AG 2019
L. T. De Paolis and P. Bourdot (Eds.): AVR 2019, LNCS 11613, pp. 220–236, 2019.
https://doi.org/10.1007/978-3-030-25965-5_17

influencing customers to buy as much as possible) or to promote a certain product or a group of products (by influencing buyer choices). Currently such merchandising research is typically performed with the use of physical mockups of real stores [3].

Virtual reality spaces offer important advantages over traditional "physical" spaces by permitting quick, easy and well-controlled rearrangement of products on the basis of customers preferences, current needs as well as the context of interaction. Also, there is no need to posses physical versions of all different kinds of products, which may quickly become unusable, because of their expiration date or changing product design.

The second purpose of building realistic immersive 3D stores are tests with customers to examine their impressions and the sense of reality. This is an important step in moving daily shopping to the virtual world – the process that has already begun with the creation of the first on-line stores and is constantly gaining importance and popularity in the modern economy. Shopping is one of the most popular on-line activities worldwide. In 2016, retail e-commerce sales worldwide amounted to 1.86 trillion US dollars and e-retail revenues are projected to grow to 4.48 trillion US dollars in 2021 [31].

2D shopping websites commonly used today, have their natural limitations, such as the lack of interaction with products, limited perception of the product size and properties, and the lack of social interaction between people, which is a particularly important element of shopping in some groups of customers and products. The use of 3D avatars to navigate and collaborate in a virtual environment introduces a social aspect into this activity, which is not achievable in traditional forms of e-commerce. These aspects can have a significant impact on the popularity of new forms of shopping [18].

Retailing in 3D virtual environments, including social virtual worlds (SVWs), is considered an evolution of the traditional web stores, offering advantages and an improved shopping experience to customers [13]. Research demonstrates that the use of virtual reality can have a positive influence on marketing communication [38]. However, VR technology is not yet ready for mass use in on-line shopping systems. In [26], authors argue that the use of virtual reality in e-commerce must rely on a mixed platform presentation to account for various levels of usability, user trust, and technical feasibility. The authors propose that e-commerce sites that implement VR commerce provide at least three layers of interaction to users: a standard web interface, embedded VR objects in a web interface, and semi-immersive VR within an existing web interface.

Currently, virtual stores and showrooms can be developed using a number of well-established methodologies, such as 3D modeling tools and game engines, programming languages and libraries, as well as 3D content formats. However, the available approaches are generic and therefore complex. Specialized solutions could enable more efficient creation of personalized virtual stores on the basis of customers' preferences and demands, resources offered by the stores, and the particular context of interaction. In addition, limitations of the available space and specific organization of work can be taken into account.

The main contribution of this paper is an approach to generating personalized virtual stores. The approach enables quick and easy reconfiguration of virtual stores and placement of particular products. It is therefore suitable for performing series of tests with different groups of users. The approach is based on the semantic web standards: the Resource Description Framework—RDF [43], the RDF Schema—RDFS [44], the Web Ontology Language—OWL [42], and the SPARQL query language [41]. The use of the semantic web in our approach has important advantages over the previous approaches to modeling 3D content. First, it enables creation of virtual stores based on general and domain knowledge, e.g., objects and properties that directly represent products, rather than concepts specific to 3D graphics and animation. This makes the content intelligible to average users and experts in marketing and merchandising. In addition, domain knowledge can also be used to describe the behavior of customers, their interest in products, and interactions with salesmen. Due to the formalism of the semantic web, the representations of stores as well as the collected data about customers can be explored with semantic queries by users, search engines and analytical tools. Furthermore, semantically represented stores and customers' behavior can be processed with standard reasoning engines to infer tacit content properties on the basis of the explicitly specified properties, e.g., to infer what are the most frequent combinations of products selected by customers with specific preferences and needs. Finally, due to the use of ontologies, which have been intended as common repositories of concepts and objects, the approach can be used in collaborative social shopping, in which multiple users jointly create and visit virtual spaces.

The remainder of this paper is structured as follows. Section 2 provides an overview of the current state of the art in the development of and interaction with VR environments. The system architecture and the process of creating personalizable virtual stores are presented in Sects. 3 and 4, respectively. In Sect. 5, different forms of interaction with virtual shopping environments are discussed. It is followed by an example virtual store in Sect. 6. Finally, Sect. 7 concludes the paper and indicates possible future research.

2   Related Works

2.1   Designing VR Environments

Creation of non-trivial interactive VR content is an inherently complex task. This results from the conceptual and structural complexity of VR content models and the variety of aspects that must be taken into account in the content creation process. A number of approaches, applicable in different scenarios and contexts, have been proposed to simplify the content creation task.

Geometry, appearance and movement of real objects can be acquired using automatic or semi-automatic 3D scanning and motion capture devices. Static 3D objects can be precisely digitized using scanners based on laser ToF measurement, triangulation, and structured light. Less precise, but more affordable, are software tools enabling reconstruction of 3D objects from series of images, such as Autodesk 123D and 3DSOM. 3D scanning can be combined with other content creation methods, allowing designers to influence the process of digitization and the created content.

Modeling of both existing and non-existing objects and places is possible with the use of visual 3D content design environments. Software packages that enable modeling or sculpting of 3D content include: Blender, 3ds Max, Modo, Maya, ZBrush and 3D-Coat. There are also programs focused on specific industries, such as Revit, Rhino and SolidWorks used mostly in architecture and product design. All these advanced professional environments offer rich capabilities of modeling various content elements, but their complexity requires high expertise. Narrowing the domain of application and the set of available operations enables development of tools that are easier to use by domain experts. Examples of such environments include AutoCAD Civil 3D, Sweet Home 3D and Ghost Productions. These tools enable relatively quick and efficient modeling without requiring users' extensive experience in 3D content creation, but this approach significantly reduces the generality of the content creation process. There are also software packages, which are quite easy to use, but their application is mostly limited to conceptual work. A good example of such software is SketchUp.

High structural complexity of 3D content, combined with the requirement of being able to adjust specific content parameters, require the development of content models – well defined structures, which describe content organization and parameterization [46]. Based on such models, the final form of 3D content can be generated by content generation software – either fully automatically or semi-automatically in an interactive process. Content models offer data structures that are better organized and easier to maintain than typical 3D content representation. They also permit automatic verification of data consistency and elimination of redundancy. Content patterns provide an additional conceptual layer on top of content models, defining roles of specific elements in a content model [21, 25].

As an alternative to fixed content models, rules of content composition can be defined. Such rules describe how different types of content elements should be combined to form the final 3D model. Rules permit flexible composition of content from predefined building blocks – components [6, 39, 45, 46]. Components may represent geometrical objects, scenarios, sounds, interaction elements, and others. Content creation based on configuration of predefined components constrains possible forms of the final 3D content. In many application domains, however, this approach is sufficient, while the process is much simpler and more efficient than creating content from scratch.

To further simplify content modeling, separation of concerns between different categories of users is required. These users may have different expertise and may be equipped with different modeling tools. A non-expert designer may use ready-to-use components and assemble them into virtual scenes. Composing a scene in such a way is relatively simple, but the process is constrained. New content creation capabilities can be introduced by programmers or 3D designers, who can add new components and new ways of combining them.

The use of semantic web techniques may further simplify the process of creating 3D content [1,30,37,50]. Semantic web techniques enable the use of high-level domain-specific concepts in the content creation process, instead of low-level concepts specific to 3D graphics. Content creation may be also supported by knowledge inference. The use of semantic content representation enables creation of content that is platform independent. Several approaches have been proposed to enable 3D content modeling with the use of semantic web techniques [4,8,10,16,48,49]. Semantic web techniques support 3D content creation in various domains, e.g., indexing movies [28], molecular visualization [33,34], underwater archaeology [9], and designing industrial spaces [22]. A detailed analysis of the state of the art in semantic 3D content representation and modeling has been presented in [11].

In the domain of architectural design, to offer the highest possible photorealism already at the design process, tools have been developed to enable VR visualization directly from within 3D modeling packages. They are mostly implemented as extension plug-ins for existing 3D modeling software packages. The design process is performed in the native application, while VR visualization requires only a push of a button. Examples of such tools include Enscape, IrisVR and Eyecad VR.

Availability of efficient and easy to use content creation methods – in particular methods based on the semantic web techniques – opens the possibility of social 3D content co-creation by users that are both producers and consumers (prosumers), similarly as in the case of the "two-dimensional" Web 2.0 [47]. 3D content sharing portals, such as Unity Asset Store [36], Highend3D [14], Turbosquid [35], 3D ContentCentral [7], and many others (e.g., CG People Network, Creative Crash, 3d Export, Evermotion, The 3D Studio, and 3D Ocean) enable access to vast libraries of 3D content.

To summarize, there are numerous approaches enabling simplification of the content creation process, but there is a lack of specific solutions intended for designing personalizable virtual stores and showrooms.

2.2   User Interaction in VR Environments

In order to build an easy-to-use configurable VR system, it is necessary to choose appropriate methods of user interaction with the virtual environment. Domain experts, who should perform the configuration task, do not need to have technical skills. For this reason, it is important that the interaction methods are intuitive and user-friendly. This is difficult because content design is a complex task and new users often find it difficult to even simply navigate and interact in immersive VR environments, such as a CAVE. This section describes different approaches to interaction of users with VR environments.

The first approach is based on classic input devices such as a mouse and a keyboard. The ability to map 2D mouse interactions to a 3D space [20] and the high degree of technological adoption make this approach preferred by many beginners in VR. However, due to the natural limitations of these devices (such as the low number of degrees of freedom and the necessity to use complex key

combinations), navigation using such devices is often non-intuitive and complicated. Moreover, this kind of interaction is not practical in CAVE systems.

Another approach is to use specific equipment: gaming input devices, such as
joysticks and pads, or dedicated VR devices, such as tracked controllers and haptic arms, to navigate and interact in virtual environments. A significant advantage of this approach is higher user comfort and good control and accuracy in
properly designed and configured environments [15]. In the case of CAVE systems, the Flystick – wireless interaction device with six buttons and an analogue
joystick – is particularly frequently used. This device meets the needs of most
users. However, the limited number of buttons and the lack of reverse communication reduce the usefulness of this device for users who are content designers.
In the case of more specialized devices, adapting them to environments other
than those for which they were originally designed is difficult or not possible at
all. Nevertheless, specialized device-based approach is often the basis for further
research [2, 32].

A quickly developing approach to users' interaction in VR environments is
the analysis of natural human behavior. It includes techniques such as motion
capture (either using marker tracking [5] or directly, e.g., using Xbox 360 Kinect
sensor system [27]), gesture recognition [17], eye tracking [24], and verbal/vocal
input [51]. All these techniques focus on providing an intuitive natural interface,
which is user-friendly even for non-experienced users. However, the problem
with using natural interaction to design content is that it requires a significant
physical effort from the user. For example, according to [23], the average time a
user can comfortably use Leap Motion (device for gesture recognition) is about
20 min. Thus, this technique is not suitable for designing VR environments, as
it is often a process that requires a long time and high accuracy.

Context-based approach is an interaction technique popular in computer
games, in particular in simulations (e.g., "The Sims" and "SimCity" series by
Maxis) and in adventure games. This approach is not in itself based on specific
input devices, but focuses on the use of available devices to navigate through a
real-time contextual interface. The content of this interface depends on the current state of the environment and its objects (e.g., time, position, current object
state). The context-based approach is also often used in modern VR environments [12]. However, users may find this approach uncomfortable due to the
mismatch between classic UI elements (buttons, menus, charts) and the 3D virtual environment. Also, this technique is not convenient for entering data (such
as text or numbers). This is a serious limitation when the interface is used for
content design.

Another approach to user interaction within VR environments is to use a
device with its own CPU for controlling the environment. Mobile devices, such as
smartphones and tablets, are often used for this purpose due to their widespread
availability and advanced user-interface features, including high-resolution touch
screen displays and various types of built-in sensors, such as gyroscope and
accelerometer. In this approach, a user has a specific predefined interface located
on the client device to control the environment. This interface can be generated

with the use of specialized software (e.g., PC Remote application by Monect [19]) or it can be dedicated to a specific environment and released in the form of an independent application. However, developing dedicated client applications is a time-consuming and costly activity, and the applicability of such an interaction interface is limited to a single VR environment.

3   System Architecture

In this section, the overall architecture of a personalizable VR store is presented. The main element of the architecture (Fig. 1) is the *Personalizable Virtual Store Application*, which integrates data from three sources:

- *Virtual Store Exposition Model* is a 3D model of a store exposition space. It may be a reconstruction of an existing shopping space, design of an intended space or a 3D model used purely for the visualization process. The exposition model may be created by a designer/architect. Any number of different virtual store exposition models may be used within the system. The only requirement for the exposition model is to provide an identification of product placeholders, i.e., locations in the model where 3D product models will be placed. The exposition model must use standard units (e.g., metric system) to enable automatic integration with 3D product models.
- *3D Product Models* is a collection of 3D models of products, which can be placed in the exposition space (e.g., on shelves). The 3D product models may be modeled or scanned by products designers/providers or retrieved from a library. All 3D product models must use the same scaling system as the exposition model.
- *Virtual Store Ontology* is a dataset that describes the virtual store, e.g., its parts, shelves, products and virtual salesmen, using the semantic web standards. A virtual store ontology may be designed by a marketing expert.

Personalization of the virtual store starts when the virtual store application receives a *personalization query*. The query may be prepared by a user or generated by a client application, which maintains user's preferences and the context of interaction. The personalization query is a tuple consisting of an ontology (OWL query) and a SPARQL query. Both the OWL query and the SPARQL query may represent customer's preferences. However, mainly the parts of the personalization queries whose fulfillment requires reasoning on numbers are expressed in SPARQL (e.g., the placement of some products relatively close to other products), whereas the other requirements are expressed in OWL (e.g., desirable classes of products). The OWL query is merged with the virtual store ontology. The merged ontology is alternately processed by the *SPARQL query engine* and the *reasoning engine* until there are no more changes introduced to the ontology. As the final result, a *virtual store descriptor* is produced, which is an ontology describing a particular personalized virtual store, including such elements as active parts of the store, interesting categories of products, favorite

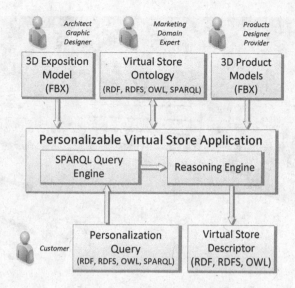

Fig. 1. Architecture of a personalizable VR store

brands, preferred salesperson, as well as the placement of the products on particular shelves. Finally, on the basis of the descriptor, the selected 3D products are imported and deployed in the 3D scene.

4   Creation of Virtual Stores

Creation of a virtual store involves two phases: *designing* and *personalization* of the store. In the first phase, a generic model is created, while in the second, a model that corresponds to specificity and preferences of a particular customer is generated as a response to a semantic personalization query.

4.1   Designing Virtual Stores

The process of designing a virtual store is divided into three separate steps. The first step of the process is *Exposition Design*. This step is performed by a highly skilled professional equipped with appropriate modeling tools. This step is performed rarely, in particular, it may be performed once, if a single exposition model is sufficient (e.g., for an existing real store). A ready-to-use virtual store model is presented in Fig. 2.

The second step is *Products Design*. It is also completed by a graphics designer equipped with a 3D modeling tool or a 3D scanner. However, in contrast to exposition design, this step is performed more frequently—every time when new types of products are introduced to the store. The process of preparing example 3D product models is presented in Fig. 3.

Fig. 2. An empty virtual store model.

The third step is *Design of the Virtual Store Ontology*. It is completed by a marketing domain expert, who uses a semantic modeling tool, e.g., Protégé. This step is repeated in case of modifying the shopping space, products or avatars that represent salesmen—depending on the ontology used in a particular store. An example of a virtual store ontology is presented in Listing 1.

Listing 1. A fragment of a virtual store ontology (RDF Turtle format)

```
1   :store rdf:type :Store.
2   :foodHall rdf:type :Hall.
3   :houseHall rdf:type :Hall.
4   :sportHall rdf:type :Hall.
5   ...
6   :store :includes :foodHall, :houseHall, :sportHall.
7   :shelf_1 rdf:type :Shelf.
8   :foodHall :includes :shelf_1, ... .
9   :houseHall :includes :shelf_11, ... .
10  :sportHall :includes :shelf_21, ... .
11  ...
12  :placeholder_1 rdf:type :Placeholder.
13  :shelf_1 :includes :placeholder_1, ... .
14  :placeholder_1 :position ... .
15  ...
16  :Food rdfs:subClassOf :Product.
17  :CleaningProduct rdfs:subClassOf :Product.
18  :Sport rdfs:subClassOf :Product.
19  :Bicycle rdfs:subClassOf :Sport.
20  :Bread rdfs:subClassOf :Food.
21  :Beverage rdfs:subClassOf :Food.
22  ...
23  :bread_1 rdf:type :Bread.
24  :bicycle_1 rdf:type :Bicycle.
25  :cleaningProd_1 rdf:type :CleaningProduct.
26  ...
27  :youngWoman rdf:type :SalesPerson.
28  ...
29  construct {:youngWoman rdf:type :SelectedSalesPerson} where {:customer :sex
        "male". :customer :age ?age. FILTER (?age > ?20 && ?age < 35)}.
```

In the ontology, different parts of the store are specified (lines 1–4). The store includes halls, which include shelves (6–10). There are placeholders on the shelves, which are empty slots for products (12–13). Every placeholder has a position specified (14). In addition to the spatial parts of the store, different classes of products are specified (16–21). Particular products, which may be put on shelves, are also specified (23–25). Moreover, avatars representing different salespersons are given (27). The SPARQL construct rule (29) selects a young woman avatar as the salesperson for customers who are males between 20 and 35. Similar rules may be specified for selecting the presented products, e.g., to present chips to customers who came to the store to buy beer.

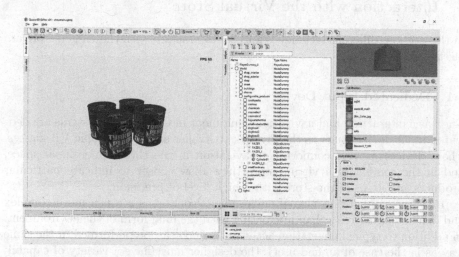

Fig. 3. Preparation of 3D products

4.2   Personalization of Virtual Store

The second phase of the virtual store creation is *Personalization of the Virtual Store*. In this phase, a personalization query is used. It may include user preferences as well as contextual data that determines the presented 3D products, e.g., only products that can be visible to the customer are imported to the 3D scene at a particular moment. An example of a personalization query consisting of an OWL ontology is presented in Listing 2. Such a query can be sent to the personalizable virtual store by the *Contextual Semantic Interaction Interface* (cf. Sect. 5.3).

Listing 2. A personalization query to a virtual store (RDF Turtle format)

```
1  :customer rdf:type :Customer , owl:NamedIndividual.
2  :customer :sex "male".
3  :customer :age "31".
4  :Bicycle rdfs:subClassOf :SelectedProduct.
5  :customer :location :sportHall.
```

The query is based on customer's characteristics: sex and age (lines 2–3), interests (line 4) and the current context of interaction (customer's location in the store—line 5). Upon receiving the query, the *Personalizable Virtual Store Application* generates a virtual store descriptor. It describes the store setup, which corresponds to the criteria provided by a customer (both explicitly and implicitly). In the example, bicycles are deployed together with associated products that may also be in the interest of the customer—food and beverages for sportsmen. Finally, a 3D virtual scene of the store is generated according to the descriptor.

5   Interaction with the Virtual Store

In this section, different techniques for user interaction with the virtual store are discussed.

5.1   Dedicated Input Device

The basic interaction and navigation equipment in CAVE-type systems are dedicated input devices such as the Flystick controller and stereoscopic VR glasses. These devices provide comfortable navigation in the three-dimensional space, and in the context of a virtual store they allow a user to indicate particular products and areas of the store with a high level of precision.

However, configuring virtual environment content is a much more complex and demanding activity than interacting with the final form of the environment. In addition to typical navigation activities (which can be performed in the same way as in the case of an end user), the designer must have a variety of capabilities for interaction with the virtual environment and the objects. Introducing the possibility of modifying the available assortment requires designing a completely new part of the user interface. This interface must allow a user to remove products currently on the shelves, add new ones (e.g., by selecting them from a list), specify the quantities or amounts, and sometimes also precisely define the options for placing the products on the shelves (e.g., whether they should be stacked, placed next to each other, at what angle they should be placed, or if they should be mixed with other products). Further development of such an environment may require additional configuration possibilities (e.g., determining the height of shelves) and improve user comfort by adding additional functions (e.g., "copying" the contents of a shelf).

As a result, in the case of configuring a virtual shopping space, dedicated input devices loose their greatest advantage – user comfort. With a small fixed number of buttons and an analog knob, the Flystick is not able to easily handle all possible activities, and using combinations of buttons significantly complicates users' interaction with the system.

Moreover, this problem cannot be solved by adding additional dedicated input devices (e.g., second Flystick), while the use of classical input devices (such as a mouse and keyboard) is typically not possible in CAVE-type systems.

All this makes it necessary to find (or develop) another method of interacting with virtual reality systems in order to enable efficient content management in virtual environments such as a virtual store.

5.2   Remote Interaction Interface

Another method that allows users to design VR content in real time is the use of a dedicated application implemented on a separate device with its own CPU. An independent application offers the convenience of having a tool specifically tailored for customization of the virtual environment.

However, a dedicated remote user interface has important disadvantages when used for configuring virtual environments. The approach requires a great deal of development work to implement the control application. If the application is generic, i.e., it has the same interface for all virtual environments and products – it cannot implement all required diversity of functions. For example, some products may have specific configuration functionality (e.g., opening a closing a laptop on the display), which is difficult to predict at the time of implementing the control application. Conversely, if the application is specific, it may offer all required functionality, but the effort, the time and the cost of development will typically not be justified by the offered advantages.

5.3   Contextual Semantic Interaction Interface

A solution to the problem mentioned in the previous subsections is the use of a new interaction method, called *Contextual Semantic Interaction Interface* (*CSII*) [29]. This method is based on a client-server architecture, where the virtual environment engine plays the role of a server, while a mobile device plays the role of a client. The client-server communication is based on WiFi, with non-sequential variant of the UDP protocol, for real-time operation.

The client displays a user interface generated dynamically based on the virtual store ontology sent by the server. The interface is adjusted to the particular VE and a particular context of interaction. Basic interface enables a user to navigate in the virtual environment. Initialization of interaction with an element of the VE is done by focusing the users' point of view on this element.

Semantic interaction metadata are included in the virtual store ontology. When a personalized virtual store is created, the metadata are assigned to particular elements of the store, and – at the runtime – are used by the CSII to generate a contextual interface. The metadata are created by a marketing specialist using a semantic editor. Different interaction metadata are assigned to different elements of the environment (e.g., shelves and refrigerators in a VR store). Also, interaction metadata can be assigned to particular products, thus implementing their specific functionality. In addition, during the use of CSII, semantic contextual information is added to personalization queries to properly adjust the store to the current context of the customer.

This type of interface may be used both by content designers configuring the space, and by end users immersed in the ready-to-use environment. This solution eliminates the standard form of navigation (such as a Flystick), so that users do not need to change the input device when switching between the content design mode and the passive customer mode. The CSII can be easily used in CAVE-type systems and other semi-immersive VR setups.

6   Example of Virtual Store

The VR environment personalization approach, presented in this paper can be used in many different application domains. This paper focuses on an application in the e-commerce industry – the personalized virtual store environment. In Fig. 4, a view of a user immersed in the VR shopping space in a CAVE system is presented.

Fig. 4. A user inside a CAVE interacting with a virtual reality shopping space.

The main element of the environment is a personalizable store shelving, which is divided into particular shelves that may contain products. The products can be food and commodity items for everyday use, such as soft drinks, corn flakes, cosmetics, etc. For the experimental evaluation, 3D models of products from open on-line repositories have been used. To perform real tests, we plan to acquire realistic 3D models of real products, for example by scanning them with the use of a photogrammetric scanner.

Apart from the shelving and the products, there are also other elements in the environment that can be typically found in real stores, such as the lighting, windows, counter and an animated model of a salesperson.

7   Conclusions and Future Works

New forms of interaction that enable convenient and intuitive use of a priori unknown and dynamically changing virtual environments, in various context and different roles, are essential to make the use of immersive VR systems simpler for non-expert users, and therefore applicable in more application domains.

In this paper, an approach to the development of personalized virtual environments has been proposed and illustrated by an application in the domain of virtual stores. The personalization can encompass product arrangement by marketing experts as well as interaction with the shopping space and the available products by customers. New forms of interaction are essential to achieve user-friendly content management by domain experts and enable merchandising research to be carried out in an efficient way.

As an initial evaluation, the usefulness and technical correctness of the system, has been verified by volunteers in the Virtual Reality Laboratory of the Department of Information Technology at the Poznań University of Economics and Business. We plan to continue the tests with marketing experts and non-expert users to evaluate the perception of virtual stores and the usefulness of the personalization process.

Future works encompass also evaluation of the performance of processing the personalization queries. In addition, the system can be extended with modules for reasoning on semantic web rules encoded in other languages than SPARQL, in particular SWRL [40].

References

1. Alpcan, T., Bauckhage, C., Kotsovinos, E.: Towards 3D internet: why, what, and how? In: 2007 International Conference on Cyberworlds, CW 2007, pp. 95–99. IEEE (2007)
2. Alshaer, A., Regenbrecht, H., O'Hare, D.: Investigating visual dominance with a virtual driving task. In: 2015 IEEE Virtual Reality (VR), pp. 145–146, March 2015
3. Borusiak, B., Pierański, B., Strykowski, S.: Perception of in-store assortment exposure. Stud. Ekon. **334**, 108–119 (2017)
4. Chaudhuri, S., Kalogerakis, E., Giguere, S., Funkhouser, T.: Attribit: content creation with semantic attributes. In: Proceedings of the 26th Annual ACM Symposium on User Interface Software and Technology, pp. 193–202. ACM (2013)
5. Cortes, G., Marchand, E., Ardouinz, J., Lécuyer, A.: Increasing optical tracking workspace of VR applications using controlled cameras. In: 2017 IEEE Symposium on 3D User Interfaces (3DUI), pp. 22–25, March 2017
6. Dachselt, R., Hinz, M., Meissner, K.: Contigra: an XML-based architecture for component-oriented 3D applications. In: Proceedings of the Seventh International Conference on 3D Web Technology, Web3D 2002, pp. 155–163. ACM, New York (2002)
7. Dassault Systèmes: 3D ContentCentral (2017). https://www.3dcontentcentral.com/
8. De Troyer, O., Bille, W., Romero, R., Stuer, P.: On generating virtual worlds from domain ontologies. In: Proceedings of the 9th International Conference on Multi-Media Modeling, Taipei, Taiwan, pp. 279–294 (2003)

9. Drap, P., Papini, O., Sourisseau, J.-C., Gambin, T.: Ontology-based photogrammetric survey in underwater archaeology. In: Blomqvist, E., Hose, K., Paulheim, H., Ławrynowicz, A., Ciravegna, F., Hartig, O. (eds.) ESWC 2017. LNCS, vol. 10577, pp. 3–6. Springer, Cham (2017). https://doi.org/10.1007/978-3-319-70407-4_1

10. Flotyński, J., Walczak, K.: Ontology-based creation of 3D content in a service-oriented environment. In: Abramowicz, W. (ed.) BIS 2015. LNBIP, vol. 208, pp. 77–89. Springer, Cham (2015). https://doi.org/10.1007/978-3-319-19027-3_7

11. Flotyński, J., Walczak, K.: Ontology-based representation and modelling of synthetic 3D content: a state-of-the-art review. Comput. Graph. Forum (2017). https://doi.org/10.1111/cgf.13083

12. Gebhardt, S., et al.: FlapAssist: how the integration of VR and visualization tools fosters the factory planning process. In: 2015 IEEE Virtual Reality (VR), pp. 181–182, March 2015

13. Hassouneh, D., Brengman, M.: Retailing in social virtual worlds: developing a typology of virtual store atmospherics (2015)

14. Highend3D: High Quality 3D Models, Scripts, Plugins and More! (2017). https://www.highend3d.com/

15. Kitson, A., Riecke, B.E., Hashemian, A.M., Neustaedter, C.: NaviChair: evaluating an embodied interface using a pointing task to navigate virtual reality. In: Proceedings of the 3rd ACM Symposium on Spatial User Interaction, SUI 2015, pp. 123–126. ACM, New York (2015). https://doi.org/10.1145/2788940.2788956

16. Latoschik, M.E., Blach, R., Iao, F.: Semantic modelling for virtual worlds a novel paradigm for realtime interactive systems? In: VRST, pp. 17–20 (2008)

17. LaViola Jr., J.J.: Context aware 3D gesture recognition for games and virtual reality. In: ACM SIGGRAPH 2015 Courses, SIGGRAPH 2015, pp. 10:1–10:61. ACM, New York (2015). https://doi.org/10.1145/2776880.2792711

18. Lee, K.C., Chung, N.: Empirical analysis of consumer reaction to the virtual reality shopping mall. Comput. Hum. Behav. 24(1), 88–104 (2008). http://www.sciencedirect.com/science/article/pii/S0747563207000155

19. Monect: Monect PC remote (2017). https://www.monect.com/

20. Nielson, G.M., Olsen Jr., D.R.: Direct manipulation techniques for 3D objects using 2D locator devices. In: Proceedings of the 1986 Workshop on Interactive 3D Graphics, I3D 1986, pp. 175–182. ACM, New York (1987). https://doi.org/10.1145/319120.319134

21. Pellens, B., De Troyer, O., Kleinermann, F.: CoDePA: a conceptual design pattern approach to model behavior for X3D worlds. In: Proceedings of the 13th International Symposium on 3D Web Technology, Web3D 2008, pp. 91–99. ACM, New York (2008). https://doi.org/10.1145/1394209.1394229

22. Perez-Gallardo, Y., Cuadrado, J.L.L., Crespo, Á.G., de Jesús, C.G.: GEODIM: a semantic model-based system for 3D recognition of industrial scenes. In: Alor-Hernández, G., Valencia-García, R. (eds.) Current Trends on Knowledge-Based Systems. ISRL, vol. 120, pp. 137–159. Springer, Cham (2017). https://doi.org/10.1007/978-3-319-51905-0_7

23. Pirker, J., Pojer, M., Holzinger, A., Gütl, C.: Gesture-based interactions in video games with the leap motion controller. In: Kurosu, M. (ed.) HCI 2017. LNCS, vol. 10271, pp. 620–633. Springer, Cham (2017). https://doi.org/10.1007/978-3-319-58071-5_47

24. Piumsomboon, T., Lee, G., Lindeman, R.W., Billinghurst, M.: Exploring natural eye-gaze-based interaction for immersive virtual reality. In: 2017 IEEE Symposium on 3D User Interfaces (3DUI), pp. 36–39, March 2017

25. Polys, N., Visamsetty, S., Battarechee, P., Tilevich, E.: Design patterns in componentized scenegraphs. In: Proceedings of SEARIS. Shaker Verlag (2009)
26. Rea, A., White, D.: The Layered Virtual Reality Commerce System (LaVRCS): an approach to creating viable VRcommerce sites. In: MWAIS 2006 Proceedings, vol. 11. AISeL (2006)
27. Roupé, M., Bosch-Sijtsema, P., Johansson, M.: Interactive navigation interface for virtual reality using the human body. Comput. Environ. Urban Syst. **43**(Suppl. C), 42–50 (2014). http://www.sciencedirect.com/science/article/pii/S0198971513000884
28. Sikos, L.F.: 3D model indexing in videos for content-based retrieval via X3D-based semantic enrichment and automated reasoning. In: Proceedings of the 22nd International Conference on 3D Web Technology, p. 19. ACM (2017)
29. Sokołowski, J., Walczak, K.: Semantic modelling of user interactions in virtual reality environments. In: Camarinha-Matos, L.M., Adu-Kankam, K.O., Julashokri, M. (eds.) DoCEIS 2018. IAICT, vol. 521, pp. 18–27. Springer, Cham (2018). https://doi.org/10.1007/978-3-319-78574-5_2
30. Spagnuolo, M., Falcidieno, B.: 3D media and the semantic web. IEEE Intell. Syst. **24**(2), 90–96 (2009)
31. Statista: Retail e-commerce sales worldwide from 2014 to 2021 (2018). https://www.statista.com/statistics/379046/worldwide-retail-e-commerce-sales/
32. Thomann, G., Nguyen, D.M.P., Tonetti, J.: Expert's evaluation of innovative surgical instrument and operative procedure using haptic interface in virtual reality. In: Matta, A., Li, J., Sahin, E., Lanzarone, E., Fowler, J. (eds.) Proceedings of the International Conference on Health Care Systems Engineering. PROMS, vol. 61, pp. 163–173. Springer, Cham (2014). https://doi.org/10.1007/978-3-319-01848-5_13
33. Trellet, M., Férey, N., Flotyński, J., Baaden, M., Bourdot, P.: Semantics for an integrative and immersive pipeline combining visualization and analysis of molecular data. J. Integr. Bioinform. **15**(2), 1–19 (2018)
34. Trellet, M., Ferey, N., Baaden, M., Bourdot, P.: Interactive visual analytics of molecular data in immersive environments via a semantic definition of the content and the context. In: 2016 Workshop on Immersive Analytics (IA), pp. 48–53. IEEE (2016)
35. TurboSquid Inc.: 3D Models for Professionals (2017). https://www.turbosquid.com/
36. Unity Technologies: Asset Store (2017). https://www.assetstore.unity3d.com/en/
37. Van Gool, L., Leibe, B., Müller, P., Vergauwen, M., Weise, T.: 3D challenges and a non-in-depth overview of recent progress. In: 3DIM, pp. 118–132 (2007)
38. Van Kerrebroeck, H., Brengman, M., Willems, K.: When brands come to life: experimental research on the vividness effect of virtual reality in transformational marketing communications. Virtual Real. **21**(4), 177–191 (2017). https://doi.org/10.1007/s10055-017-0306-3
39. Visamsetty, S.S.S., Bhattacharjee, P., Polys, N.: Design patterns in X3D toolkits. In: Proceedings of the 13th International Symposium on 3D Web Technology, Web3D 2008, pp. 101–104. ACM, New York (2008). https://doi.org/10.1145/1394209.1394230
40. W3C: SWRL. On-line (2004). http://www.w3.org/Submission/SWRL/
41. W3C: SPARQL query language for RDF (2008). http://www.w3.org/TR/2008/REC-rdf-sparql-query-20080115/
42. W3C: OWL. http://www.w3.org/2001/sw/wiki/OWL

43. W3C: RDF. http://www.w3.org/TR/2004/REC-rdf-concepts-20040210/
44. W3C: RDFS. http://www.w3.org/TR/2000/CR-rdf-schema-20000327/
45. Walczak, K.: Flex-VR: configurable 3D web applications. In: Proceedings of the Conference on Human System Interactions, pp. 135–140. IEEE (2008)
46. Walczak, K.: Structured design of interactive VR applications. In: Proceedings of the 13th International Symposium on 3D Web Technology, Web3D 2008, pp. 105–113. ACM, New York (2008). https://doi.org/10.1145/1394209.1394231
47. Walczak, K.: Semantics-supported collaborative creation of interactive 3D content. In: De Paolis, L.T., Bourdot, P., Mongelli, A. (eds.) AVR 2017. LNCS, vol. 10325, pp. 385–401. Springer, Cham (2017). https://doi.org/10.1007/978-3-319-60928-7_33
48. Walczak, K., Flotyński, J.: On-demand generation of 3D content based on semantic meta-scenes. In: De Paolis, L.T., Mongelli, A. (eds.) AVR 2014. LNCS, vol. 8853, pp. 313–332. Springer, Cham (2014). https://doi.org/10.1007/978-3-319-13969-2_24
49. Walczak, K., Flotyński, J.: Semantic query-based generation of customized 3D scenes. In: Proceedings of the 20th International Conference on 3D Web Technology, Web3D 2015, pp. 123–131. ACM, New York (2015). https://doi.org/10.1145/2775292.2775311
50. Zahariadis, T., Daras, P., Laso-Ballesteros, I.: Towards future 3D media internet. In: NEM Summit, pp. 13–15 (2008)
51. Zielasko, D., Neha, N., Weyers, B., Kuhlen, T.W.: A reliable non-verbal vocal input metaphor for clicking. In: 2017 IEEE Symposium on 3D User Interfaces (3DUI), pp. 40–49, March 2017

Towards Assessment of Behavioral Patterns in a Virtual Reality Environment

Ahmet Kose(✉), Aleksei Tepljakov, Mihkel Abel, and Eduard Petlenkov

Department of Computer Systems, Tallinn University of Technology,
Ehitajate tee 5, 19086 Tallinn, Estonia
ahmet.kose@taltech.ee

Abstract. Virtual Reality (VR) is a powerful modern medium for immersive data visualization and interaction. However, only a few research efforts targeted the issue of complementing VR applications with features derived from real-time human behavior analysis in virtual environments. This paper addresses an interactive application for analysis of user behavior in a VR environment. In this work, real-time data communication is employed to collect data about the VR user's location and actions in the virtual environment. To ensure the authenticity of interactions in the virtual environment, the VR application aims at achieving complete immersion. Our findings pertaining to behavioral patterns in immersive environment suggest that there is a potential in applying knowledge of user behavior models to improve the interactivity between the user and the virtual environment. Analysis of VR users' behavioral models also complements studies typically performed by traditional survey techniques.

Keywords: Virtual Reality · Real-time communication ·
Data analysis · Tracking systems · Interactions · Human behavior

1 Introduction

Virtual Reality (VR) is a computer generated audio-visual environment that strives to immerse users completely in a simulated interactive environments [1]. Therefore, present advancement in VR technology allows inducing a persistent effect of presence in visual world. This technology has become available for a broader audience in recent years with the advent of low-cost head-mounted display (HMD) devices such as Oculus Rift [2] and HTC Vive [3]. These plug and play devices are also featured with user virtual-world position and orientation tracking that complement the presented work. Virtual environments (VE) have proven effective in many scientific, industrial and medical applications [4–8] and are applied in, including but not limited to: computer-simulated environments, visualizations of complex data, joyful learning tools, etc.

© Springer Nature Switzerland AG 2019
L. T. De Paolis and P. Bourdot (Eds.): AVR 2019, LNCS 11613, pp. 237–253, 2019.
https://doi.org/10.1007/978-3-030-25965-5_18

VR is also a favored tool to utilize interactions in various fields. The interactivity usually encourages users to be more operative, complements their learning process and assists them in exploring the virtual environment and complex data visualizations [9]. Furthermore, one of attractive extension of interactive applications is to examine behavioral aspects mainly for research purposes [10]. In other words, VR provides an opportunity to evaluate human performance and perception in a specific task orientated VE by eliminating distractions and granting them with artificial skills. Accordingly, we aim to provide a solution for analysis of users' behavior to complement interactive VR applications. Consequently, the analysis part of the research into user's behavior will provide more advanced knowledge compared to traditional survey techniques.

In our earlier work [11], we have introduced the fully immersive environment and we followed a similar process to create the presented application. In this work, we describe our approach to analyze behavioral and attentional aspects by delivering a VR based application. The aim is to investigate success rate of the player and the efficiency of the application in different categories by taking into account the player's profile. Besides that, we try to answer the question whether the subjective self-evaluation of the player correlates the objective measurement data by monitoring movements and actions in immersive environment. Previously implemented real-time data communication [12] is also improved to collect and store necessary data for this work. As a remarkable benefit of this framework, interactions by users are stored in real-time and further feedback with realistic perspective can be provided based on the user's actions.

The main contribution of the present paper as follows: First, we present the framework used to conduct the experiments. In following, we explain a customized User Interface (UI) and we present a system configuration in order to enable real-time data communication. We also investigate the employed communication protocol to minimize latency during the VR experience. Next, we introduce created 3D models and used the game engine to create the application. In the following section, we describe preferred methods to provide novel findings and the outcome of the conducted experiment respectively. We also evaluate the application based on collected data and survey results. Finally, we report and address the conclusion towards behavioral aspects while the user experiences the application and outline some related items for future research.

The structure of the paper is as follows. In Sect. 2, the recreation process of the application, the game play and UI are described. The system configuration including technical specifications is presented in Sect. 3. In Sect. 4, the reader is introduced to human-computer interaction conducted with the immersive environment. The statistical findings towards behavioral aspects and the complementary questionnaire are addressed in Sect. 4. Finally, conclusions are drawn in Sect. 5.

1.1 Related Work

There exist a number of studies about interaction techniques considering human behavior, motion and actions. Many of such studies have focused on training with interactive VR applications. For instance, VR based application [13] is employed to train early medical professional by interactions. Participants who performed in another interactive VR training [14] increased retrieval success almost two times. Similar applications have also proven their positive impact in education. The steady decrease of students' failure rate is observed by training students with VR based application in [15]. Game-based learning using VR in education is presented in [16]. As the result, the learning efficiency is increased in terms of the time and the amount of content. Overall, VR based applications can provide a novel approach towards interactions which can be claimed one of the key factors for effectiveness [17]. Researchers have also introduced a few methods and applications towards human behavior [18], actions [19] and emotional states [20] in real-time in application-specific VEs. The user behavior and emotional state models are also used in design of user interfaces [21,22] linked to VR based applications. Furthermore, the location of participants in virtual-world is recorded in [23] to measure the engagement of participants in VR based application. It is claimed in [24] that head orientation recorded by inertial sensors may be sufficient to prediction purposes with reasonable accuracy.

2   The Framework for Analysis of Human Behavior

It is apparent that a complete analysis of human behavior in VR requires a significant amount of time and effort. In order to elaborate the framework of this research, we created a prototyping platform to acquire data necessary for assessing a user behavior in VR based applications. Replication of large scale building for interactive VR based application [25] can be considered the first stage of the framework. We gained technical experience and knowledge as well as remarkable feedback from hundreds of participants during the early stage. The gain gave rise to create another VR based application-Swedbank VR Experience [26]. The first version of the gaming-oriented VR application was completed at the beginning of 2018 in cooperation with a private bank in Estonia. The bank is now using the first version of the application for marketing purposes on various fairs, in their headquarters among Baltic countries to exhibit the innovative and entertainment sides of their business.

The first version of the application consists of three main components: starting screen with a custom UI form, tutorial level and main game level. The main level is situated in the hill-top virtual environment and the screenshot of the level is depicted in Fig. 1. It was designed to let the users perceive in a borderless environment. Secondly, the tutorial level was designed to keep the player in boundaries in order to direct the player to complete the specific tasks. The tutorial level thus took place in a modern circle shaped hall with a dome. The tutorial environment remained simple in order to induce the player to take desired actions leaving only minimal possible distractions. The primary motivation of tutorial level is devoted to giving an understanding of employed functions in the application such as teleportation, interactions etc. Additionally, users benefit from the warm-up session before the main game, precisely who does not have previous a practical VR experience.

Fig. 1. The game play view of the presented application

The tutorial level is excluded in the presented work since it would have given the player a better notion about the game-play. The dedicated experiment was designed to gather data at two different situations: before the players had some experience in VE and after. Therefore, in order to succeed with further analysis, the application is configured to provide a set of data in real-time. As we have primarily been dealing with setting up the stage for further experiments, it will be too early to draw any statistically relevant conclusions from the experiments described herein. In other words, the experiment is in early stage and the aim now is to design the experiment including the development of technical tools enabling the experiment. The experiment itself is on-going process and the number of participants is being extended constantly with the goal of obtaining statistically relevant data based on a large number of experiments.

2.1   Data Processing and Game Play

The data set is collected in three phases of the experiment. In the first phase, the customized UI is in charge of handling the collection of initial data set. The primary purpose of created UI is to collect the information about the participant's background.

In the first phase, the data is categorized based on age group, gender, education, previous gaming experience and hours spent weekly on physical activities. Users are asked to choose the best fit option based on multiple choice questions. The data collection is only activated by filling up the form and accepting to share the data for research and development purposes for each user. Options provided to the user are listed in Table 1.

Table 1. Given options in UI

Options	Age	Gender	Educational level	VR experience	Activity level
1	3–7	Female	Primary school	None	None
2	7–14	Male	High school	Low	1–2 h
3	15–24		Bachelor's degree	Moderate	3–4 h
4	25–49		Master's degree	High	4–7 h
5	50+		Doctor of philosophy		7+ h

The second phase of process is allocated into game sessions, the player's goal in the game is to collect a maximum number of points possible during the two-minutes game time. There are four different items in the game for the defined goal; yellow leaf, green leaf, orange leaf, and coin. Each item rewards the player with 1, 2, 3, 10 points respectively.

In game level, the items are randomly generated and scattered but only around the defined play area. Those items are placed in the level by the algorithm for each game. Players are motivated to use teleportation feature to move around the play-area in VE since the distance between collectable objects are random and often the objects are far apart from each other. The objects spawn

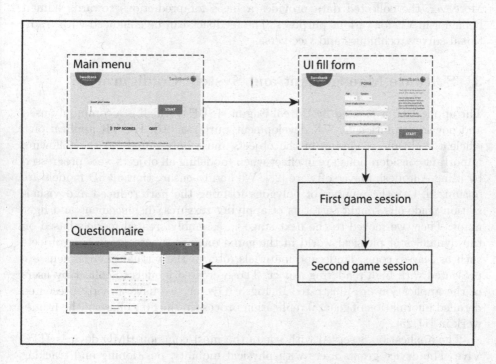

Fig. 2. Data collection process for the experiment

in different locations and at a varying distance from the ground, so the player must still move physically.

The last phase is placed after the game sessions. Each player is also asked to fill the survey that consists of seven questions by rating 1 (very low) to 5 (very high). The process of the application for experimental purposes is depicted in Fig. 2. To sum up, collected data with classified types for the framework is presented in Table 2.

Table 2. Classified types of collected data

Data type	Collected information
1	Age, gender, education level and experience in VR
2	Teleportation usage, amount of collected items
3	User location in VR, head movement of user
4	Traditional survey

On the whole, understanding human behavior in VE might also encourage joyful learning approaches. Therefore, existing difficulties to ensure rewarding feedback loop for creating more efficient and compelling VE can be alleviated. Moreover, the collected data provides a basis for predictions towards human behavior in VE for various purposes. The feedback can be enhanced with traditional survey techniques and vice-versa.

3   Software Management and System Configuration

The application is created by Unreal Engine 4 (UE4) which is a commonly used and powerful solution for VR development purposes [27]. For VR applications, efficient real-time rendering of the objects must be ensured, so the following important considerations are in effect when modeling all objects were progressed by using Autodesk Maya software [28]. We had to ensure that all 3D models are optimized, i.e., the number of polygons forming the part reduced and visualization trade-offs sought in terms of applying textures, displacement and light maps. Then, we moved to the next stage to assembly created models based on the dynamics of physical world in the game engine. Whereas, existing objects such as leaves, coins, landscape materials, etc. in the virtual environment are presented with high rendering aspect. Those created models to collect by users in the application are illustrated in Fig. 3. Overall, we have already presented detailed information of general replication process in the first phase of the framework in [11, 25].

The application is served with one of the most common HMD devices–HTC Vive. The device grants users with physical mobility, positioning and tracking based on real time with high display resolution. The HMD provides 1080×1200 resolution per eye, and the 9-DOF with 2 lighthouse base stations for tracking.

(a) 3D models of green, orange and yellow leaves

(b) 3D coin model by both side

Fig. 3. Existing objects to collect in VR based application

The two sets of controller features 24 sensors, dual-stage trigger and multi-function trackpad [29]. Launching the application and running the experiment require specifications in terms of hardware and software. The application runs on a PC equipped with a 4.00 GHz Intel i7-6700 processor, 32 GB RAM, and an NVidia GTX 980 graphic card.

The real-time data communication part of this work is remarkable. The visual scripting system in UE4 typically referred to as *Blueprints* is a complete game-play scripting system based on the concept of using a node-based interface to create game-play elements from within Unreal Editor [30]. The Blueprint system allowed us to create a User Datagram Protocol (UDP) plugin [31] to communicate between local database via python made scripts and the physics engine based on a custom C++ class. In order to ensure smooth experience with minimum latency possible for the user, UDP socket is employed to collect data with $f_s = 1\,\text{Hz}$ sampling rate. Nevertheless, UDP plugin as well as exported modules can also be reused in any project. Since HTC Vive is used, the corresponding UE4 VR template is employed and thus the user can navigate inside of the virtual environment. The main idea here is to synchronize the user location in VE and head movements tracked via HMD device.

The UDP plugin requires custom level blueprints for communications, meaning that the reference to UDP actor in visual level can be given only into a corresponding game level. Thus, it is required to gather all the needed variables that contained desired data with "get" function from game instance blueprint. All the form input data was translated into numeric values to be stored in the local database if necessary. For instance, "Gender" is the part of the required form to fill by the user via UI and selected output is defined as 0 or 1 meaning male or female to keep the database structure steady and organized.

The game instance in the blueprints of the game engine acts as a collection of global variables. The collection can exchange values of global variables between all the blueprints within the project via "get" function. Furthermore, new values can be written into global variables by other blueprints via the "set" function. "get" function is used to get the value of certain variable and "set" function was used to forward new value to the global variable [30]. UI, teleportation and interactions with four collectable objects are composed by this approach to provide real-time data communication.

Consequently, we implemented the application and the devoted experiment with three software packages: The game engine is employed as the visualization platform while the user experiences the virtual facility with interactive equipment. Motion capture is achieved using present HMD technology which provides access to user movement data. UDP serves as a communication plugin between the game engine and software environment that makes real-time data communication and simulation possible. To proceed with modeling of the user's head and controller movement, the data of user is sent out to Python and local database to store and apply statistical methods. The screenshot while the user is playing and all software packages are working is depicted in Fig. 4.

Fig. 4. The screenshot while experiencing VR application with timer and score

4   Human Computer Interaction in Immersive Environment

Human-Machine Interaction (HMI) is one of the relevant research topics. The interactivity directly impacts the efficiency whether HMI applications encourage users to operate, explore and manipulate. Merging interactions in VR brings different perspective to HMI. In other words, HMI can be advanced in artificial reality conditions [32]. One of the primary attitudes of VR development is to allow the user to interact as much as they desire. It gives them independent movement, thinking and improving themselves with support of VE. Moving in the scene of VR can be operated through HMD controller using teleportation, physical movement etc. The presented VR based application also allows users to interact by handles are provided with a VR device. Users can direct any locations where navigation maps are implemented. Previously, we provided an questionnaire to participants who had experienced the replication of the architecture [25]. According to results, we experienced that there are no major differences between real building and its digital twin. Basic interactions in the application also engaged their attention. However, we also observed that functionalities of virtual environment were limited. They would like to demonstrate practical activities. Therefore, presented VR based application concentrate to grant users with interactions while minimizing limitations.

The virtual environment is employed for multiple purposes such as self-learning activities, psychological and behavioral aspects while users experience the application. We conduct our experiment with twenty volunteers. After the intro session and the UI phase, we asked to them play the game two times in a row and each time frame is restricted to 120 s. The motivation for that is to observe the difference from user perspective between first and second attempt for the VR based application. The provided results may benefit to have better understanding of behavioral aspects considering the first type of collected data-background of participants. According to collected data, most diversified information is gathered by physical activity level and VR experience. While experience in VR levels raised in all categories almost equally, activity levels of participants remain coincidental. Participants who selected low VR experience is referred to none of physical experience with HMD devices previously. Further classification is illustrated by drawing pie charts in Fig. 5. The ultimate goal of the presented framework is to endow the VE with the additional quality of intelligence derived from a thorough study of the users' behavior and the way they interact with VE.

4.1   Assessment of Interactions

One of the main attitude of VR applications is to allow users to interact. VR based applications are often delivered with the interaction that may also allow an evaluation of the engagement of participants [23]. In what follows, we present our findings by measuring user's location and action with a case study; Teleportation and Collected Items.

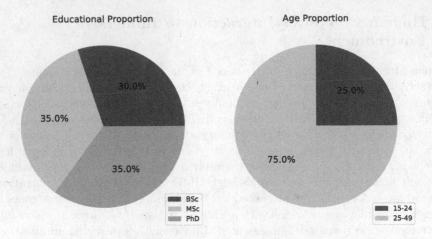

(a) Education and age division

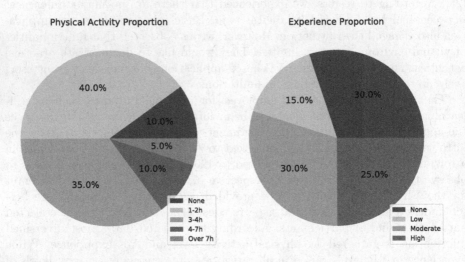

(b) Physical activity and experience division

Fig. 5. Collected background profiles of participants

Experimental Study-Teleportation Within Immersive Environment.
The teleportation is a type of locomotion and desired feature for the presented
application like other large scale immersive environments [33]. Although the
feature is only artificial and is unfamiliar to users in terms of actual experience
(with the exception of users having previously used VR), it also allows users
to move independently and effortless in immersive environment. Furthermore,
independent movement may also grant users to sense self-learning activities in
VE. Thus, the part of this work is devoted for investigating the teleportation
feature which let the user interact with objects during continuous movements.

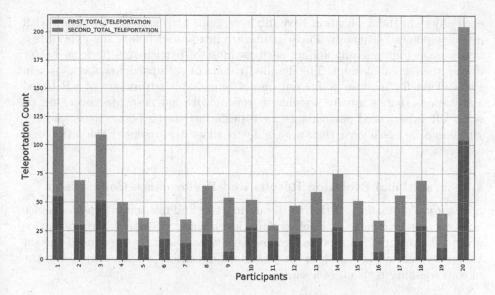

Fig. 6. Statistical view of interactions into the teleportation

Studying user behavior while using the previous version of the application, some difficulties while attempting to use teleportation such as insecure emotions were observed. Thus, tracking teleportation usage of participants seemed relevant for further research analysis. Participants who claim to have VR experience in high and moderate levels used teleportation feature at most in both times. Additionally, the rest of participants who had less VR experience have used the teleportation feature much more frequently in the second time of the experience. As a conclusion based on collected data, the teleportation feature is easy to adapt and very convenient to employ even for users who might not have previous VR experience with HMD device. Detailed information about teleportation usage for participants including first and second time is presented in Fig. 6.

Experimental Study-Decision Making in Limited Time to Collect Maximum Points. We have already described in Sect. 2.1 that the game session has a goal to maximize performance in terms of collecting items within limited time-frame. From the user perspective, such interactions linked to the gamification based engagement are of curiosity driven experimental nature. Moreover, the user can adapt to the gaming environment by interacting with it, and the perceives tasks to be completed [34]. Once the motivation is obtained, an accomplishment of attentional aspects by delivering a VR based application matters of given objects and tasks. Training for real-world scenarios with strict time limitations can be form effectively by following similar methodology [35,36]. However, applying interactions, teleportation and physical movements at once are challenging for users without previous VR experience. Hence, we let participants to try second time to observe the differences. As it is described previously in that each object in the application have different color and users are rewarded with different points adequately. Participants are introduced before the experiment and we conclude relevant finding such as most of participants perform better during the second attempt. The detailed results of each participant is shown in Fig. 7. That finding also points out one of remarkable advantages of VR based applications that is usually possible to repeat with low cost. Moreover, low or none VR experienced participants have performed much better in second try which can be considered that were able to adapt to presented the VR based application.

4.2   Traditional Evolution Results and Performance Comparison

Statistical Terminology. Prior to discussing the findings towards performance analysis of participants, a brief explanation of used statistical terminology is given. The mean ($\bar{x}$) is used to summarize interval data by the sum of the set values divided by the length of set. Standard deviation (σ): An estimate of the mean variability (spread) of a sample [37].

Table 3. Survey results

Questions	Mean ($\bar{x}$)	Std (σ)
Did you enjoy the VR application?	4.15	0.7
How much mental and perceptual activity was required?	2.75	0.9
How much physical activity was required?	2.3	1.2
How satisfied were you with your performance?	3.55	1.1
Did you perform better for the second time?	4.15	1.1
Did you have the immerse feeling while experiencing the app?	3.85	0.7
Would you like to try the VR application again?	4.3	0.7

According to survey results in Table 3, participants claim that the performed better in second time. The outcome of collected data in Table 4 also confirms the statement of participants. Utilizing basic statistical methods to support collected data has also shown that low or none VR experienced participants have performed almost four times better in the second try. The second time performance of the participant who achieved the highest difference compare first and second times is illustrated in Fig. 8. Moreover, the second attempt results indicates collected leaves do not have significant difference among participants based on VR experience level. However, participants with lower level VR experience did not pay the same level of attention for points to compare with amount of items. In other words, even though the point rewarding mechanism was known, it was not practical significantly.

Nevertheless, the results based on conducted survey can reveal subjective assessment to evaluate such as whether the application is joyful, easy, immersive, inconvenient etc. On the other hand, tracking the behavioral as well as

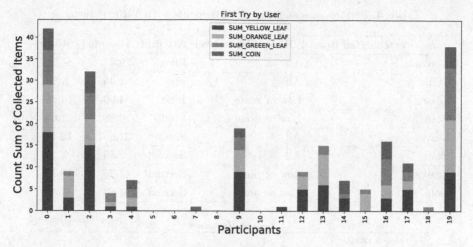

(a) Collected objects for first time

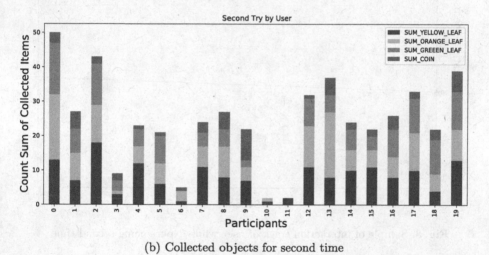

(b) Collected objects for second time

Fig. 7. Collected objects by users in both attempts

attentional aspects by given tasks and enabled interaction in VR based applications can validate the survey results. The method might also provide remarkable, objective and accurate feedback loop to improve and assess the application.

Table 4. Performance analysis of participants with VR experience

Parameter/collected items	VR experience level	Attempt	Mean ($\bar{x}$)	Std (σ)
Leaves	All	First	9.4	11.25
Coin	All	First	1.45	1.94
Leaves	Low or none	First	4.00	4.47
Coin	Low or none	First	0.88	1.65
Leaves	All	Second	21.65	12.17
Coin	All	Second	2.85	2.25
Leaves	Low or none	Second	17.25	10.12
Coin	Low or none	Second	1.63	1.40

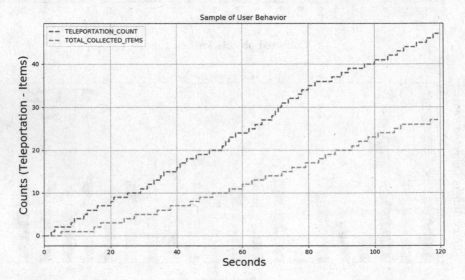

Fig. 8. Sample of interaction track of user while experiencing second time

5   Conclusions

In this paper, the general framework towards developing an interactive VR based application for analysis of user behavior was described. First, the background of the application was presented. Next, we have described complete system configuration including software management, 3D modeling, UI and integration of hardware devices. In this contribution, we have two main goals to employ the real-time streaming and collecting of datasets: to track accurate location and interaction of the user in immersive environment and orientation of HMD device worn by the user. As a result, although the experiment is in early stage, the application already illustrates the potential of using VR technology in practice. We conducted the application with two concrete experiments into assessments of locomotion and interaction features. These experiments were accomplished and provided important insights into the development of interactive VR based application. Furthermore, our findings may be beneficial for development of user experience assessment tools for VE. The data collection and automatic analysis may work in background during demonstrations when people are playing the game. Thus, the statistical relevance of the results is improved as the number of players increases. All the participants are informed in advance regarding collecting the data for research purposes. The data collection is only activated by accepting to share the data for research and development purposes. The unified and plug and play solution to integrate the other VR applications is yet to be investigated. Collected data sets of head orientation as well as user location in immersive environment will also be presented in future.

References

1. Zhu, L., Wang, J., Chen, E., Yang, J., Wang, W.: Applications of virtual reality in turn-milling centre. In: 2008 IEEE International Conference on Automation and Logistics. IEEE, September 2008
2. Oculus VR, LLC. Oculus Rift (2017). Accessed 10 Feb 2019
3. HTC Corporation. HTC Vive (2017). Accessed 20 Jan 2019
4. Cordeil, M., Dwyer, T., Klein, K., Laha, B., Marriott, K., Thomas, B.H.: Immersive collaborative analysis of network connectivity: CAVE-style or head-mounted display? IEEE Trans. Visual Comput. Graph. **23**(1), 441–450 (2017)
5. Beidel, D.C., et al.: Trauma management therapy with virtual-reality augmented exposure therapy for combat-related PTSD: a randomized controlled trial. J. Anxiety Disord. **61**, 64–74 (2019)
6. Valmaggia, L.R., Day, F., Rus-Calafell, M.: Using virtual reality to investigate psychological processes and mechanisms associated with the onset and maintenance of psychosis: a systematic review. Soc. Psychiatry Psychiatr. Epidemiol. **51**(7), 921–936 (2016)
7. Yiannakopoulou, E., Nikiteas, N., Perrea, D., Tsigris, C.: Virtual reality simulators and training in laparoscopic surgery. Int. J. Surg. **13**, 60–64 (2015)
8. Tepljakov, A., Astapov, S., Petlenkov, E., Vassiljeva, K., Draheim, D.: Sound localization and processing for inducing synesthetic experiences in virtual reality. In: 2016 15th Biennial Baltic Electronics Conference (BEC), pp. 159–162, October 2016
9. Dormido, R., et al.: Development of a web-based control laboratory for automation technicians: the three-tank system. IEEE Trans. Educ. **51**(1), 35–44 (2008)
10. Tsaramirsis, G., et al.: Towards simulation of the classroom learning experience: virtual reality approach, October 2016
11. Kose, A., Tepljakov, A., Petlenkov, E.: Towards assisting interactive reality. In: De Paolis, L., Bourdot, P. (eds.) AVR 2018. LNCS, vol. 10851, pp. 569–588. Springer, Cham (2018). https://doi.org/10.1007/978-3-319-95282-6_41
12. Kose, A., Tepljakov, A., Astapov, S., Draheim, D., Petlenkov, E., Vassiljeva, K.: Towards a synesthesia laboratory: real-time localization and visualization of a sound source for virtual reality applications. J. Commun. Softw. Syst. **14**(1), 112–120 (2018)
13. Rajeswaran, P., Hung, N.-T., Kesavadas, T., Vozenilek, J., Kumar, P.: AirwayVR: learning endotracheal intubation in virtual reality. In: 2018 IEEE Conference on Virtual Reality and 3D User Interfaces (VR). IEEE, March 2018

14. Schöne, B., Wessels, M., Gruber, T.: Experiences in virtual reality: a window to autobiographical memory. Curr. Psychol. **38**, 715–719 (2017)
15. Pereira, C.E., Paladini, S., Schaf, F.M.: Control and automation engineering education: combining physical, remote and virtual labs. In: International Multi-Conference on Systems, Signals and Devices. IEEE, March 2012
16. Rozinaj, G., Vanco, M., Vargic, R., Minarik, I., Polakovic, A.: Augmented/virtual reality as a tool of self-directed learning. In: 2018 25th International Conference on Systems, Signals and Image Processing (IWSSIP). IEEE, June 2018
17. Ragan, E.D., Bowman, D.A., Kopper, R., Stinson, C., Scerbo, S., McMahan, R.P.: Effects of field of view and visual complexity on virtual reality training effectiveness for a visual scanning task. IEEE Trans. Visual Comput. Graph. **21**(7), 794–807 (2015)
18. Moeslund, T.B., Granum, E.: A survey of computer vision-based human motion capture. Comput. Vis. Image Underst. **81**(3), 231–268 (2001)
19. Pham, D.-M.: Human identification using neural network-based classification of periodic behaviors in virtual reality. In: 2018 IEEE Conference on Virtual Reality and 3D User Interfaces (VR). IEEE, March 2018
20. Suhaimi, N.S., Yuan, C.T.B., Teo, J., Mountstephens, J.: Modeling the affective space of 360 virtual reality videos based on arousal and valence for wearable EEG-based VR emotion classification. In: 2018 IEEE 14th International Colloquium on Signal Processing and Its Applications (CSPA). IEEE, March 2018
21. Sutcliffe, A.G., Poullis, C., Gregoriades, A., Katsouri, I., Tzanavari, A., Herakleous, K.: Reflecting on the design process for virtual reality applications. Int. J. Hum.-Comput. Interact. **35**(2), 168–179 (2018)
22. Wang, W., Cheng, J., Guo, J.L.C.: Usability of virtual reality application through the lens of the user community. In: Extended Abstracts of the 2019 CHI Conference on Human Factors in Computing Systems. ACM Press (2019)
23. Merino, L., Ghafari, M., Anslow, C., Nierstrasz, O.: CityVR: gameful software visualization. In: 2017 IEEE International Conference on Software Maintenance and Evolution (ICSME). IEEE, September 2017
24. Sitzmann, V., et al.: Saliency in VR: how do people explore virtual environments? IEEE Trans. Visual Comput. Graph. **24**(4), 1633–1642 (2018)
25. Kose, A., Petlenkov, E., Tepljakov, A., Vassiljeva, K.: Virtual reality meets intelligence in large scale architecture. In: De Paolis, L.T., Bourdot, P., Mongelli, A. (eds.) AVR 2017. LNCS, vol. 10325, pp. 297–309. Springer, Cham (2017). https://doi.org/10.1007/978-3-319-60928-7_26
26. Tallinn University of Technology. Official website of Re:creation Virtual and Augmented Reality Laboratory (2018). Accessed on 01 Mar 2018
27. Epic Games. Unreal Engine. Accessed 03 Jan 2019
28. Autodesk Maya Software. Features (2017). Accessed 25 Aug 2018
29. Dempsey, P.: The teardown: HTC vive virtual reality headset. Eng. Technol. **11**(7), 80–81 (2016)
30. Epic Games. Blueprints Visual Scripting (2019). Accessed 18 Jan 2019
31. Madhuri, D., Reddy, P.C.: Performance comparison of TCP, UDP and SCTP in a wired network. In: 2016 International Conference on Communication and Electronics Systems (ICCES), pp. 1–6. IEEE, October 2016
32. Combefis, S., Giannakopoulou, D., Pecheur, C., Feary, M.: A formal framework for design and analysis of human-machine interaction. In: 2011 IEEE International Conference on Systems, Man, and Cybernetics. IEEE, October 2011

33. Coomer, N., Bullard, S., Clinton, W., Williams-Sanders, B.: Evaluating the effects of four VR locomotion methods. In: Proceedings of the 15th ACM Symposium on Applied Perception - SAP 2018. ACM Press (2018)
34. Begg, M., Dewhurst, D., Macleod, H.: Game-informed learning: applying computer game processes to higher education. Innovate **1** (2005)
35. Piromchai, P., Avery, A., Laopaiboon, M., Kennedy, G., O'Leary, S.: Virtual reality training for improving the skills needed for performing surgery of the ear, nose or throat. Cochrane Database Syst. Rev. (2015)
36. Ingrassia, P.L., et al.: Virtual reality and live scenario simulation: options for training medical students in mass casualty incident triage. Crit. Care **16**(Suppl. 1), P479 (2012)
37. Christopoulos, A., Conrad, M., Shukla, M.: Increasing student engagement through virtual interactions: how? Virtual Real. **22**(4), 353–369 (2018)

Design and Implementation of a Reactive Framework for the Development of 3D Real-Time Applications

Francesco Scarlato[1,2], Giovanni Palmitesta[1,2], Franco Tecchia[1,2],
and Marcello Carrozzino[1,2(✉)]

[1] Dipartimento di Informatica, Università di Pisa, 56127 Pisa, PI, Italy
[2] PERCRO, Scuola Superiore Sant'Anna, 56127 San Giuliano Terme, PI, Italy
m.carrozzino@santannapisa.it, istituto-tecip@sssup.legalmail.it
https://www.di.unipi.it/it,
https://www.santannapisa.it/it/istituto/tecip/
perceptual-robotics-percro-lab

Abstract. In this paper we present Bestbrau, a reactive programming framework for the development of 3d real-time applications. We show its most relevant features, explaining how it differs from the systems used nowadays in commercial game engines and underlining which new functionalities it would provide to a 3D engine integrating it. Our framework is entirely written in C++17 and has been tested on the SAM Engine, a work-in-progress 3D game engine that we have designed and developed in parallel with the reactive framework itself. Bestbrau features are also compared with those offered by Unreal Engine 4 and Unity 2019, two of the most important commercial game engines today available. After a brief introduction, in the second section we briefly introduce the generalities of reactive programming, in the third section we focus, instead, on the FRP (Functional Reactive Programming). In the latest sections we make an accurate and exhaustive description of the peculiarities of such a framework and of the opportunities it provides to 3D and VR programming.

Keywords: FRP · C++ · Game engine

1 Introduction

During the development of a system that models a virtual environment, expecially in the complex case of video games and virtual reality, a programmer has to face a lot of challenges. One of them is represented by the definition of the behaviours of the entities in the virtual world, in order to make the simulation as imperfection-free (and bug-free) as possible. There are indeed tasks that can be assigned to an entity whose declarative formulation is pretty straight forward, but whose transposition in actual source code may result annoying and repetitive. This leads to writing ad-hoc systems or facilities that, case by case, allow to express concepts such as: "Wait ten seconds, then start to shoot at regular

© Springer Nature Switzerland AG 2019
L. T. De Paolis and P. Bourdot (Eds.): AVR 2019, LNCS 11613, pp. 254–273, 2019.
https://doi.org/10.1007/978-3-030-25965-5_19

intervals of 20 ms and then restart from scratch", or: "Assign the player a bonus if he hits more than N flying rockets with a gun".

The main limit of the todays available 3D engines event systems, such as Unreal and Unity, lays in the impossibility to compose events to generate new ones. Therefore, in order to derivate information from the payload of an old event sequence or in order to recognize an event sequence itself, it is needed to register and manage a lot of callbacks; these callbacks will raise new events with the information coming from the previous ones. This process leads to error prone code, and suffers from lack of locality. Errors directly arise from the difficult management of the execution, activation and deactivation of those callbacks that recognize event patterns and throw new events.

In this article we investigate and experiment how functional reactive programming principles can be successfully integrated in a 3D engine using C++ as host language. For the purpose of this study we chose SAM (State Against The Machine) Engine, a work-in-progress game engine written in C++ as a master graduate thesis.

This is the first official attempt to integrate functional reactive programming features written in C++ in a 3D game engine. Nevertheless, it is not the first time that someone implements functional reactive programming in C++, as the reader can see in [16, 18] and previous work about FRP and 3D game engines already exists, although implemented in other languages (usually functional ones), such as Haskell [17].

2 Reactive Programming

There are many different reactive programming models; what truly characterizes this paradigm is definitely the way expressions are evaluated. Reactive programming is, in fact, based on **streams of values changing in time**. An exaustive survey about reactive programming can be found in [15]. Let's now show a significant example:

```
1  1 :  a = 4;
2  2 :  b = 3;
3  3 :  c = a + b;
4  4 :  a = 1;
5  5 :  c;
```

In a purely functional paradigm, as well as in an imperative one, the last expression would return 7. In a reactive programming model the returned value of the expression at line five is, instead, 4.

Expressions, instead of building a classical syntactic evaluation tree, build a dataflow graph, where functional dependencies are defined between nodes and these functional dependencies are kept even at runtime.

The assignment done at line 4 triggers the re-calculation of all values depending from the variable a and, among these values, there is c value, that functionally depends from both a and b. As a matter of fact, there is a substantial difference between this evaluation technique and the classical functional and imperative methods. This difference becomes even bigger if we consider that this **dataflow graph** can be executed in a **multi-threaded** context (Fig. 1).

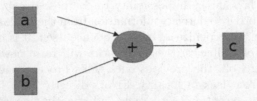

Fig. 1. Schematic representation of the dataflow graph describing the calculations presented in the previous code snippet

The example reported in Fig. 2 allows to deeply catch the nature of the reactive programming paradigm. There are many different implementations and semantics of reactive programming, but all of them are based on the same concept: "values that change through time and that depend on other values". Models differ one from another on the basis of various features, such as:

1. **Synchrony**: updates to the nodes can propagate synchronously or asynchronously way. The synchronous model provides the instant propagation of the updates (stored in a transaction), while the asynchronous one implies the propagation of the updates from a node to be subject to the end of the computation of the node itself. Microsoft Rx extensions are a remarkable case of asynchronous model [27].
2. **Tools**: refers to the available tools for the composition of the dependency graph; it also deals with the possibility of the graph to evolve and to modify its own structure in time; it even refers to the possibility of the programmer to create pieces of reactive code and to modify the graph at will.
3. **Determinism**: the evaluation order can be both deterministic or nondeterministic. We refer to the process of propagation and consequent computation of those nodes depending from a father node whose value has changed. In the first case a **Direct Acyclic Graph** (**DAG**) has to be deduced and a topological visit has to be established. We could, instead, rely on a model that evaluates nodes of the graph in a non-deterministic way, therefore without foreseeing an order in the evaluation of those nodes depending from a father node whose value has changed at a certain time.
4. **Push/Pull** [4,5]: another fundamental feature deals with the way the updates are propagated on the graph. In a push model, as soon as a portion of the graph is invalidated (after the value of a node changes), that whole part has to be re-calculated. A pull model, on the other hand, has a **lazy** nature: in this case we re-calculate the invalidated portions of the graph only when we need to recall the value of a node. This, of course, generates problems such as having to save the whole story of the "seen but not processed values" and not knowing when they will actually be necessary. A trully correct implementation should, in fact, memorize all of them.

3 Functional Reactive Programming

While a reactive calculation model is based on the declarative composition of dataflow graphs, used to specify purely functional runtime dependencies between the nodes and these nodes can also represent the state of the application, we are in the case of functional reactive programming.

The key ideas of functional reactive programming are its notions of **behaviour** and **event**. Behaviours are values changing in time in a continuous way and represent the state of the application; events, on the other hand, are ordered sequences of values emitted in time in a discrete way. We notice that the semantics of both behaviours and events depends on time; it becomes an integrating part of both semantics and implementations of the FRP, enabling the concepts of **succession of values** and of coincidence in time.

We can classify the various FRP implementations considering the possibility to modify the structure of the graph and how this can actually be done.

1. **First order FRP**: once built, the dependency graph is immutable [22–24].
2. **Higher order FRP**: the nodes of the graph model infinite signals and graphs become first class citizens; this means that we have nodes that can emit even reactive graphs as values. In the semantics of the language we can find the primitives **flatten** and **switch**, that allow parts of a graph to reconfigure so that they become the graph they received as input [25,26].
3. **Async/Dataflow**: this family of models allows the composition of **asynchronous** and **event-based programs** in an agnostic way relative to the kind of programming paradigm (functional or imperative) of the programs themselves. The composition of the programs is, instead, functional, returning streams of values asynchronous and discrete in time. Actions on the system take place thanks to callbacks, registered on the emission of these values. In every moment, in the program, it is possible to define new streams and compose already existing ones, as well as to register new callbacks [27,28].
4. **Arrowized FRP**: this model is probably the most complex among those presented, but it is also the most expressive one. Just like the higher order model, arrowized FRP allows to define a set of graphs, with the possibility to switch among them. The number and the type of these graphs however, is finite and is used statically (at compilation time). The name "arrowized" comes from the theory on which the combinators used for creating functional dependencies that originate graphs is built. These combinators are different from the so called monadic combinators, typical of the functional paradigm, such as: **filter**, **map**, **fold** etc... [3,8–10,29].

Our framework can be considered a higher order, synchronous and push-propagated FRP model.

In order to better explain FRP and the opportunities its provides, we introduce some **tipical tasks** we may encounter during the everyday programming activity in **3D real-time computer graphics** and in **virtual reality**, showing how we can deal with them using this different programming paradigm:

1. We want to model the behaviour of a military tower that has a 5 s cooldown and then fires with its machine gun.
2. We want to assign a bonus to the player if he catches more than 20 gems in less than 10 s, always keeping a 120 km/h minimum speed.

Our goal is to simplify writing behaviours; we now show how to achieve this goal.

3.1   The Events

If we analyze the first case, we may not find anything exceptional in it at a first glance: this problem requires, for its solution, the coordination between a boolean variable (in order to know if we are firing or not) and two time accumulators (in order to count the **cooldown** and the **fire rate**) For each entity showing a behaviour with a similar pattern, however, we are forced to repeat the same three variables solution. According to the relevance of the considered functionality, it may be useful to write an ad-hoc facility (a state machine), configurable in many different ways (configuration file, interfaces to implement whose object life cycle is managed by an extern system).

We looked after a programming model that offers the possibility to face these situations in a native way, i.e. without the necessity to build complex ad-hoc structures. We also wanted this model to be generic enough. Thanks to the FRP tools we can write a function that, given two intervals (cooldown and fire rate), returns an event that has the desired behaviour: during the cooldown it is never raised, then it starts to raise events at the fire rate, then it returns to cooldown and then, eventually, everything is repeated. The following function, implementing the above mentioned functionality, is fully valid C++ code, used in both our framework and our engine

```
auto delayed_interval(float t1, float t2) {

    return interval(t1) | evswitch([t2](int c) {

        if ( c % 2 == 0 )
            return interval(t2);
        else
            return event<int>();
    });
}

event<int> i1 = delayed_interval(10, 1);

auto l1 = i1.listen([](int c) { shoot_target(PLAYER1); });
auto l2 = delayed_interval(5, 0.01).listen([](int c)
{
    shoot_target(PLAYER2);
});
```

- *interval(t)*: returns an event that produces over time the sequence $0, 1, 2, 3, 4, 5...$ where the emission of the single values takes place every t seconds.
- *event<int>*: returns an event with *int* payload, that is never raised (evento *never*).
- *evswitch*: when a value comes from *interval(t1)* the argument function is executed; this function returns an event itself: we see that in even cases we return an event that goes with the fire rate, while in odd cases we return an event that is never raised.

The callbacks *l1* and *l2*, registered on the event built by the *evswitch* are called in reply to the event returned by its argument function. Here we show how the implementation of a task that would have otherwise been repetitive and error prone was made generic, immediate and error-free.

We redefined the operator "|" for the event type, this way allowing the **monadic composition** of those primitives that transform events into events. This allows to write expressions like:

```
1  auto  a = map(...)  |  filter(...)  |  reduce(...)
```

instead of:

```
1  auto  a = reduce(filter(map(...), ...), ...);
```

The second style is, without doubt, more verbose and confusing. Equivalent to the first, instead, is the "chained" style, typical of object oriented programming:

```
1  auto  a = map( ... ).filter( ... ).reduce( ... );
```

Names such as *map*, *filter* and *reduce*, were not chosen randomly; there is, indeed, a deep bond between FRP and **streams** from the functional world. An FRP event is meant as an infinite sequence of events emitted over time, while, dually, a stream is a sequence of values generable on demand. Therefore, if applying the *filter* operator on a stream, we obtain a new stream whose values are those of the first one, following a particular condition, analogously the *filter* primitive, applied to an event, returns a new event that emits all the values belonging to the first event that satisfy a particular condition.

The **elementary events**, i.e. those events we do not obtain through composition (**compound events**), are the *never* event (an event that is never raised) and a *sink*.

We mention, for completeness sake, the signatures of the operators for mapping, filtering and accumulating events:

- **evmap**: returns an event that emits values of the first argument transformed following the function of the second argument.

```
1  evmap :: event<T> -> ( T -> T1 ) -> event<T1>
```

- **evfilter**: returns an event that emits values of the first event that satisfy a certain condition.

```
1 evfilter :: event<T> -> ( T -> bool ) -> event<T>
```

- *evreduce*: returns an event that accumulates values received from the first argument following the function of the second argument.

```
1 evreduce :: event<T1> -> ( T1 -> T2 -> T2 ) -> event<T2>
```

The signature of the elementary events are:

- *never*: an event that never emits any value.

```
1 never :: event<T>
```

- *evsink*: an event that provides the *push(T value)* value, that allows to raise an event inside the FRP system and to raise all the other compound events.

```
1 evsink :: event<T>
```

The following program, printing even multiples of three, is self explanatory.

```
1  int  c = 0;
2  evsink<int> source;
3
4  auto compound = source  | evmap([](int c) { return c * 3; })
5                          | evfilter([](int c) { return c % 2 == 0;
       });
6
7  auto listener = compound.listen([](int v) { cout << v << "it's
       even!"; });
8
9  while ( true ) {
10      count.push(c++);
11 }
```

OUTPUT:

- 0 is even!
- 6 is even!
- 12 is even!

3.2   The Cells

Discrete functional reactive programming describes the world and its evolution by means of **events**, just like those we have already seen, and **cells**. Cells build up the state of the system and are obtainable through functional composition, such as the events.

```
1  cell<int> a = 4;
2  cell<int> b = a | map([](int v) { return v * 3 });
```

The value of the b cell functionally depends on that of the a cell, so that, when the assignment $a = 2$; is done, the value of the b cell is updated to 6.

FRP semantics integrate the concept of time and events, and cells can be thought as generators of signals. Events are discrete generators, since their value exists only in the infinitesimal instant in which they appear; cells, on the other hand, are continuos generators, since in every moment they have/present/pulse a value.

In order to deal with the second example presented above (determine if the player catches more than 20 gems in less than 10 s, keeping a 120 km/h minimum speed) we need a cell that accumulates the amount of gems picked up during a certain time interval that respects some specific limits and conditions. In the following a possible code is presented:

```
1  // Here we accumulate the number of picked up gems
2  cell<int> gems;
3
4  // When the ring_collected event is raised, sDelta pulses with
       value 1 (the increment for the counter)
5  event<int> sDelta = ring_collected | evmap( [](int _)
6      { return 1; } );
7
8  // When sDelta is raised, the value of counts is summed to 1 (the
       value of sDelta) and this value is pulsed by incrs
9  event<int> incrs = gems | snapshot(sDelta, +);
10
11 // Every ten seconds emits the value, that in this case, sets the
       counter to zero
12 event<int> sClear = interval(10) | evmap( [](int _)
13      { return 0; } );
14
15 // In this event we merge the values for updating the counter
16 event<int> updates = evmerge(sClear, incrs);
17
18 // The value of the counter holds the values emitted by the
       updates event, starting from value 0
19 gems = updates | hold(0);
20
21 // The final event is obtained by sampling the value of counts at
       every frame. If it is bigger than 20, the double damage
       event is raised
22 event<PowerUp> to_give = gems | snapshot(game_timer)
23      | filter( c -> c >= 20 );
24      | map ( _ => DOUBLE_DAMAGE );
```

Introducing the FRP instrument often means to change the way we face and describe a problem; this requires the programmer to make a significant adaptation effort. The main benefits lay in the **implementative correctness** of the solutions. As a matter of, since we do not have to manually keep track of a mutable state, the number of errors (and bugs) is drastically lowered. Moreover, modifying or extending the existing logic is often simple and the compositional nature allows to make the written expressions both **generical** and **versatile**.

For example, if we want to introduce a minimal speed constraint, assuming we have access to speed by means of a cell, it is sufficient to modify the definition of the *sClear* event:

```
1  cell<float> speed;
2
3  event<int> sClear = speed | snapshot(game_timer)
4      | evfilter([](float s) { return s < 120; })
5      | evswitch([](float speed) {
6
7          return interval(10)
8
9          }, interval(10) )
10     | evmap( [] (int _) { return 0; } );
```

This way we emit the value that resets the counter to zero every time we sample a speed lower than 120 Km/h.

Going more in depth about the generic nature of this approach, we could also abstract from the number of events to count (*ring_collected*): instead of the collection of some gems, or the number of bullets that hit, or other more complex conditions, we could even generalize on the time limit that we need to obtain the bonus, or also on the event that causes the resetting of the counter to zero (go under minimum speed). A simple function would be a factory for this behaviour:

```
1  event<PowerUp> build_powerup_event(event<T> to_count, event<T>
       denier, int duration, int how_many, PowerUp powerup) {
2
3      cell<int>  counter;
4      event<int> sDelta = to_count | evmap( [](int _)
5          { return 1; } );
6      event<int> incrs = counter | snapshot(sDelta, +);
7
8      // The second argument of the switch is the initial event
9      event<int> sClear = denier
10
11          | evswitch( [](float speed) { return interval(duration)
       }, interval(duration))
12
13          | evmap( [] (int _) { return 0; } );
14
15      event<int> updates = evmerge(sClear, incrs);
16
17      counter = updates | hold(0);
18
19      return counter | snapshot(game_timer)
20          | filter([](int c) { return c >= how_many });
21          | map ([](int _) { return powerup });
22
23 }
```

New primitives met in this example are:

- **snapshot**: snapshot(c, e) produces an event that, when the *e* event rises, samples the value of the cell in that instant and pulses with that value.

```
1  snapshot ::  cell<T> -> event<T1> -> event<T>
```

- **snapshot**: a second version of the snapshot, used for building the *incrs* event and that allows to combine the value of the event and that of the cell.

```
1  snapshot ::  cell<T> -> event<T1> -> ( T -> T1 -> T2 ) -> event<T2>
```

- **hold**: given an event and an initial value, it builds a cell that holds the most recent value emitted by the source event.

```
1  hold  ::  event<T> -> T -> cell <T>
```

We also notice that *counter* variable is first declared and then defined in terms of events that depend from *counter* itself, thus creating a cycle between the functional dependencies of our terms.

4   Distinctive Features of the Bestbrau Framework

In this section we present the main analogies and differences between our model and those implemented by the two most successful game engines today on the market: **Unity 2019** and **Unreal Engine** 4. We decided to organize the evolution of the system without a strict life cycle such as that of Unity. In that case the *Update()*, *UpdateFixed()* and *LateUpdate()* functions are delegated to most of the actions, with the event system grafted inside the life cycle itself.

We tried to make the model uniform, proposing a system in which the events are the most important means that the entities have to trigger updates and actions. In this model, instead of a classical game loop in which, for every component of an entity, methods delegated to the system update are called, we have, instead, a **reactive system**. Inside this system the components **listen to** messages (the events) and **react to** them using proper functions; every function registered by a component acts on the state of the component itself, and it is also able to raise new events, and therefore to send new messages towards the system. This difference shows up, for instance, in having a **tick event**, raised every n milliseconds, instead of a generic *Update* method. Registered components react to that tick, executing actions and making changes to their state.

This is a model that shows deep similarities with the actor-based one; **Carl Hewitt**, in [14], defines an **actor** as:

"a computational entity that, in response to a message it receives, can concurrently: send a finite number of messages to other Actors, create a finite number of new Actors, designate the behavior to be used for the next message it receives."

In the **actor model** there is no intrinsic order for the execution of the actions described above. They can, therefore, be executed **concurrently** and **asynchronously**; moreover, two messages sent simultaneously can be processed in any order.

A synchronous propagations of the effects is what, at a first analysis, distinguishes our reactive framework from an actor-based model. The entities, furthermore, at this point of the evolution of our framework, are not provided with addresses, but they listen to events published inside the system; also these events are not messages directed to a specific entity.

Functional game engines, such as the **Nu Engine** [34], define the concept of **address**, giving it to both events and entities. This way, sending a message to an event coincides to raising that event. The message itself, instead, could be routed to an entity, thus causing a reaction from it.

The introduction of addresses can be considered as a possible evolution of our reactive framework; in any case we remark the current difference with the actor model that is, by nature, asynchronous and that is largely used in the writing of big distributed systems, in which it is necessary to ensure location transparency, automatic scale out, persistence, failure recovery, resilience, strong isolation and back pressure functionalities.

Our design is still a **component-based** one, analogous to that of most of the present game engines. While building it we followed a data-driven model, defining in proper data structures the representation part (data) and in other structures the logical part (i.e. the part oriented to the definition and generation of events and actions in response to the events themselves). This way it becomes easier, among entities that have the same representation, to reuse portions of code to express different behaviours, a problem in which the inheritance system, typical of the object oriented paradigm, shows its limits.

The three most important objects of our reactive framework are: the **entities**, the **behaviours** and the **events**.

Entities host the data part and are provided with tools to manage behaviours that act on each of them. For instance, at a certain moment, a *bullet* entity can have an active behaviour that makes it follow a target just like if it was remote-controlled. In the meantime, if the target is suddenly removed from the scene, the *bullet* entity can listen to that event, activating a new behaviour (target-seeking) in response and deactivating the first behaviour.

The activation of a behaviour consists in the definition of new events inside the system and in the registration of callbacks. Events were implemented with the tipical functional programming constructs, such as *filter*, *map*, *reduce*, thus with a semantics that can be related to that of the stream calculus and to that of the functional reactive programming. Events are, for all purposes, streams emitting values over time and that, potentially, support the dual of the operations on classical streams.

4.1   SAM Engine

The functionalities of our reactive event handling system were tested on the **SAM Engine**, a work in progress game engine we have designed and implemented as part of a master degree thesis. SAM Engine is currently endowed with the following features:

1. **Graphics**: 3D models import, 3D transformations, scene lighting, normal mapping of 3D objects, tessellation of 3D objects, application of displacement maps to tessellated surfaces, simultaneous, practical and efficient draw

of many different objects, provided with different shaders and OpenGL states, application of filters and graphical effects to a scene rendered through render to texture, lighting through deferred shading, obtained with the multiple render target technique, draw of text and bidimensional figures using the orthogonal projection;

2. **Physics**: collisions handling through a bounding spheres system;
3. **Memory**: automatic memory management at all levels of the use of the engine (RAM, VRAM), automatic deduction, through introspection, of the structure of the uniform blocks inside the shaders, with reflection in RAM of this data structure, lack of memory leaks;
4. **Resources**: resource handling through serialization and deserialization from and to JSON;
5. **Events**: reactive event handling, implemented with a framework of functionally composable events;
6. **Portability**: compatibility of SAM exports with both Windows and Linux systems on witch OpenGL version 4.2 or superior is installed.

4.2   Discussion About the Event System

In this section we highlight the major strengths, the potential and the distinctive traits of our framework, implemented inside the SAM Engine.

Functionally Composable Events and Cells. Functionally composable events allow to easily collect information to support the behaviours.

The task of an entity inside our system is that of orchestrating the execution of callbacks registered on the events of the system, as well as hosting the data that these functions are going to transform. A behaviour is made up of a list of callbacks, together with an activation method (*awake()*), a deactivation method (*sleep()*) and a termination event (*complete*).

We consider the **termination event** as the key element to enable the **compositon of behaviours**, i.e. the possibility to define sequences of actions and animations in a declarative way. All this with the goal of writing code such as the one presented below.

```
1 attack_sequence = do_rotate(speed, angle) |
2 do_bounce(10) |
3 play_sound("sword_clash") |
4 do_damage(inventory->weapon) |
5 do_bounce(1);
```

This would also allow tho write **recursive behaviours**.

```
1 behaviour do_chase;
2 do_chase = find_target("Player") |
3 shoot_target |
4 do_chase;
```

The *shoot_target* behaviour fires towards the target, ending when the latter is no longer visible. Once it looses the sight of the target, it restarts to seek the target and then it shoots again.

This perspective wants to make it possible to build new mechanisms that allow an entity to pass from a set of active behaviours to another one in a native, symbolic and descriptive way. By the term "in a native way" we mean at code level. Unreal offers functionalities similar to those ones we described, thanks to tools such as the event graph and the blueprints system (usable through the GUI). We believe, however, that the lack of such expressive power at code level is, in the long run, a limiting factor for an event system and for a game engine in general. Behaviours expressed through the Unreal GUI must be compiled in a code that does not respect the compositional nature, that, instead it means to implement. This makes it difficult and expensive (in terms of time) for the programmer to introduce and maintain new features at GUI level because, for each of them, ad-hoc code has to be written by specialists, taking care not to create uncorrect behaviours in a non composable system by nature. The same argument applies to Unity too, that, furthermore, has a much less expressive GUI than Unreal's.

First Order FRP. Our framework supports first order FRP. Its expressive power in respect to the Higher Horder variant must not be underestimated. As a matter of fact there are programming languages that integrates natively the FRP, like Elm [22], with a large use of applications. A major use case is found in writing user interfaces, where it is useful to have a machinery which binds in a reactive way the view values with the model ones, automatically updating the view values at each change in the model. The Bestbrau reactive system was conceived to fully support First Order FRP and higher order concepts. We cannot declare a fully FRP compliant implementation though, since we allow the firing of events in the event listeners themselves, as a way of updating the FRP logic and switching behaviours.

Native Multi-threading. In the Unreal and Unity approach to multithreading programming, most of the components lifecycle, as the event system and callbacks are executed in the main thread. Renderer and physics are executed in their own thread. When users need to perform a heavy task, for example an A* search executed on behalf of a player needing to do the next move, it is possible to define jobs (Unity) or tasks (Unreal). These computations are then executed asynchronously and the programmer can later retrieve the value of the computation. Constructs similar of those offered by Fastflow [35] or OpenMP [36], like map-reduce and parallel-for are also offered as a way of using the various contexts offered by the CPU, making the execution of specific honerous tasks explicitly asynchronous and distributed.

Nevertheless we believe that more than this can be achieved, aiming at making the code written by the user-programmer to be inherently parallel. Our event system, with callbacks listening to the events, fullfills the requirements for achieving those results. The first step for making this real, consists in the use of the N contexts provided by the CPU for the propagation of the effects in the graph in a purely dataflow fashion. Achieving this does not present too many difficulties as the nodes dependencies do not allow race conditions, since they are purely functional and each event node operates only on its state. **Lock-Free queues** [37] are needed if one wants to achieve a good scalability at the graph interpreter level.

The second step, clearly more challenging is the parallelization of the callbacks registered on the events. For each entity (actor) there are multiple callbacks registered at the same time and race conditions can arise. A clean approach is given by the Nu Engine where Actors do not directly hold their representation, but just Attribute Tags. The representation -attribute values- are instead stored in a highly optimized functional map. Immutable/functional data structures [38] are used for ensuring a transactional way of using memory, thus avoiding read and write race conditions.

Other solutions could make use of assumptions typical of the actor model, that is, an actor can access and mutate only its state while reading external values; changing the state of other actor can be obtained, instead, by means of message passing. By using these assumptions we can treat each entity as a task, serialize all the callbacks relative to a single entity, and execute in parallel the callbacks of other entities. Otherwise, one can rewrite more and more logic in a FRP way, thus achieving multi-threading and thread safety for free.

One of the reason behind the creation of SAM and its reactive system, other than all the points made in this paper, lies in the awareness that an increasing number of developers and corporations are building and adopting reactive models when writing their software systems. The same goes for **virtual reality** and **game development** in general: we believe that currently available 3D/ Virtual Reality development frameworks will be probably augmented or replaced by reactive programming models. We do cite here, a Tim Sweeney's (creator of the Unreal Engine) tweet, dated January 2017, in which he exposes his interest in this field:

– *"Currently exploring purely functional data structures and algorithms based on futures to ensure maximum async progress. It's early, but so far the problem space feels "fortuitously constrained" to yield particularly clean and efficient solutions. Specifically: minimizing dependencies on futures tends to force container designs that minimize combinatorial complexity. It's very similar to functional reactive programming, which is simpler than imperative and is dual to lazy evaluation."*

Moreover, because of the poor presence of reactive frameworks implemented in C++ (such as RxC++ and Voy C++ [39]), we wanted to give our contribution in this field, making an implementation of a first order FRP and offering, at the same time, a programming model suitable for our host language, C++.

We did not talk nor implemented any kind of support towards entity and behaviour serialization, since we are currently exploring programming models like CSP and idiomatic ways for C++ for composing SAM behaviours. Only after that we will focus on the serialization of those data structures. We also have the feeling that pairing the FRP system with a library like Evreact [32], can provide a boost in the expressivity of an actor system.

Evreact is a library written in F# which allows to define Petri nets capable of recognize event sequences and execute actions in response to those events. This nets are built in a declarative way and are just F# expressions defined by a simple grammar. In this library we can define a net that tracks the drag and drop process just like this:

```
1 let net =
2     +(
3         (!!md |-> fun e -> printfn "Mouse down @(%d,%d)" e.X e.Y)
4         - (+(!!mm) |-> fun e -> printfn "Mouse move @(%d,%d)" e.X
           e.Y) / [|mm; mu|]
5         - !!mu |-> fun e -> printfn "Mouse up @(%d,%d)" e.X e.Y
6     )
```

For any clarification about the code, refer to the evreact documentation [32].

Easing Functions. Our framework has been enriched with a bunch of useful easing functions. The easing functions (a.k.a. interpolation functions) are one of the building blocks of the procedural animations, specifying for a generic animation, its acceleration, thus making it more realistic.

In facts, in the real world, an object does not start its movement at its full velocity, nor the speed stay constant in time. Taking those examples from the real life:

- When a drawer is opened, it is first given an acceleration an then it slows down until it stops.
- When something falls of, it first accelerates towards the ground, then it bounces after it hits the floor.

These behaviours can be described by functions of one variable with the following signature:

```
1 float ease(float t);
```

With $t \in [0, 1]$

In Fig. 2 it is depicted a plot of a linear easing function, the simpler examples of this kind of functions, while in Fig. 3 are depicted plots of many other possible easing functions which can describe more complex animations.

Fig. 2. Linear easing function

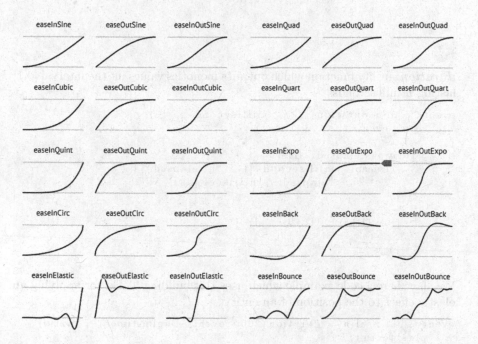

Fig. 3. Various kinds of easing functions

We show how an easing function has been implemented in our framework, introducing some utility functions for building the base events.

– *elastic_in*: easing function.

```
float elastic_in(float t) {
    return sin(13 * t * PI / 2) * pow(2, 10.0 * (t − 1));
}
```

- **ms_elapsed**: utility function that takes a timer in input, accumulates the passing times publishing it's accumulated value at every timer tick

```
event<float> ms_elapsed(const event<float>& timer)
{
    return timer
        | evreduce([](float accum, float dt) {
            return accum + dt*1000;
            }
            , 0.0f
        );
}
```

- **duration**: utility function which outputs monotics values int the interval [0,1] lasting n milliseconds

```
event<float> duration(float milliseconds,
    const event<float>& timer)
{
    return ms_elapsed(timer)
        | evmap( [milliseconds](float elapsed) {
            return elapsed / milliseconds;
        })

        | evfilter( [](float d) {
            return d <= 1.f;
        });
}
```

We show here a code example which uses this building blocks for applying an elastic effect to the position of an entity.

```
event<float> elin = duration(2000,events->anim_timer) | evmap(
    elastic_in);

auto l1 = elin.listen(
    [_this=shared_from_base<player>() ](float t) {
        _this->bounding_coords.translate( 0, 0, t );
});
```

5 Conclusions

We consider the introduction of an FRP system to support a 3D/VR engine a novelty in the realm of frameworks and libraries available for the C++ language. System integration was successful and in a first test, we built an isometric shooter game in the forty hours of a game jam. Further explorations will be done about how we can pair the FRP system with an actor model, evaluating if the FRP graph is suitable for the message transport of an actor system, which is, by definition, asynchronous.

The fact is: an FRP system is Turing-complete and capable of modeling far more complex scenarios than the ones depicted in this paper. The entire VR simulation can be delegated to an FRP system. We wanted, instead, to add a functional framework to a programming model that stays imperative, and familiar to the C++ users. One of the purpose of the event listeners in Bestbrau is, in facts, to create new FRP logic and to push new events into the system, something that's forbidden in most of the FRP frameworks.

An interesting case of study is how the CSP (Communicating Sequential Process) [20] can give a compositional model capable to describe the actor's behaviour evolution with a defined semantics.

Our work allowed us, fundamentally, to design and build a framework inspired to the functional reactive programming model of computation, with push propagation, written entirely in C++17 and tested on our engine called SAM. The framework exposes the following features:

1. functionally composable events:
   - *map, filter, reduce, accum, sink, never, merge, switch* nodes.
2. functionally composable cells:
   - *map, hold, snapshot, reduce, switch* nodes.
3. implementation by means of a dataflow graph:
   - synchronous propagation by means of transactions.
   - transactional context that implements the FRP concept of time.
4. support to native multithreading.

Our framework's particular structure is pretty suitable to the future implementation of various forms of multithreading:

1. **Ad-hoc multi-threading**: individuation and parallelization of critical zones of our code, trying to use FastFlow, OpenMP, or simple pthread pools.
2. **Event multi-threading**: implementation of the propagations of the events on the graph in a parallel way. The idea consists in assigning the execution of all the nodes of the graph executable during a single transaction to a thread pool. It is a pretty straight forward implementation, given the structure of our framework, because, not having a mutable shared state, all the threads can concurrent to the simultaneous computation of the nodes.
3. **Callback multi-threading**: implementation of the execution of the callbacks in a parallel fashion. Much more complex than event multi-threading, since, in this case, we have a mutable and shared state. There can be, in facts, two or more threads that simultaneously execute callbacks of the same entity, modifying its state. This fact can generate either race conditions or invalid states. The solution we are thinking to adopt, is based on the use of persistent data structures, following a transactional atomicity logic similar to that of databases. A persistent data structure remembers the state it had before the beginning of the transaction and definitively modifies its state only at the end of the transaction itself.

We wait with enthusiasm the future introduction, in C++20, of the coroutines, that will represent an enabling factor for the introduction in our framework of the constructs for asynchronous programming.

References

1. Elliot, C., Hudak, P.: Functional reactive animation
2. Wan, Z., Hudak, P.: Functional reactive programming from first principles
3. Hughes, J.: Generalizing monads to arrows
4. Amsden, E.: Push-pull signal-function functional reactive programming
5. Elliott, C.: Push-pull functional reactive programming
6. Krishnawami, N.R., Benton, N., Hoffman, J.: Higher order functional reactive programming in bounded space
7. Krishnawami, N.R.: Higher order functional reactive programming without space-time leaks
8. Wadler, P.: Comprehending monads
9. Wadler, P.: The essence of functional programming
10. Wadler, P.: Monads for functional programming
11. McKee, H., White, O.: Akka A to Z - an architect's guide to designing, building, and running reactive systems
12. Bruni, R., Melgratti, H., Montanari, U., Sobocinkski, P.: Connector algebras for C/E and P/T nets interactions
13. Czaplicki, E.: Controlling time and space: understanding the many formulations of FRP
14. Hewitt, C.: Actor model of computation
15. Bainomugisha, E., Carreton, A.L., van Cutsem, T., Mostinckx, S., de Meuter, W.: A survey on reactive programming
16. Čukić, I.: Functional and imperative reactive programming based on a generalization of the continuation monad in the C++ programming language
17. Cheong, M.H.: Functional programming and 3D games
18. Al-Khanji, L.: Expressing functional reactive programming in C++
19. Blackheath, S., Jones, A.: Functional Reactive Programming, 1st edn. Manning, New York (2016)
20. Hoare, C.A.R.: Communicating sequential processes. In: Hansen, P.B. (ed.) The Origin of Concurrent Programming. Springer, New York (1978). https://doi.org/10.1007/978-1-4757-3472-0_16
21. Author, A.-B.: Contribution title. In: 9th International Proceedings on Proceedings, pp. 1–2. Publisher, Location (2010)
22. Elm Homepage - A delightful language for reliable Web Apps. https://elm-lang.org/
23. React Homepage - A JavaScript Library for building user interfaces. https://reactjs.org
24. Angular Homepage - One Framework Mobile and Desktop. https://angular.io/
25. Reactive-Banana Homepage. https://wiki.haskell.org/Reactive-banana
26. Sodium Homepage. https://github.com/SodiumFRP/sodium
27. Reactive Extension Homepage. https://reactivex.io
28. Bacon Homepage. https://baconjs.github.io/
29. Yampa Homepage. https://wiki.haskell.org/Yampa
30. LINQ Homepage. https://msdn.microsoft.com/en-us/library/bb308959.aspx
31. GestIt Homepage - A framework for reactive programming. http://gestit.github.io/GestIT/
32. Evreact Homepage - A lightweight framework for reactive programming. http://vslab.github.io/evreact/

33. Reactive Extensions for C++ Homepage - A lightweight framework for reactive programming. https://github.com/ReactiveX/RxCpp
34. Nu Game Engine Homepage. https://github.com/bryanedds/Nu/blob/master/Nu/
35. FastFlow (FF) Homepage. http://calvados.di.unipi.it/
36. OpenMP Homepage. https://www.openmp.org/
37. An Introduction to Lock-Free Programming. https://preshing.com/20120612/an-introduction-to-lock-free-programming/
38. Persistent Data Structures. https://www.geeksforgeeks.org/persistent-data-structures/
39. Voy: Message Passing Library for Distributde KRunner. https://cukic.co/2016/05/31/voy-message-passing-library-for-distributed-krunner/
40. Easing.net Homepage. https://easings.net/it

A Real-Time Video Stream Stabilization System Using Inertial Sensor

Alessandro Longobardi[(✉)] ⓘ, Franco Tecchia ⓘ,
Marcello Carrozzino ⓘ, and Massimo Bergamasco ⓘ

PERCRO, Scuola Superiore Sant'Anna, 56127 San Giuliano Terme, PI, Italy
alessandro.longobardi@santannapisa.it,
istituto-tecip@sssup.legalmail.it
https://www.santannapisa.it/it/istituto/
tecip/perceptual-robotics-percro-lab

Abstract. This paper has the purpose to show a stabilization video-streaming methodology feasible to low-power wearable devices. Thanks to an Inertial Measurement Unit (IMU) mounted together with the camera we are ready to stabilize directly on video-stream without the delay and the complexity due to image processing used by classic software stabilization techniques. The IMU gives information about the angle rotation respect to the three main orthogonal axes of the camera; the wearable device transmits the video along with the IMU data synchronized frame per frame then a base station receives and stabilizes while renders the video. The result is that the shaking and the unwanted motions of the human body wearing the system are compensated giving a clear and stable video. Numeric results prove that the video is more stable: we cut-off the half of the motion noise in the scene.

Keywords: Video-stabilization · Wearable device · Video-streaming

1 Introduction

The main purpose of real-time video stabilization is to limit or completely eliminate the delay due to stabilization process.

Classical stabilization techniques use sophisticate algorithms that try to find correlated points between frames and calculate the motion; this requires a lot of computational power and processing time making it unfeasible for video-streaming. The method proposed in this paper uses data from a tiny inertial measurement unit (IMU) tightly mounted with the camera to retrieve angular position (referred to the three standard camera axes) of each frame in order to perform stabilization in real-time during the streaming.

The stabilization process is pretty simple: each frame is rotated in the direction and with the magnitude sampled by the IMU and a rectangle is cropped from the frame. The final effect is the removal of shaking and noise due to small movement, preserving great and desired motions.

Application fields vary from sports to recordings on drones.

L. T. De Paolis and P. Bourdot (Eds.): AVR 2019, LNCS 11613, pp. 274–291, 2019.
https://doi.org/10.1007/978-3-030-25965-5_20

In sport could be helpful: thanks to the limited invasiveness of modern camera and IMU sensor it is possible to stream directly from a referee (or players) removing disturbances given by running, jumping, sprinting etc.... making an enjoyable experience for viewers.

We can achieve stable recordings on all kind of drones without mounting mechanical stabilizer, leaving the drone structure lighter and less cumbersome. Moreover, it gives the possibility to position the camera in places that normally are unfeasible due to vibrations (like near motors) and expands the vision robustness (e.g., suppose having a flying drone in heavy ventilated area or a naval drone in rough sea, the real-time stabilization is a powerful tool in these cases).

2 State of the Art

A video captured with a hand-held device (e.g., a smartphone or a portable camcorder) often appears remarkably shaky and undirected. Digital video stabilization improves the video quality by removing unwanted camera motion. It is of great practical importance because the devices capable of capturing video have become widespread and online sharing is so ubiquitous.

In this section we present an overview of the most important approaches to the topic of video-stream stabilization. It is usually achieved by means of computer vision approaches or inertia/rotation compensation using electronic sensor we will subdivide this part in the same two broad categories:

- Video-streams stabilization using image-based approach.
- Real-time video stabilization approach.

2.1 Video-Streams Stabilization Using Image-Based Approach

The technique post-processes the video to determine how the camera is moving and what part of that motion is unwanted. The advantage digital stabilization holds over mechanical or optical stabilization is that it requires no additional hardware and can be performed after the video is taken, making it possible to stabilize video that has been archived.

Digital video stabilization workflow (see Fig. 1) is generally discussed in three main sections: global motion detection, motion filtering and motion compensation.

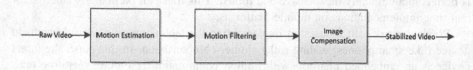

Fig. 1. Digital video stabilization workflow

Global motion detection is the process of finding matching features between frames and forming a model that describes the motion in the image plane. Motion filtering separates the wanted motion (zoom, pan…) from unwanted motion (shake, jitter…) and outputs a model representing the unwanted motion. The process is completed with frame compensation, which moves the image plane opposite the direction of the unwanted motion model.

A first group of image-base techniques is based on black on block matching: they use different adaptive filters to refine motion estimation from block local vectors [1–3]. These algorithms do generally provide good results but are more likely to be misled by video content with moving objects. This happens because they do not associate a unique descriptor to a block and neither track blocks along consecutive frames. This results in an inefficient motion estimation, since such objects will generate incorrect matching blocks and therefore wrong estimated motion.

Feature-based algorithms extract features from video images and estimate inter-frame motion using their location. Some authors present techniques [4, 5] combining features computation with other robust filters; these methods have gained larger consensus for their good performances.

In the motion filtering stage several techniques such as Kalman filtering [6] and Motion Vector Integration [7] have been proposed and modified and enhanced to correct translational and rotational jitters [8] according to real systems constraints. A video stabilization system based on SIFT (Scale-invariant feature transform) features [9] has been presented in [10].

Searching for feature points in a sequence of frames is a classic form of video stabilization. Prior video stabilization, those methods synthesize a new stabilized video by estimating and smoothing 2D camera motion [11, 12] or 3D camera motion [13, 14]. Briefly, 2D methods are more robust and faster but may sacrifice quality (e.g., introducing unpleasant geometrical distortion or producing less stabilized output), while 3D methods can achieve high-quality results but are more fragile.

2.2   Real-Time Video Stabilization Approach

This approach achieves stabilization in real-time using data from an inertia sensor attached to the camera, it retrieves absolute position and angular velocity respect to the canonical three axes for the image compensation directly. Thus, there is no need to detect key-points or to perform motion estimation stage.

Specifically, an inexpensive microelectromechanical (MEMS) gyroscopes to measure camera rotations is employed. The approach uses these measurements to perform video stabilization (inter-frame motion compensation). This kind of approach is both computationally inexpensive and robust. This makes it particularly suitable for real-time implementations on mobile platforms.

Previous work [18] focuses on implementing the technique on powerful handheld device (like smartphones) adding roller-shutter compensation. In this paper the target device is an embedded platform with limited computation resources, therefore optimizations and model simplifications have been taken into account.

An approach mixing inertial data and optical flow has been proposed [19], it employs a series of processing stages: template matching, feature detection and optical flow stage analysis. Each stage has a certain complexity and introduces delay that makes it not suitable for streaming application (except with low frame rate).

The technique presented in the following section uses only geometry rotations provided by OpenGL, reducing at the minimum the processing time required by the stabilization process and rendering high framerate video; the lightweight and portable software guarantees compatibility across different devices.

3   System Model

In our system there are two components:

- The wearable device: in charge of streaming the video along with inertia data associated to each single frame.
- The base station: in charge of stabilizing the video.

Components communicate using a single direction channel (the video channel) (Fig. 2).

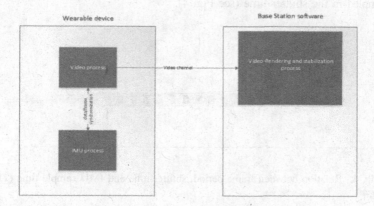

Fig. 2. The two components

3.1   The Wearable Device

The wearable device is a low-power and low-consumption device, designed to be small and light. Due to its constrains must perform simple tasks:

- Sample inertia measurements.
- Synchronize each video frame with the associated measurement.
- Stream the video frames with their associated inertia data.

The camera and the IMU must be attached together: the camera sensor axes must coincide with the IMU ones (Fig. 3).

Each sample (called *IMU data*) from the IMU is the angular position variation from the previous frame along the three main axes. It means that if we consider a frame i, the associated *IMU data* on x axis is given by $\theta_{x,i} - \theta_{x,i-1}$, where θ_x is the angular position on x axis.

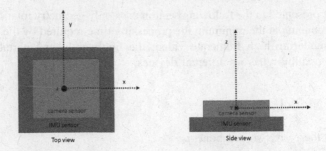

Fig. 3. IMU and camera sensor placement: axes coincidence

Therefore, each packet being streamed is a couple *<Video frame, IMU data>*.

How to synchronize the *IMU data* and the video frame depends on the device. The frame is taken at the end of the frame-period (1/(video frequency)) in general and time required is called shutter-time. The IMU data associated to the frame should be the latest sampled in the shutter time (see Fig. 4).

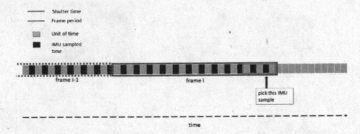

Fig. 4. Relation between frame-period, shutter-time and IMU sample time (1)

Hence the IMU sampling time must be lower than the shutter-time in order to be safe:

$$T\_IMU \leq T\_shutter$$

Alternatively, if the software and camera electronics assure that there is not jitter in shutter-time (i.e., the frame is acquired always at the same time units in frame-period) we can relax the above constraint and set the IMU sampling time to the frame-period (Fig. 5):

$$T\_IMU = T\_frame$$

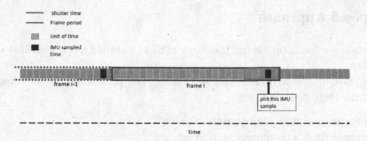

Fig. 5. Relation between frame-period, shutter-time and IMU sample time (2)

3.2 The Base Station

In this paper, we suppose that the wearable device has not GPU capabilities then the base station performs the stabilization process.

Details will be given in Sect. 4. The interaction between the wearable device and the base station is described in Fig. 6.

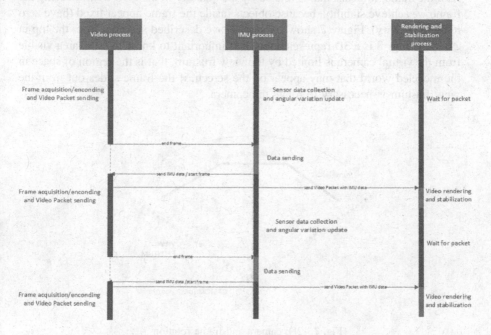

Fig. 6. Software modules interaction, video and IMU process are on wearable device, rendering and stabilization on base station.

The video process signals the IMU process when a frame is acquired, the IMU replies with the angular position variation from the previous frame. The video process starts acquiring the next frame and, at the same time, encodes the actual frame. After the encoding, the video-packet along with the IMU data is sent to the base station.

4   Proposed Approach

In this section we focus on the methodology used to stabilize the frame. It is described step by step.

The input is the video frame and *IMU data*. The parameters to set (described in details during the discussion) are:

- Angle range on x axis (compensation angle x).
- Angle range on y axis (compensation angle y).
- Angle range on z axis (compensation angle z).
- Crop size.

1. We render every image-frame in a rectangle (that keeps the original video aspect ratio) at the video frame frequency (e.g., 30 fps). The rectangle rotates around a pivot-point (that indicates the camera position) and with a certain distance from this one (that recalls the focal-camera length). The rotation follows the pose associated with the frame, it means that is rotated by the quantity in the *IMU data*. Moving the frame we achieve stability, because objects inside the frame appear fixed (have zero relative velocity). Figure 7 shows what we have described seeing it from the top in 2D view, Fig. 8 is a 3D representation. It is important to know that the area visible from the virtual camera is limited by the view frustum, that is the region of space in the modeled world that may appear on the screen, if the frame slides out from the view frustum is necessary to rotate the camera.

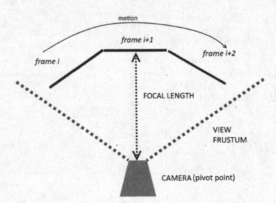

Fig. 7. 2D: camera and frame rotation

2. The next step is to limit the frame rotation angles along the three axes, then we force the camera to be fixed and frame can move in a range delimited by the camera view frustum (setting an upper-bound in rotations range). Using this solution preserve the frame to slide out.
3. The first parameter to set is how much we want to stabilize by defining the angle variations that must be compensated, in other words: how much we want to move

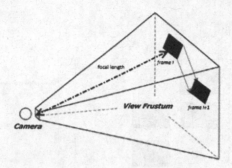

Fig. 8. 3D: camera and frame rotation

the frame in order to reduce the shaking and the undesired motion. Keep in mind that large and desired motion must be preserved. To do that we must set rotation angle ranges on the three axes; those ranges delimit a "*rotation area*" (smaller than view frustum, see Fig. 9) where the frame is rotated within range limits. Larger rotation area gives more compensation power. Summarizing, we must set:

- $[min_x, max_x]$ delimits how much the frame can rotate along x axis, must be symmetric $min_x = -max_x$.
- $[min_y, max_y]$ delimits how much the frame can rotate along y axis, must be symmetric $min_y = -max_y$.
- $[min_z, max_z]$ delimits how much the frame can rotate along z axis, must be symmetric $min_z = -max_z$.

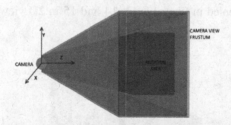

Fig. 9. Camera view frustum and rotation area (front view)

Example: suppose we start with a frame in the center of the view frustum and then it rotates along x axis until it reaches max_x degree, at this point the frame is hold fixed as long as the variations are positive (x angle is increasing) and the frame starts to move back when variations becomes negative (x angle is decreasing). This solution allows to compensate small movements because, until max_x, it could be a simple shaking (frame can go "up and down" quickly), whereas if it is a wanted large motion (so the x angle is still increasing) when frame reaches max_x it is not more compensated (frame fixed).

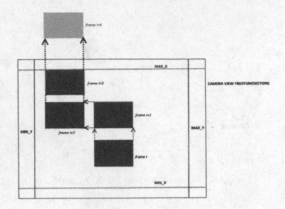

Fig. 10. Frame motion

Figure 10 shows an example (note: only a projected section on a plane of the frustum is depicted in figure and frame inclination effect due to rotation around the camera is ignored for simplicity).

4. In order to avoid seeing the entire frame motion, we crop the frame to obtain a view only on a fixed area. This gives us the feedback of stability. Thus, we must "carve out" a rectangle (a sort of window) inside area rotation in order to see only the fixed part of the frames; the size of cropped and rotation area (whose size is defined by the rotation ranges defined in the previous step) must be set properly: a trade-off has to be defined. The objective is to find balance on how much we want to compensate and how much we want to see of the frame.

In Fig. 11 is shown a 3D view of the cropped area, instead various possible frame motions are represented in Figs. 12, 13, 14 and 15 in 2D view (the cropped area is the grey one).

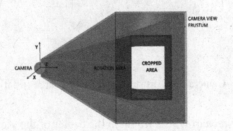

Fig. 11. 3D view of cropped area

The more compensate the more the scene appears fixed, but as a drawback we have to decrease the cropped area and consequently lose visual information. On the contrary less compensation leads to instability, but we can focus on a bigger part of the frame.

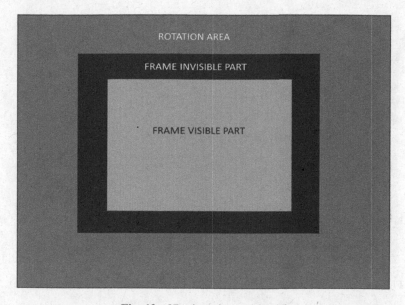

Fig. 12. 2D view: frame centered

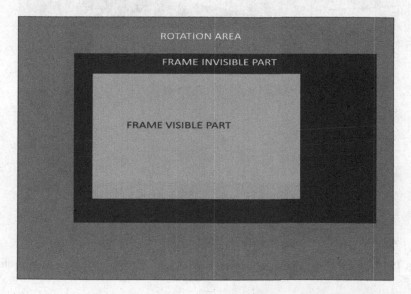

Fig. 13. 2D view: positive rotation around y axis (frame motion towards right)

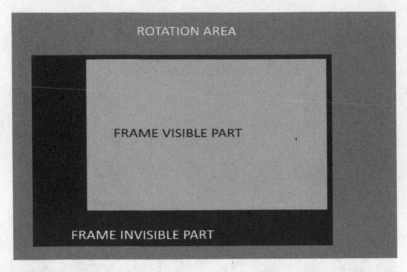

Fig. 14. 2D view: negative rotation around x axis (frame motion towards down)

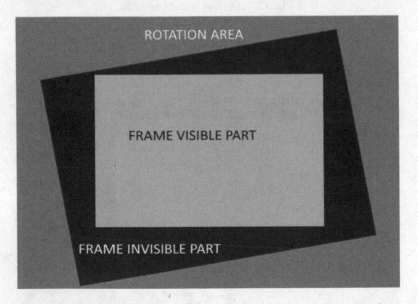

Fig. 15. 2D view: positive rotation around z axis (frame inclination respect to its center)

4.1 Considerations

The stabilization process illustrated has been designed to be as general as possible, in order to adapt to specific cases we have to set the proper cropping and rotation parameters seen before.

We ignore translations because they are difficult to measure accurately using IMU. Also, accelerometer data must be integrated twice to obtain translations. As a result, translation measurements are significantly less accurate than orientation measurements. Even if we could measure translations accurately, this is not sufficient since objects at different depths move by different amounts. Therefore, we would have to rely on stereo or feature-based structure from motion algorithms to obtain depth information. Warping frames in order to remove translations is non-trivial due to parallax and occlusions. These approaches are not robust and are currently too computationally expensive to run in real-time.

5   Experimental Results

In this chapter we discuss how we achieve analytical results to prove our stabilization method. Basically, we compare if the amount of motion in the stabilized video is less than in the not-stabilized one. In order to evaluate the motion we calculate the Optical Flow of the video and displacement of a maker through Object Tracking Technique.

We stream a video at the resolution of 1280×960 at 30 fps.

We set the wearable recording system mounted around the head (like a headband). Concerning the stabilization parameters we set the compensation angle on three axes to the same value. Since the head shaking is limited and small in general, we choose a small compensation angle value; starting from a qualitative result, we set $4°$ compensation angle, thus the rotation area on all three axes is in range $[-4°, +4°]$. The numerical results on the goodness of such angle and stabilization improvements are given in the following sub-sections.

We set as cropping area the 80% of the original image.

The stabilization technique described in Sect. 4 has been implemented on XVR [17].

5.1   The Optical Flow

Optical Flow is the distribution of the apparent velocities of objects in an image. By estimating Optical Flow between video frames, you can measure the velocities of objects in the video. Optical Flow estimation is used in computer vision to characterize and quantify the motion of objects in a video stream, often for motion-based object detection and tracking systems.

We use the Lucas-Kanade method [15] because is the least sensitive to image noise.

The Optical Flow output (Fig. 16) is a matrix $N \times 1$, where N is the frames number of the video. Each element is another sub-matrix of size $H \times W$, where H is the height and W is the width in pixel of the video; each element (i, j) of this sub-matrix is the Magnitude (the modulus) of the Optical Flow of the pixel (i, j).

In our measurements we always use the mean of Magnitude in frame i as unit base: that is the mean value among all pixel in the frame i.

$$\text{Mean\_Magnitude}_i = \sum_{k=1}^{H} \sum_{j=1}^{W} \text{magnitude\_pixel}_{kj} \qquad (1)$$

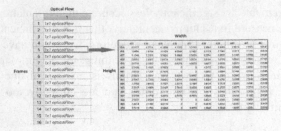

Fig. 16. Optical flow output

Methodology and Results

Video setting

In order to perform our measurements, we made a video with a black background where only a black-cross marker on a white rectangle is visible. This simple setting gives the possibility to have the clearest possible results reducing image noise. Hence we shake the camera causing the marker motions: in stable video the marker must move as little as possible while in the not-stable video (the original) it can move freely (Fig. 17).

Fig. 17. Video test setting: a cross on a white rectangle in black background

Results

We perform the Optical Flow in both videos and compare the results: in the stabilized video must be less than in the original.

The graph in Fig. 18 is referred to a video of 2:15 min (4050 frames) where shakings are performed in all directions and with different amplitudes. As you can see the Magnitude per frame, on average, is lower in the stabilized video. If we calculate the mean of the Magnitude (1) across all the frames in both videos we can have an index on how globally motion there is on average per frame:

$$\text{Mean\_Stabil} = \frac{\sum_{i=1}^{numFrames} \text{Magnitude\_Stabil}_i}{\text{numFrames}} \qquad (2)$$

$$\textit{Mean\_Not\_Stabil} = \frac{\sum_{i=1}^{numFrames} \textit{Magnitude\_Not\_Stabil}_i}{\textit{numFrames}} \qquad (3)$$

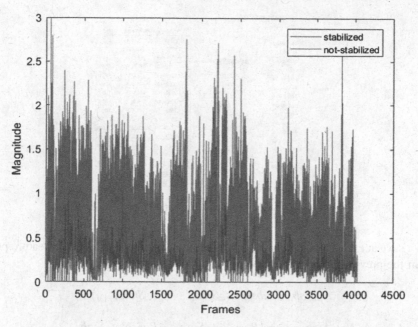

Fig. 18. Optical flow comparison between stabilized video and the original

Then we estimate an index on how much is more stable than the original:

$$Performance\_Stabil = 1 - \frac{Mean\_Stabil}{Mean\_Not\_Stabil} \qquad (4)$$

In our test the value is **46.82%**. This means that almost the half of entire motions have been compensated.

5.2 Marker Displacement Through Object Tracking

Another way to infer an index on stabilization performance is to calculate how much the black-cross marker position changes during the video. Using the same video setting and the two videos we track the marker and verify that in the stabilized video the displacement between one frame and the subsequent is less on average than in the original.

Methodology and Results

The cross is tracked using feature points extracted by the Shi and Tomasi algorithm [16]. For each frame the tracker gives as output the new positions (x, y) of feature points and a confidence score indicating how much are reliable them, the range of score values is in [0, 1], where 0 means unreliable and 1 totally reliable (Fig. 19).

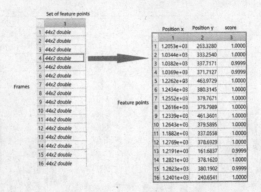

Fig. 19. Tracking output

We save at each frame i the absolute difference position (delta) of each feature point j from the previous frame $i - 1$:

$$\Delta\_Stab_{ij} = abs\left(Pos\_Stab_{ij} - Pos\_Stab_{(i-1)j}\right) \tag{5}$$

$$\Delta\_Not\_Stab_{ij} = abs\left(Pos\_Not\_Stab_{ij} - Pos\_Not\_Stab_{(i-1)j}\right) \tag{6}$$

Therefore, we have collected all delta of all feature points of all frames. We calculate the weighted average of delta of all feature points in every frame i using scores as weights:

$$Mean\_FP\_Stab_i = \frac{\sum_{j=1}^{numFP} \Delta\_Stab_{ij} * Score\_Stab_{ij}}{\sum_{k=1}^{numFP} Score\_Stab_{ik}} \tag{7}$$

$$Mean\_FP\_Not\_Stab_i = \frac{\sum_{j=1}^{numFP} \Delta\_Not\_Stab_{ij} * Score\_Not\_Stab_{ij}}{\sum_{k=1}^{numFP} Score\_Not\_Stab_{ik}} \tag{8}$$

In Figs. 20 and 21 are reported how (7) and (8) vary along y and x respect to frames.

In both graphs the marker in the stabilized video has less displacement. In numerical results we can calculate a stabilization performance in a manner similar to the one seen in the Optical Flow: starting from (7) and (8) we infer (2), (3) and the index (4). In our test the value of (4) is **49.15%**. Very close to the one given by the Optical Flow (**46.82%**).

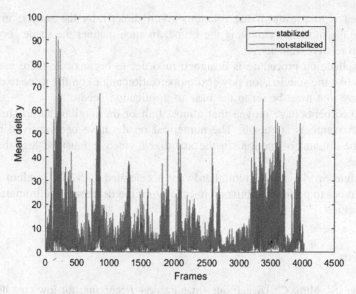

Fig. 20. Feature points mean delta on y axis

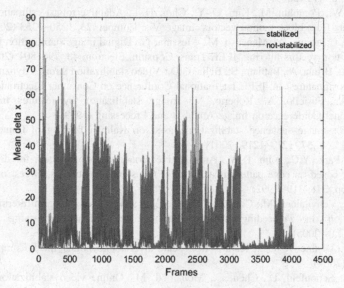

Fig. 21. Feature points mean delta on x axis

6 Conclusions

In this paper a methodology to achieve real-time video stabilization using inertial sensor is presented. This technique is suitable for wearable devices, where processing power is limited and motion estimation based on complex algorithm is unfeasible.

The idea is to use the frame rotation, given directly by the sensor, in order to compensate the motion, by moving the frame. In such manner the scene seems fixed, having zero relative velocity.

The stabilization procedure is designed in order to be flexible: there are a set of parameters like the stabilization power (compensation angle) on the three axes and the cropping size that must be set to the base of application needs.

The experiments have proven that almost half of the motion due to shaking and unwanted movements is cut off. The numerical results have been derived analyzing index of the amount of motion: in the stabilized video it must be less than in the original one.

The future work is to exploit hardware-accelerated GPU of modern wearable devices in order to perform stabilization directly on the devices and eliminate the need of a base station.

References

1. Auberger, S., Miro, C.: Digital video stabilization architecture for low cost devices. In: Proceedings of the 4th International Symposium on Image and Signal Processing and Analysis, p. 474 (2005)
2. Jang, S.-W., Pomplun, M., Kim, G.-Y., Choi, H.-I.: Adaptive robust estimation of affine parameters from block motion vectors. Image Vis. Comput. **23**, 1250–1263 (2005)
3. Vella, F., Castorina, A., Mancuso, M., Messina, G.: Digital image stabilization by adaptive block motion vectors filtering. IEEE Trans. Consum. Electron. **48**, 796–801 (2002)
4. Bosco, A., Bruna, A., Battiato, S., Bella, G.D.: Video stabilization through dynamic analysis of frames signatures. In: IEEE International Conference on Consumer Electronics (2006)
5. Censi, A., Fusiello, A., Roberto, V.: Image stabilization by features tracking. In: International Conference on Image Analysis and Processing (1999)
6. Erturk, S.: Image sequence stabilization based on Kalman filtering of frame positions. Electron. Lett. **37**, 1217–1219 (2001)
7. Paik, J., Park, Y.C., Kim, D.W.: An adaptive motion decision system for digital image stabilizer based on edge pattern matching. In: Consumer Electronics, Digest of Technical Papers, pp. 318–319 (1992)
8. Tico, M., Vehvilainen, M.: Constraint translatiònal and rotational motion filtering for video stabilization. In: Proceedings of the 13th European Signal Processing Conference (EUSIPCO) (2005)
9. Lowe, D.: Distinctive image features from scale-invariant keypoints. Int. J. Comput. Vis. **60**, 91–110 (2004)
10. Yang, J., Schonfeld, D., Chen, C., Mohamed, M.: Online video stabilization based on particle filters. In: IEEE International Conference on Image Processing (2006)
11. Matsushita, Y., Ofek, E., Ge, W., Tang, X., Shum, H.Y.: Full-frame video stabilization with motion inpainting. IEEE Trans. Pattern Anal. Mach. Intell. **28**, 1150–1163 (2006)
12. Grundmann, M., Kwatra, V., Castro, D., Essa, I.: Calibration-free rolling shutter removal. In: Proceedings of the ICCP (2012)
13. Liu, F., Gleicher, M., Jin, H., Agarwala, A.: Content preserving warps for 3D video stabilization. In: ACM Transactions on Graphics (Proceedings of SIGGRAPH), vol. 28 (2009)

14. Liu, S., Wang, Y., Yuan, L., Bu, J., Tan, P., Sun, J.: Video stabilization with a depth camera. In: Proceedings of the CVPR (2012)
15. Patel, D., Upadhyay, S.: Optical flow measurement using Lucas Kanade method. Int. J. Comput. Appl. **61**, 6–10 (2013)
16. Shi, T.: Good features to track. In: IEEE Conference on Computer Vision and Pattern Recognition (1994)
17. Tecchia, F., Carrozzino, M., Bacinelli, S., et al.: A flexible framework for wide-spectrum VR development. Presence Teleoperators Virtual Environ. **19**(4), 302–312 (2010)
18. Karpenko, A., Jacobs, D., et al.: Digital video stabilization and rolling shutter correction using gyroscopes. Stanford Tech Report CSTR (2011)
19. Smith, M.J., Boxerbaum, A., Peterson, G.L., Quinn, R.D.: Electronic image stabilization using optical flow with inertial fusion. In: 2010 IEEE/RSJ International Conference on Intelligent Robots and Systems, Taipei, pp. 1146–1153 (2010)

Using Proxy Haptic for a Pointing Task in the Virtual World: A Usability Study

Mina Abdi Oskouie$^{(\boxtimes)}$ and Pierre Boulanger

Department of Computing Science, University of Alberta, Edmonton, AB, Canada
{mina.abdi,pierre.boulanger}@ualberta.ca

Abstract. Virtual Reality (VR) applications are getting progressively more popular in many disciplines. This is mainly due to (1) technological advancement in rendering images and sounds, and (2) immersive experience of VR which intensifies users' engagement leading to improved joy, sense of presence and success. Although VR creates a realistic experience by simulating visual and auditory sense, it ignores other human senses. The absence of haptic feedback in Virtual Environments (VE) can impair users' engagement. Our objective is to provide haptic feedback in VE using the most affordable method, proxy haptics. Proxies are physical objects which are co-registered with virtual objects in a way that by touching the proxies, users see their virtual hand touching the virtual objects. So, the haptic feedback will be applied to users' hands by touching the proxies. The important question which arises here is: what are the influences of providing haptic feedback on users' performance and usability? In this work, we have designed an experiment in Virtual Reality to answer this question. In the virtual environment, we presented a keyboard with touchable keys and asked our participants to type three words using the keyboard. They repeated the experiment twice: once they used their own hands and pressed the keys and once they used VR controllers for typing. Our results show that participants can type words with fewer errors when they are using their own hands while they found VR controllers easier to use.

Keywords: Virtual Reality · Sense of touch · Passive haptic · Usability test

1 Introduction

Significant improvements in audio, visual, rendering, and motion tracking technologies have enabled VR systems to create high-quality and low-cost immersive environments that can be used to create an intense sense of presence.

Unfortunately, haptics - the sense of touch - remain relatively undeveloped in virtual environments. In current VR systems, the sense of presence disappears when the user's hand goes through an object. Being able to touch objects in VR is critical in order to make the experience significantly more engaging and natural.

© Springer Nature Switzerland AG 2019
L. T. De Paolis and P. Bourdot (Eds.): AVR 2019, LNCS 11613, pp. 292–299, 2019.
https://doi.org/10.1007/978-3-030-25965-5_21

This is especially important for simulations like virtual surgery or mechanical assembly. Coupling the sense of touch with other senses (audio and visual) not only enables medical students to believe that they are operating on a real-patient but also give them the unique chance to practice important skills such as hand-eye coordination.

Many researchers have tried to enrich VR experience with the sense of touch. One can find in the literature two main categories, *Active Haptic* and *Passive Haptic*.

In Active Haptic, the haptic feedback is provided by small mechanical robots called haptic interfaces that are controlled by a haptic rendering computer simulating the physical interaction between a virtual tool (registered with a real tool) and the virtual environment. The haptic interface actuation is calculated based on the user's actions in the virtual environment and is returned to the user as a force and a displacement.

Active haptic interfaces can be designed in wearable or non-wearable forms. Most of the available devices are non-wearable [3,10,11]. The problem with this approach is that the user is obliged to hold the devices at all time making it hard to create a real sense of touch.

The other approach to Active Haptic is to use wearable robots to simulate the sense of touch by sending tactile signals to the skin. These robots have a wide range of sizes and shapes. From the small ones just covering fingertips to the huge robots which cover the whole upper part of the body. Prior work by [1,4,9] have proposed wearable haptic devices which create the illusion of touching virtual objects by applying appropriate forces to fingers and fingertips. Other researchers [7] provide the sense of touching walls or lifting heavy objects by sending electrical signals to the arms or hands muscles.

Being forced to wear a robot exoskeleton or receive small electric shocks are not always pleasant for the users. Plus, feeling that there are devices on their skin makes the users aware that they are holding a haptic device hence losing the sense of presence one wants to achieve. Additionally, these haptic robots are very expensive and hard to program.

Passive Haptic, on the other hand, uses physical objects, called *Proxy haptic objects* or *props*, to simulate the sense of touch [5]. When the user touches a virtual object, its co-registered proxy haptic object will provide feedback about the texture, weight or temperature of the virtual objects.

Using passive haptic, one can make the tactile VR experience more affordable. In this approach, users are actually interacting with real physical objects called "proxy haptics objects" that are spatially co-registered with virtual objects that may have different appearances and shapes. In other words, proxy haptics objects are providing haptic feedback without the need to build complex robots and making users wear or hold them.

The question that requires more investigation is: how providing haptic feedback influences users' performance? We have designed an experiment to investigate these influences. Since this is one of the first step in this research direction, we simplified the usability experiment to a pointing task.

2 Experiment

The pointing task is presented as typing in which a word is shown to the participants and they should type the word by pointing and selecting keys on a keyboard and press the OK key to finish the task.

2.1 Apparatus

We used tracked HTC Vive head-mounted display (HMD) to represent the virtual environment and head tracking. Our goal was to keep the experiment as simple and affordable as possible so we employed the most affordable equipment for tracking hands, the Leap Motion Controller. With this device attached on the HMD, one could see a representation of their hand in the virtual environment without wearing any marker.

The experiment took place in front of a sponge attached to a wall. The sponge is used as a proxy for keys. In the Virtual Environment (VE), designed using Unity 3D version 5, participants saw a keyboard in front of them. Figure 1 shows the physical and virtual environments.

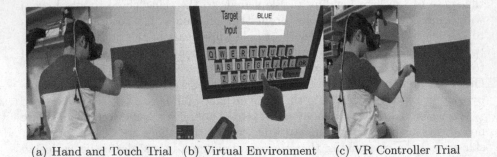

(a) Hand and Touch Trial   (b) Virtual Environment   (c) VR Controller Trial

Fig. 1. Physical and virtual environment of the experiment

The proxy was co-registered with their virtual representations. Therefore, when participants touched them, they were actually touching a physical object in the real world. The co-registration is done based on the position of the virtual and the physical hands, in a way that when we saw our virtual hand touching a virtual object, its corresponding proxy should be touched by our physical hand.

2.2 Task

In this experiment, we provided users with a standard computer keyboard and a word that the user needed to type on the keyboard. The experiment was repeated with 3 different words, "BLUE", "CHAIR", and "FAMILY". We wanted to have 3 words with different lengths and also make sure that none of the two consecutive characters in these words are neighbours in the keyboard, so users should aim and reach each character separately. In order to have a baseline to compare the results with, we conducted this experiment twice. First, we asked the participants to type the word using their hands. In this case, the keys were touchable and their hand was detected using Leap Motion controller (No makers were used). Then we repeated the experiment with a the HTC Vive VR controller. Figure 1 shows a participant during each of the trials.

2.3 Participants

Based on the common practice, by setting alpha $= 0.5$ and effect size $= 0.4$, ANOVA test showed that we need 33 participants to achieve power equal to 90%. Thirty-three participants, thirty-two of them aged between 22 and 34 and one was 53 years old ($M = 27.03$; $SD = 5.52$), took part in the study. All of them were graduate students at the University of Alberta, Department of Computing Science. We asked them how frequently they used video games, VR, and VR controllers on a scale of 1 to 4. As the video games generally are more available, we interpret its scales as 1 (not frequently play) to 4 (play every day). For VR and VR controllers 1 means "Never worked with it" and 4 means "Frequently working with it". Our sample consisted of a group of in-frequent gamer ($M = 1.94$; $SD=1.10$) with little experience with VR and VR controllers (VR: $M = 2.33$; $SD = 1.09$, VR controller: $M = 2.21$; $SD = 1.17$).

2.4 Ethical Consideration

This work has received research ethics approval from the University of Alberta Research Ethics Board, Project ID, Pro00081301.

3 Measurements

In order to be able to compare the two trials, using hands or controllers, we need to define our measurements. The experiment is not designed as a game and also users work with it in a very short time, so we cannot measure users' engagement, excitement or joy as a measure of success.

What we mainly looking to answer: (1) does using bare hands, compared to controllers, make the task easier and more comfortable for users? (2) does providing haptic feedback make VR more believable? To find answers to these questions, we used some quantitative measures and a questionnaire. In what follows these measures will be explained:

3.1 Quantitative Measures

With quantitative measures, we are mainly focusing on measuring how flawless users perform in the task. The less number of errors the easier the task is. The other important factor is the time they consume for typing the words. The measures we collected data for are "Time", "Number of Deletes" and "Accidentally Pressing OK key".

3.2 Questionnaire

We asked our users to answer 27 usability questions. The questions are a mixture of 4 questionnaires each of which will be explained below. We have eliminated 3 questions from the questionnaires, 6 in total for both trials because we did not want our participants to get bored and answer the questions carelessly. We considered 7 response options for each question; from "Not at all True" to "Very True", and ask each questionnaire twice; once after each trial.

- **Perceived Competence Scales**
  Self-determination theory (SDT) [8] states that perception of competence can facilitate goal attainment and make people feel effective in the activity. If users feel competence doing a task in VR that mean the task was doable and they are engaged. Perceived Competence Scales (PCS) is a short 4-item questionnaire designed to assess the participants' feeling of competence about doing a task or following through on some commitment. In this work, we used 3 questions of it.
- **System Usability Scale**
  System Usability Scale (SUS) is an instrument to measure the usability [2]. Originally, it consists of 10 questions. We only used 8 of them.
- **Believability**
  The last factor that we want to assess was about the believability of trials. As we could not find any established instrument to measure it, we have designed 2 questions for this matter. The statements are: "I felt like I am dealing with the real world" and "I felt strange using the system".
- **Preference**
  In the end, we asked users which trial they prefer and why.

3.3 Observation

In order to fully understand how users interact with the physical and virtual environments, we employed *Observer as Participant* observation method. In this method, the researcher, who is known for the participants, can communicate with them while researcher's main role is to collect data [6].

Table 1. Experimental results - quantitative measures

	Hand Trial			Controller Trial		
Words	BLUE	CHAIR	FAMILY	BLUE	CHAIR	FAMILY
Time	$17.73_{\pm10.89}$	$18.52_{\pm12.13}$	$17.17_{\pm8.27}$	$13.04_{\pm5.37}$	$10.37_{\pm4.26}$	$11.94_{\pm6.89}$
Avg #Errors	$0.39_{\pm0.81}$	$0.61_{\pm1.04}$	$0.67_{\pm1.17}$	$0.97_{\pm1.38}$	$0.93_{1.86}$	$1.15_{\pm1.67}$
Accidental OKs	1	0	3	4	4	3

4 Experimental Results

4.1 Quantitative Measures

- **Time**

  We measured the time consumed to type the words in both trials. We ignored the trials in each users' accidentally press the OK key because the consumed time in them are shorter but they could not finish typing the words.

  Table 1 shows the average and standard deviation of the time consumed to type the words completely in each of the trials. Users took a longer time to type the word when they were using their hands. Observation notes show that is because the Leap Motion Controller was not always accurate and it sometimes showed the wrong hand gesture or lost the hands completely. This made it impossible to press keys. Users had to wave their hand in front of the Leap Motion Controllers and gave it some time to restart. While the VR controller was always responsive.

- **Number of Deletes**

  The number of times participants pressing delete key is an indicator of how many time they pressed a key by mistake or accuracy of their performance. Table 1 shows the average and standard deviation of the number of deletes for the hand and Controller trials.

  Based on the Table, people can type more accurately when they are using their hands comparing to controllers. Based on observations, that is mainly because pressing the trigger button on the controller make the controller shake and that leads to pressing neighbour keys by mistake. While in the Hand Trial users have more control over the movement of their hand. As mentioned before, the inaccuracy of Leap Motion Controller in detecting hand sometimes cause a problem as well. Using a more accurate system to detect hand gesture and position can help us achieve even more flawless performance in the Hand Trial.

- **Accidentally Pressing OK**

  Accidentally Pressing OK is another indicator of the accuracy of users' performances. Table 1 shows the number of people who accidentally pressed OK during typing each word. Fewer users accidentally pressing OK key in the Hand Trial. Just like the "Number of Deletes", based on observations, it is because people have generally better control over their fingers comparing to the VR controllers.

4.2 Questionnaire

Table 2 shows the results of the questionnaires separately. Based on the Table, users felt more competence in the Controller Trial. This trial has also gained a better usability score. While participants found Hand Trial more believable.

Most of the participants preferred Controller Trial. Their reasons for this choice shed light on the advantages and disadvantages of these two systems. In what follows we will summarise users' feedback:

Table 2. Experimental results - questionnaire

	Hand Trial	Controller Trial
PCS	5.41	5.8
SUS	5.03	5.64
Believability	5	4.93
Preference	34.38%	65.63%

- The Hand Trial is more immersive and natural. Some users felt more control over it because they could do the task by their hand instead of a tool.
- Inaccurate hand tracking and hand gesture detection made this trial time consuming, cumbersome and hard.
- The size of the keys on the keyboard were the same for both trials, while the fingers were much bigger than the laser came out of the controller. This made it harder for the users to press keys by their fingers.
- Participants found the Controller Trial to be faster and easier to use. Some of them expressed that this is because they are used to work VR controller or tools like it.
- The Controller trial was more unrealistic and erratic. It was hard for users to avoid pressing the neighbours of a key accidentally.

5 Discussion

Although the Hand rial was more realistic it failed to improve the system usability and ease of use. That was mainly because of the poor performance of the Leap Motion interface we used in this work. Providing passive haptic feedback needs the virtual objects to be exactly aligned with their proxies. In the typing task, presenting the keyboard deep into the physical wall made users push the sponge, keyboard's proxy, too hard which was not pleasant for them. On the other hand, showing the virtual keyboard outside the wall cannot provide any haptic feedback because users cannot touch the sponge anymore. The co-registration was too fragile and we needed to re-do it for each user. It is because the Leap Motion controller was not accurate in showing the position of users' hand and also since the sizes of fingers are different for each individual, the system needed to be calibrated for each user separately.

6 Conclusion and Future Work

We added haptic feedback to a Virtual Reality experiment of a simple task of typing. With the goal of keeping the experiment as affordable as possible, we have employed HTC Vive HMD for representing the Virtual Environment and Leap Motion Controller for hand tracking and hand gesture detection. The results show that the Hand Trial is less error-prone and more believable. On the other hand, Controllers Trial is faster and easier to use. This is mainly because of the inaccuracy in hand tracking and hand gesture detection. Using better devices, e.g. OptiTrack and active markers, for this matter can improve the usability of the experiment significantly while making the experiment more expensive.

References

1. Bortone, I., et al.: Wearable haptics and immersive virtual reality rehabilitation training in children with neuromotor impairments. IEEE Trans. Neural Syst. Rehabil. Eng. **26**(7), 1469–1478 (2018)
2. Brooke, J., et al.: SUS-a quick and dirty usability scale. Usability Eval. Ind. **189**(194), 4–7 (1996)
3. Choi, I., Ofek, E., Benko, H., Sinclair, M., Holz, C.: CLAW: a multifunctional handheld haptic controller for grasping, touching, and triggering in virtual reality. In: Proceedings of the 2018 CHI Conference on Human Factors in Computing Systems, p. 654. ACM (2018)
4. Gu, X., Zhang, Y., Sun, W., Bian, Y., Zhou, D., Kristensson, P.O.: Dexmo: an inexpensive and lightweight mechanical exoskeleton for motion capture and force feedback in VR. In: Proceedings of the 2016 CHI Conference on Human Factors in Computing Systems, pp. 1991–1995. ACM (2016)
5. Hinckley, K., Pausch, R., Goble, J.C., Kassell, N.F.: Passive real-world interface props for neurosurgical visualization. In: Proceedings of the SIGCHI Conference on Human Factors in Computing Systems, pp. 452–458. ACM (1994)
6. Kawulich, B.B.: Participant observation as a data collection method. In: Forum Qualitative Sozialforschung/Forum: Qualitative Social Research, vol. 6 (2005)
7. Lopes, P., You, S., Cheng, L.P., Marwecki, S., Baudisch, P.: Providing haptics to walls & heavy objects in virtual reality by means of electrical muscle stimulation. In: Proceedings of the 2017 CHI Conference on Human Factors in Computing Systems, pp. 1471–1482. ACM (2017)
8. Ryan, R.M., Deci, E.L.: Self-determination theory and the facilitation of intrinsic motivation, social development, and well-being. Am. Psychol. **55**(1), 68 (2000)
9. Scheggi, S., Meli, L., Pacchierotti, C., Prattichizzo, D.: Touch the virtual reality: using the leap motion controller for hand tracking and wearable tactile devices for immersive haptic rendering. In: ACM SIGGRAPH 2015 Posters, p. 31. ACM (2015)
10. Strasnick, E., Holz, C., Ofek, E., Sinclair, M., Benko, H.: Haptic links: bimanual haptics for virtual reality using variable stiffness actuation. In: Proceedings of the 2018 CHI Conference on Human Factors in Computing Systems, p. 644. ACM (2018)
11. Whitmire, E., Benko, H., Holz, C., Ofek, E., Sinclair, M.: Haptic revolver: touch, shear, texture, and shape rendering on a reconfigurable virtual reality controller. In: Proceedings of the 2018 CHI Conference on Human Factors in Computing Systems, p. 86. ACM (2018)

Medicine

Surgeries That Would Benefit from Augmented Reality and Their Unified User Interface

Oguzhan Topsakal[1($\boxtimes$)] and M. Mazhar Çelikoyar[2]

[1] Department of Computer Sciences, Florida Polytechnic University,
4700 Research Way, Lakeland, FL 33805, USA
otopsakal@floridapoly.edu
[2] Department of Otolaryngology, Istanbul Florence Nightingale Hospital,
Abide-i Hürriyet Cad. No: 166, 34381 Şişli, Istanbul, Turkey

Abstract. Augmented Reality (AR) has a high potential to help healthcare professionals in the operating room and researchers have been working on bringing AR to the operating room for many different types of surgeries so far. However, there are thousands of surgery types that might benefit from the AR. Our goal is to identify which surgeries could benefit the most from AR and identify the common functionality desired in an AR app for surgery and then come up with a unified user interface (UI) design for these surgeries. To achieve this goal, we analyzed the historical number of surgeries performed at a teaching and research hospital to find the most frequently performed surgeries. Starting with the most frequently performed surgeries, we collaborated with the experienced surgeons to find out the surgeries that could benefit from AR the most. For each surgery, we identified the functionality and the information that could be presented on an AR app before, during and after surgery. We prepared the list of surgeries that might benefit from AR through a series of thorough interviews with experienced surgeons. We also identified the functionalities desired in AR apps for surgery and prepared a unified UI design for the common functionality.

We believe that the unified UI that we designed for the AR App of surgical procedures is the first step towards an AR app that might be beneficial for many types of surgeries. We believe surgeons equipped with AR apps will achieve better results; the success rates of surgical procedures will increase while the number of errors decreases.

Keywords: Augmented Reality · Mixed reality · Surgery ·
Surgical procedures · AR in medicine · AR functionality for surgery ·
User interface for AR surgery app · Computer assisted surgery

1 Introduction

Augmented Reality (AR) is mostly achieved by providing overlapping information on the user's field of view [1]. The user can make more effective, more accurate and better decisions with the help of the information provided. AR applications for medicine are expected to help healthcare professionals (doctors, patients, nurses, etc.) and improve the quality of service and decrease the number of errors.

L. T. De Paolis and P. Bourdot (Eds.): AVR 2019, LNCS 11613, pp. 303–312, 2019.
https://doi.org/10.1007/978-3-030-25965-5_22

The third leading cause of death in the United States is medical errors and about 50% of these errors happen during surgery operations resulting much pain to the families of the patients and to the economy [2] Even if it is not a deadly mistake, errors caused during a surgery can cause injuries that put financial and emotional burden to the families. Improvements made through AR technologies would lead to minimizing the costly errors and improve healthcare satisfaction benefitting both medical practitioners and the patients. Moreover, improvements in the quality and success of surgeries would lower the cost of the healthcare system as a whole.

However, to achieve the benefits, AR apps should provide an easy to use, intuitive interface [3]. The usability of the apps will help them to be easily adapted by healthcare professionals [4]. If not designed well, an AR app can become hard to use or can be rejected by surgeons who are already working in a crowded space and are processing much information in a short amount of time [5]. To design a usable AR app for surgery, we need to know the details of the requirements and the usage. However, each surgery has different requirements and setup. It is important to know more about the environments, users and the details of the procedures so that solutions can be tailored to solve the specific problems [6].

Designing AR Apps is already a challenging task due to limited and unconventional methods for user-device interaction and due to limited space on the see-through screen for placing information [7–9]. Moreover, the apps can only be used for a limited time due to ergonomics and battery life. Using an AR app becomes even more challenging in the operating room because touch-based interaction is often not possible since surgeon's hands are busy or covered with gloves, and surgery room is mostly a crowded interactive space with other surgeons, nurses, etc. working together. Besides, it is difficult for the surgeon to quickly and accurately process information coming from many sources. It is important not to cognitively overload the surgeon with unnecessary information.

In this study, we review the surgeries and identify which surgeries could benefit the most from AR. During these reviews, we identify the information that can be presented through an AR app to healthcare professionals. We present a unified user interface design that can provide most of the information needed before, during and after surgery. In the following phases of the research, our goal is to identify UI/UX design principles of AR apps to achieve the best results in operating room environments.

2   Methodology

Instead of reviewing thousands of surgical procedures from the perspective of AR, which would require excessive resources, we wanted to focus on the most common surgical procedures. To do so, we got the three years of statistics of surgeries performed at the research and teaching hospital of Uludag University in Turkey. Research and teaching hospital of Uludag University is serving an area of 6 million populations with its over 300 faculty members and specialist doctors. We calculated the average number of times each surgical procedure performed in a year. We identified around 50 frequently performed surgery types from around 2,500 surgical procedures in the following specialty fields; Otolaryngology, Urology, Neurology, Orthopedics, Ophthalmology, Gastroenterology, Obstetrics, and Gynecology.

With the names of the most frequent surgical procedures at hand, we scheduled interviews with specialty doctors and scholars to discuss these procedures from the perspective of AR. We started our conversation by educating the surgeons about the AR technology without giving any boundaries about what can or cannot be done. After reviewing and discussing the most frequently performed surgeries in their specialty field, we also opened a discussion about the surgeries that they think AR might help based on the challenges they are facing during the surgical procedures.

In general, each surgery has three phases; pre-surgery preparation and planning, in-surgery execution of the plan (with adjustments if needed), and post-surgery evaluation of the surgery results and the patient status [10]. During our discussions, we investigated the information sought in these three phases of surgery. We asked what kind of information is needed before, during and after surgery and whether utilizing AR to deliver this information would make sense or not. We have discussed if there are any difficulties to access the information especially during surgery and if the current techniques to access information cause any errors. We have also investigated if a wearable computer could do any computations or measurements instantaneously to help the surgeon perform his/her tasks more conveniently.

We have interviewed 12 surgeons from 4 hospitals. The names of the surgeons are listed in the acknowledgements section. The hospitals that surgeons are affiliated with are Uludag University Teaching & Research Hospital, Bursa Sevket Yilmaz Post-graduate Teaching & Research Hospital, Istanbul Bilim University Florence Nightingale Hospital, Bursa Inegol Arrhythmia Private Hospital.

During our selection of the surgeons, our main criterion was the experience in their specialty. Almost all of the surgeons we interviewed had at least 20 years of experience in their fields. Other criteria for selection were being interested in technology and being open to new advancements. Based on these criteria, we have received a list of sug-gested academicians and specialty doctors from the Uludag University Teaching & Research Hospital. We reached out to these surgeons and explained our study and asked if they would like to participate. As we have reached to these surgeons, we have also requested suggestions about surgeons who might be interested in participating in our study and then tried to reach the suggested surgeons as well.

For some of the specialties, such as General Surgery, Orthopedics, Neurosurgery, we have interviewed two surgeons. For some of the specialties, such as Urology, Otolaryngology, and General Surgery we interviewed more than once with the same surgeons. During an interview with a surgeon, we have shared suggestions/ideas from other surgeons to spark new ideas/suggestions.

3   Results

We have identified 10 surgeries from 4 different specialties that AR would help to surgeons. We need to emphasis that we are not providing the complete list of surgeries that AR might help. However, we have come up the list of top 10 surgeries based on the through interviews with the experience surgeons. We not only list the names of these surgical procedures in Table 1 but also discuss how AR would help to those 10 surgeries in Table 2.

3.1   Results - Surgeries that Can Benefit from AR

Based on the interviews with experienced surgeons, we were able to identify the below surgical procedures that can possibly benefit from AR:

Table 1. Surgeries that would benefit from AR grouped by specialties.

Specialty	Surgical procedure name
Otolaryngology	Rhinoplasty, Thyroidectomy, Neck Dissection
Urology	Kidney Transplantation, Percutaneous Nephrolithotripsy
Orthopedic	Knee Arthroplasty, Shoulder Arthroplasty, Hip Arthroplasty
Neurosurgery	Hydrocephalus, Intradural and Intramedullary Spinal Cord Abscess Drainage

For the surgical procedures listed above, based on the experiences of the surgeons, the following suggestions were made regarding how AR could help the surgeons:

Table 2. How AR would help in selected surgical procedures.

Surgical procedure	Desired augmented reality functionality
Rhinoplasty	Rhinoplasty (i.e. nose reshaping) is one of the most common plastic surgery procedures. One of the biggest challenges of rhinoplasty is to achieve a perfect tip and nostril rim/shape symmetry. Currently, no device is used for the measurements and the success of this operation is mainly based on the surgeon's experience. During a procedure where millimeters can make a difference, AR can provide instant measurements regarding the size, angle, and symmetry helping the surgeon to achieve better results
Thyroidectomy	Zoom In/Out features can be utilized dissecting the critical structures, i.e. laryngeal nerves, parathyroid glance as well as vascularity of these glances. Applying filters on the image/video feed of the working area might also be helpful in distinguishing critical structures
Neck dissection	Zoom In/Out features can be utilized dissecting the critical structures, i.e. carotid vessels, internal jugulars vein, cranial nerves. As well as help differentiate different tissues such as a nerve from a vessel, a lymph node from fatty tissue
Kidney Transplantation	Kidney transplantation is a life-saving procedure. Zoom In/Out features can be utilized while sewing the blood vessels to successfully achieve the procedure

(*continued*)

Table 2. *(continued)*

Surgical procedure	Desired augmented reality functionality
Percutaneous Nephrolithotripsy	Percutaneous Nephrolithotripsy is a procedure for removing large kidney stones. Wearable AR glasses can be utilized to find exact locations of the stones and to make sure no pieces of the stone were left in the kidney
Knee Arthroplasty, Shoulder Arthroplasty, Hip Arthroplasty	Knee/shoulder/hip arthroplasty is a surgical procedure to replace the weight-bearing surfaces of the knee/shoulder/hip joint to relieve pain and disability. Alignment is very important for arthroplasty surgeries. If measurements can be done via a wearable device to guide the surgeon for the alignments, the success rate of knee, shoulder and hip arthroplasty procedures increases
Hydrocephalus	Hydrocephalus is an excessive accumulation of cerebrospinal fluid (CSF) in the brain. Hydrocephalus is most often treated surgically by inserting a shunt system. This system diverts the flow of CSF from the Central Nervous System (CNS) to another area of the body where it can be absorbed as part of the normal circulatory process. A shunt is a flexible but sturdy plastic tube. AR wearable devices can be very beneficial in accurately placing the shunt in the brain. Accurately locating the shunt of vital importance for the success of the procedure
Intradural and Intramedullary Spinal Cord Abscess Drainage	A spinal cord abscess (SCA) can cause permanent damage to the spinal cord. Accurately locating the abscess is very important for the drainage operation. Magnetic Resonance (MR) compatible barcodes can be used while taking the MR so that AR device can be used to overlap the MR on the back of the patient during the procedure for more accurate and successful operation

3.2   Results - Common Functionality Needed for Surgeries

Based on the discussion over 50 specific types of surgeries, surgeons frequently needed the following information for surgeries:

- **Access Patient Information:** Surgeons think it would be beneficial to have quick access to the Computerized Tomography (CT), Magnetic Resonance (MR), Xrays, Lab results and the surgery plan via an AR app.
- **See Warnings for the Procedure:** For each patient, there might be notes and warnings from healthcare professionals based on lab results and patient history. These can be presented through the AR app as well.

- **Do Measurements & Image Manipulations:** For several surgeries such as Knee Arthroplasty or Rhinoplasty, measurement of angles and distances of the surgical area is important for accuracy. Moreover, applying filters on the image/video feed of the working area might also be helpful in distinguishing critical structures.
- **Zoom In/Out:** Zoom in/out to see the working area more precisely has been requested for some of the surgical procedures.
- **Record & Share:** A wearable head-mounted device can be used to record and share the surgeon's point of view for training, research, archiving, teaching, and consultation purposes during or after the surgery [11].

3.3   Results - Unified UI Design of AR Apps for Surgery

We designed a unified user interface that presents the common information and provides the common functionality needed by surgeons. On every screen, the 'eye' icon indicates that user can gaze on the area to have a 'click' effect on the UI. On the AR Surgery App, user and UI interaction will be achieved by eye gaze detection and voice recognition. We present and explain the UI design below:

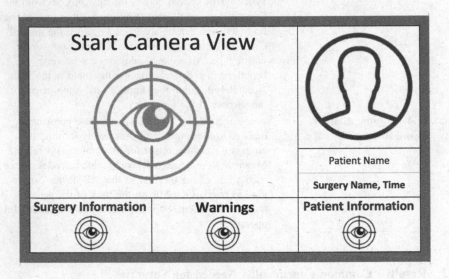

Fig. 1. Start screen

1. **Start Screen:** Start screen is depicted in Fig. 1. It has brief information (patient name, gender, surgery name and time) about the patient and the surgery and four sections to navigate to other screens. These sections and the related screens are as follows:
   a. **Start Camera View Section:** This section becomes active only after user visits 'Warnings' and 'Patient Information' screens. When active, user gazes on the icon for a couple of seconds to start the 'Camera View' screen described below.

b. **Warnings Section**: User gazes on the icon to go to the 'Warnings' screen described below.

c. **Patient Information Section:** User gazes on the icon to go to the 'Patient Information' screen described below.

d. **Surgery Information:** User gazes on the icon to go to the 'Surgery Information' screen. Surgery information screen provides general information, videos, and techniques about the procedure if the surgeon wants to review.

2. **Warnings Screen:** Warning screen is depicted in Fig. 2. It lists the warnings about the patient in three groups: allergies, co-morbidities, and warnings specific to the patient. The warnings must be reviewed before the surgery. If this screen is not reviewed and not marked as 'read and understood', camera view does not get activated on the start screen.

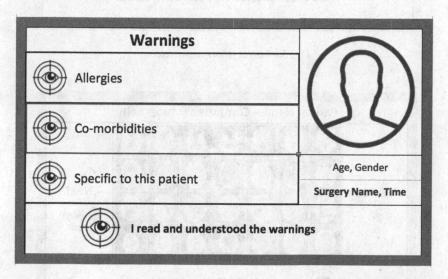

Fig. 2. Warning screen

3. **Patient Information Screen**: Patient Information screen is depicted in Fig. 3. It has 3 sections to navigate to sub-screens. These sub-screens are 'Surgery Plan', 'CT, MR, XRays' or 'Lab Results' to get further information about the patient. 'CT, MR, XRays' screen is described below. All the sub-screens are supposed to be reviewed before the surgery. If all patient information screens are not reviewed and not marked as 'read and understood', camera view does not get activated on the start screen.

a. **'Surgery Plan' Screen:** This screen has the details of the surgery plan.

b. **'CT, MR, XRays' Screen:** This screen is depicted in Fig. 4. The user can navigate through CT, MR, and XRays to get further information about the patient.

c. **'Lab Results' Screen:** This screen has the details of the lab results.

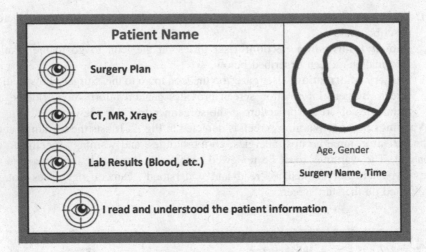

Fig. 3. Patient information screen

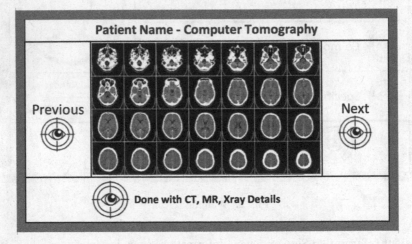

Fig. 4. CT, MR, XRays screen

4. **Camera View Screen:** Camera View screen is depicted in Fig. 5. The screen is used during the surgery. It provides zoom in/out (+/−), record, share live feed via wifi functionalities as well as an option to switch to measurement mode. In the measurement mode, the user can make measurements of angles, distances by utilizing markers. As the technology advances, there will be no need for markers and the app will be able to create the 3D mesh of the surgical area to make measurements between detected points.

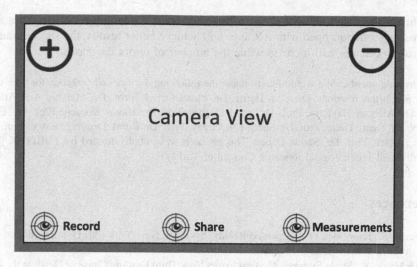

Fig. 5. Camera view screen

4 Future Directions

We have identified surgical procedures that can benefit from AR and the functionalities needed for each surgery. We then listed the common functionalities and designed a unified UI. From now on, we plan to continue this research in two steps;

1. We plan to develop a prototype of an AR app utilizing the unified UI design to provide the functionalities needed for one of the specific surgeries: Rhinoplasty. As we develop the prototype, we plan to provide the desired functionality for Rhino-plasty listed in Table 2.
2. We plan to enrich the above with more functionalities to make it applicable for other types of surgeries listed in Tables 1 and 2.
3. As we enrich the AR app, we plan to conduct studies to assess how success rate of surgical procedures increases and the number of errors decreases as surgeons utilize the AR technology.

For the prototypes, we plan to use Magic Leap One (Magic Leap Inc.) and Microsoft Hololens (Microsoft Corporation). The AR apps will be developed utilizing Unity 3D (Unity Technologies).

5 Conclusion

In this research, we have listed 10 surgical procedures that would benefit from AR. We believe, AR might be beneficial for more than 10 surgeries however we have identified the top 10 surgeries through a series of thorough interviews with experienced surgeons. We were able to identify the common functionalities that should be included in an AR app for surgeries. We believe that the unified UI designed for AR apps would provide an intuitive interface that would encompass the needs of many types of surgeries. We believe surgeons equipped with AR apps will achieve better results; the success rate of surgical procedures will increase while the number of errors decreases.

Acknowledgements. We would like to thank the following doctors and surgeons for their time and input in this research; Dr. Cem Heper, Dr. Murat Öztürk, Prof. Dr. Mehmet Aral Atalay, Prof. Dr. Mehmet Baykara, Prof. Dr. Şeref Doğan, Prof. Dr. Hasan Kocaeli, Prof. Dr. Ersin Öztürk, Dr. Deniz Tihan, Prof. Dr. Sadık Kılıçturgay, Prof. Dr. Burak Demirağ, Assoc. Prof. Dr. Burak Akesen, Prof. Dr. Serhat Özbek. This research was initially funded by TUBITAK (The Scientific and Technological Research Council of Turkey).

References

1. Furht, B.: Handbook of Augmented Reality. Springer, New York (2011). https://doi.org/10.1007/978-1-4614-0064-6
2. McMains, V.: Study Suggests Medical Errors Now Third Leading Cause of Death in the U.S. https://www.hopkinsmedicine.org. Short url https://bit.ly/2iOcECc. Accessed 14 Feb 2019
3. Zhou, F., Dun, H.B.L., Billinghurst, M.: Trends in augmented reality tracking, interaction and display: a review of ten years of ISMAR. In: Proceedings - 7th IEEE International Symposium on Mixed and Augmented Reality, ISMAR, pp. 193–202 (2008)
4. Behringer, R., Christian, J., Holzinger, A., Wilkinson, S.: Some usability issues of augmented and mixed reality for e-health applications in the medical domain. In: Holzinger, A. (ed.) USAB 2007. LNCS, vol. 4799, pp. 255–266. Springer, Heidelberg (2007). https://doi.org/10.1007/978-3-540-76805-0_21
5. Livingston, M.A.: Issues in human factors evaluations of augmented reality systems. In: Huang, W., Alem, L., Livingston, M. (eds.) Human Factors in Augmented Reality Environment, pp. 3–9. Springer, New York (2013). https://doi.org/10.1007/978-1-4614-4205-9_1
6. Ma, M., Jain, L.C., Anderson, P.: Virtual, Augmented Reality and Serious Games for Healthcare 1. Springer, Heidelberg (2014). https://doi.org/10.1007/978-3-642-54816-1
7. Ganapathy, S.: Design guidelines for mobile augmented reality: user experience. In: Huang, W., Alem, L., Livingston, M. (eds.) Human Factors in Augmented Reality Environments, pp. 165–180. Springer, New York (2013). https://doi.org/10.1007/978-1-4614-4205-9_7
8. Huang, W., Alem, L., Livingston, M.A.: Human Factors in Augmented Reality Environments. Springer, New York (2013). https://doi.org/10.1007/978-1-4614-4205-9
9. Gabbard, J.L., Swan, J.E.: Usability engineering for augmented reality: employing user-based studies to inform design. IEEE Trans. Vis. Comput. Graph. **14**(3), 513–525 (2008)
10. Kalkofen, D., et al.: Integrated medical workflow for augmented reality applications. In: International Workshop on Augmented Environments for Medical Imaging and Computer-Aided Surgery (AMI-ARCS), Copenhagen (2006)
11. Celikoyar, M.M., Aslan, C.G., Cinar, F.: An effective method for video recording the nasal–dorsal part of a rhinoplasty – a multiple case study. J. Vis. Commun. Med. **39**(3–4), 112–119 (2016)

Assessment of an Immersive Virtual Supermarket to Train Post-stroke Patients: A Pilot Study on Healthy People

Marta Mondellini[1]([✉]), Simone Pizzagalli[1], Luca Greci[2], and Marco Sacco[1]

[1] Institute of Intelligent Industrial Technologies and Systems for Advanced Manufacturing – STIIMA, Via Previati 1/A, 23900 Lecco, Italy
{marta.mondellini, simone.pizzagalli, marco.sacco}@stiima.cnr.it
[2] Institute of Intelligent Industrial Technologies and Systems for Advanced Manufacturing – STIIMA, Via Alfonso Corti 12, 20133 Milan, Italy
luca.greci@stiima.cnr.it

Abstract. Technological improvements have contributed to the development of new innovative approaches to rehabilitation. The use of virtual reality (VR) in physical and cognitive rehabilitation brings several advantages, such as greater motivation and patient enjoyment, allowing for the performance of everyday life tasks in a safe and controlled environment. This work presents a virtual supermarket to be experienced with the HTC Vive Head Mounted Display, aiming at the training of post-stroke patients. The immersive virtual environment was validated on a sample of healthy subjects; a comparison was made with the previous software version of the virtual supermarket developed to train cognitive capabilities in elderly users. System usability, cyber-sickness and sense of presence have been investigated to highlight desirable modifications before a final assessment on clinical patients.

Keywords: Virtual Reality · Usability · Rehabilitation

1 Introduction and Related Works

Stroke is one of the most common causes of death and one of the main causes of acquired adult disability [1]. The effects of stroke can be both physical and cognitive, with consequent behavioral and emotional problems. Up to 42% of stroke survivors require assistance for daily living activities (ADL) due to a combination of physical, cognitive and perceptual problems [2].

Several longitudinal studies indicate that in almost 66% of the cases of post-stroke patients with a paralyzed arm the limb is still immobile after 6 months from the event, while only a small percentage of the subjects recover the functionality of the upper limb [3]. This is one of the reasons why medical treatment of stroke patients still heavily relies on rehabilitation interventions [4].

In this scenario the introduction of gaming elements and immediate feedback on performance appears to be really important as it enhances motivation, thereby

© Springer Nature Switzerland AG 2019
L. T. De Paolis and P. Bourdot (Eds.): AVR 2019, LNCS 11613, pp. 313–329, 2019.
https://doi.org/10.1007/978-3-030-25965-5_23

encouraging higher numbers of repetitions of the exercises. An immersive Virtual Environment could be useful to this purpose because a repetitive exercises, salient at the motor level, is associated with an entertaining and motivating feedback, thus optimizing the motor learning [5]. For example, a work by Cho et al. [6] confirmed that virtual reality (VR) therapy is capable of stimulating new motor and sensory abilities responsible for maintaining balance.

Furthermore, by simulating real-life activities in VR, stroke patients are able to work on self-care skills in a setting and scenarios that are usually impossible to reproduce in a hospital environment [7]. Despite being full of potential, the application in the field of post stroke treatment and rehabilitation is quite recent and needs to be further explored [8].

In most of the reported cases focusing on rehabilitation, Virtual Reality for post-stroke patients has been employed in the rehabilitation of the upper limb, with the aim of re-acquiring specific motor skills and daily living tasks dexterity. This is a very important goal as people that lost the use of an arm will almost certainly see the quality of their life heavily reduced [5]. Other examples show how Virtual Reality has been used to exercise cognitive functions [9], such as orientation and navigation [10]. Researchers designed rehabilitation interventions with the aim of retaining specific ADLs, such as preparing a hot drink [11] or driving a scooter [12].

Shopping is certainly one of the instrumental activities of daily living (IADL) [10] and it presupposes both the use of cognitive and physical capacities [10]. For this reason, performing tasks in a Virtual Environment (VE) reproducing a supermarket scenario is a good opportunity to provide post-stroke patients with an effective cognitive and physical training in a familiar and controlled environment. A virtual supermarket scenario is employed in several studies to improve specific deficits in post stroke patients as, for instance, in [7, 13].

Rand et al. [7] designed an intervention focused on improving multitasking while participating in a virtual shopping task. 4 post-stroke patients received ten 60-min sessions over a 3 weeks training in a Virtual Mall. Subjects had to plan and activate multitasking and problem solving skills while performing an everyday shopping task. All subjects made fewer errors during the latest training sessions, supporting the hypothesis that the V-Mall was an effective tool for the rehabilitation of post-stroke multitasking impairments during the performance of daily tasks.

Farìa et al. [4] include a virtual supermarket in a much larger virtual space - Reh@City, with the aim of exercising memory, attention, visuospatial abilities and executive functions in post stroke patients. Results show significant improvements in global cognitive functioning when training in Virtual Reality compared to conventional therapy. Lee et al. [13] built a rehabilitation tool for shopping IADL; despite the promising results, the use of the joystick was problematic for patients with hemiplegia, because perceived as unstable and heavy.

This article describes the Virtual Supermarket –VS-, an immersive environment initially designed to train visuospatial capabilities in MCI patients [14], and, in this stage, adapted and modified for the specific needs of post-stroke patients presenting reduced movement ability of the upper limbs. The interactions with the VR navigation and visualization hardware system, HTC VIVE, have been simplified by reducing the use of the controllers to only one and a single button to push during the tasks. This solution is regarded as greatly improving the usability of the VS by hemiplegic people or users with reduced mobility of the arms.

A pilot study is necessary to asses this VE on a sample of healthy subjects before using it with post-stroke patients. The goal of this evaluation is to confirm the results of the tests conducted with the first version of the VS over the acceptability of the VE in terms of collateral effects, and to evaluate the effectiveness of the simplification of the interactions and use of controls, as previously explained, on the general usability of the system, in such a way that a more engaging, pleasant and intuitive rehabilitation experience could be offered to post stroke patients in the near future.

2   Methods

2.1   Participants

Fifteen subjects (3 females and 12 males) recruited within the Italian National Research Council participated in this study. The inclusion criteria were age >18 years old and good general health status. Subjects with deficits preventing the correct visualization of the Virtual Environment or impeding the correct performance of the task were excluded from the pilot study.

Subjects with glasses or corrective lenses (11 out of 15) choose freely whether to use them or not in combination with the HMD (Head Mounted Display).

The subjects are asked about their scholastic level, if they have already used HMDs in the past and which specific model, if they have theoretical or practical knowledge about Virtual Reality and, since the test involved a shopping assignment, how often they perform this task in everyday life. These questions are asked in order to assess differences in the subjective experience of interaction with the system due to personal characteristics or previously acquired knowledge about VR.

Table 1 shows all these sample characteristics.

2.2   Equipment

The HTC Vive head mounted display is employed in this study both as visualization device to render the VE and to interact with VE [15]. The headset is the same employed in the previous study. Noises and audio feedbacks are also unchanged from the previous software version.

HTC VIVE is equipped with two infrared cameras operating as room tracking units, and two controllers allowing the interaction with objects inside of the virtual environment. This specific setup allows the user to move freely in a physical space of approximately 4 by 3 m, while being immersed in the virtual scene or play area. The Virtual Supermarket is developed with Unity 3D game engine [16] and shows two specific areas of the supermarket, being an aisle with different products (Fig. 1), and the cash desk payment area (Fig. 3).

The first VS scenario has been modified and/or simplified in order to use one HTC VIVE controller, allowing the potential patient with hemiplegia to use only one limb, to handle the interaction with the virtual environment.

Table 1. Characteristics of the sample

Age	32.87 ± 5.99
Sex (M/F)	12/3
Type of vision impairment	
• Myopia	4
• Astigmatism	1
• Squinting	1
• Myopia and Astigmatism	4
• No visual problems	5
Level of education	
• High School Diploma	3
• Bachelor Degree	4
• Master Degree	8
Past experience with HMD	
• No	5
• Yes	10
Kind of HMD used in the previous experience	
• Samsung Gear VR	2
• HTC Vive	1
• Others	1
• More than one type	6
• None	5
Familiarity with VR	
• None	5
• Enough	8
• Good	1
• Excellent	1
Frequency of grocery shopping	
• Never	0
• Occasionally (once or twice a month)	2
• Often but less than 50% of the days in a month	11
• More than 50% of the days in a month	1
• Everyday	1

In the first scene (Fig. 1) the list of products to be purchased has been attached to the cart positioned against one of the walls and not displayed above the second VIVE controller.

To perform the object grabbing action, the trigger button (Fig. 2) needs to be held pressed by the user while the controller avatar is colliding with the virtual object in the VE. The object grabbing action on the wrong product is not possible and returns an error sound.

The cash desk payment scene (Fig. 3) is loaded automatically after the shopping task is completed. This differs from the previous VS software version where the

Fig. 1. The aisle scene

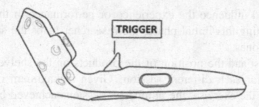

Fig. 2. The HTC Vive controller trigger button

payment task had to be started manually by the user through the interaction with a virtual interface, absent in this version. In the cash desk scene, the virtual scenario and the task to be performed are the same of the previous version of the VS. The same control as in the shopping scene is employed in this scene. The pressure on the trigger button activates a pointer enabling the user to highlight the desired banknotes and coins visualized on the cash monitor and select them by releasing the button.

After performing the payment, the exercise starts again with a new shopping list of eight products to be purchased.

The interaction and task performance data of each user, together with their errors, named in the same way as in the previous VS version ("Wrong item Error"-WE, "Drop Error"-DE, "Fall Error"-FE, "Payment Error"-PE) are counted and saved in a XML file.

2.3 Study Protocol

All subjects of the sample follow the same protocol. Before starting the exercise and similarly to our previous study, subjects complete an anamnestic questionnaire (see Methods Section). Moreover, users need to fill two other questionnaires with the aim of evaluating the tendency to get involved in the present experience and the personal attitude to motion sickness (see Measures Section). These two questionnaires were not included in the first study while in this case they are used to assess if any characteristics

Fig. 3. The cash desk payment scene

of the subject would influence the experience or performance in the virtual environment. After completing this initial phase, the researcher helps the subject to wear the headset and headphones.

The shopping list and the position of the products on the shelves are identical for every subject and for each different session. Given a maximum of 10 min to each subject to complete the exercise, the number of sessions achieved by each user is not fixed but depends on subjective performance and dexterity.

In order to compare the results between the two studies, at the end of the test, subjects are required to complete the same questionnaires employed in our previous work, evaluating the usability of the system, the physical side effects of being immersed in virtual reality, and the sense of presence experienced (see Measures Section).

As in the other study, spontaneous user comments are transcribed by the experimenter to evaluate if differences in the qualitative data appear between this version of VS and the previous one.

2.4   Measures

The user experience of using HTC Vive in the Virtual Supermarket application is evaluated taking into account mostly quantitative but also qualitative data.

Subjects complete the Motion Sickness Susceptibility Questionnaire - revised version (MMSQ-long) [17], an useful instrument in the prediction of the motion sickness caused by a variety of environments [18]. Moreover they fill out the Immersive Tendencies Questionnaire (ITQ-rev) [19], measuring the individual tendency to "get lost" in the present experience and that is correlated with the construct of "Sense of Presence" [19].

In order to compare qualitative data between the two studies, user's spontaneous comments and questions are transcribed by the researcher, as well as behaviors that indicated difficulties in using the system.

At the end of the task subjects complete the System Usability Scale (SUS) questionnaire [20], the Sense of Presence Inventory (ITC-SOPI) [21] and the simulator sickness questionnaire (SSQ) [22].

The information saved in the XML file for each subject in the shopping scene (WE, DE, FE) and the payment scene (PE) is treated as objective performance data.

2.5 Statistical Analysis

All the statistical analysis is performed with IBM SPSS v.25. The usability of this system is evaluated following the scoring indications in [20]. Due to the small sample, the nonparametric test of Kolmogorov-Smirnov is run to establish if SUS scores are generalizable to the population.

Sample characteristics and performance are analyzed using descriptive and frequency analyses.

Asymmetry and kurtosis are observed for all the evaluated variables to verify any variation from their normal distribution. Total MMSQ score is the only variable with an asymmetry higher than $|2|$; in this case, a log transformation is applied to the values.

Multiple linear regression is performed with the purpose of investigating whether usability, sense of presence and cyber-sickness would influence performance.

Simple correlation is performed to evaluate association between SUS scores and number of completed shopping lists and product payment tasks with the aim of establishing the relationship between usability and the other user experience variables, and between usability and personal features.

Similarly, a simple correlation is employed in establishing the relationships between all types of execution times, the number of completed shopping tasks and the different types of errors. In order to exclude the influence of external factors in the relationship between two variables, partial correlations are run whenever a significant simple correlation between more than two variables is highlighted by the analysis.

The time needed to take all the products and the time necessary to put them on the register's tape in each session is compared by within-subjects ANOVA. Post-hoc tests using Bonferroni correction are performed in case of significance. Assumptions for ANOVA and linear regression are verified before running the tests.

T-test are run to compare shared results among questionnaires related to this and to our previous study, in particular results of the SSQ, SOPI and SUS questionnaires.

Only the statistically significant results will be reported in the following section.

3 Results

3.1 Participants

The initial sample consists of 15 subjects, of which one was removed from the statistical analysis due to the not completed questionnaires.

All subjects declare to have computer expertise and 10 subjects out of 15 have already experienced VEs through HMDs. Everyone already experienced 3D computer generated images, even if with a very low frequency, between never (N = 10) and occasional use (N = 5).

More than half of the sample already experienced Virtual Reality and has enough knowledge about its working principles; this is probably due to the individuals' common working context.

Other characteristics of the sample are reported in Table 1.

3.2 Usability

The minimum sample required for usability studies is 5 subjects [20].

SUS scores ($M = 85.71 \pm 8.23$ out of 100) indicate excellent usability, as reported in [23]. The minimum score given is 70, which represents a good usability [24]. The Kolmogorov Smirnov test is not significant ($p > 0.2$), indicating that the result is generalizable to the population.

On average, each subject completed 4 and a half times the shopping sessions (min 3, max 6 times).

In a model that assumes UX variables (values reported in Table 2. Values in the brackets are related to the previous VS version) and the total time necessary to complete three sessions as dependent variables, usability does not affect performance.

Table 2. User experience raw scores: usability, sense of presence, side effects. Previous VS version scores in the brackets

	Mean	Std. Dev.	Min.	Max
Usability (SUS)	85.71 (81.56)	8.23 (7.19)	70 (70)	97.5 (92.5)
Spatial Presence (SOPI)	58.93 (64.25)	12.24 (7.2)	25 (56)	73 (79)
Realness (SOPI)	18.29 (20.75)	3.47 (2.12)	9 (18)	23 (25)
Engagement (SOPI)	48.86 (49.5)	10.1 (4.31)	33 (43)	63 (57)
Side Effects (SOPI)	9.71 (10)	4.03 (3.74)	6 (6)	19 (17)
	Median	Mode		
Cybersickness (SSQ)	5 (5)	5 (1)	0 (0)	15 (13)

Furthermore, usability is unrelated to any kind of number of errors recorded by the system during the shopping sessions.

Usability appears to be correlated with the engagement shown during the experience ($R = 0.57$, $p < 0.05$) only running simple correlation; this association disappears when removing the influence of the "spatial presence" variable (which is highly related to "engagement").

A positive correlation is found between ITQ total score and usability ($r = 0.56$, $p < 0.01$).

Execution Times. The average time required to complete the first 5 sessions is shown in Table 3. Values of the previous VS version are shown in the brackets.

Table 3. Comparison between the average time necessary to complete the first five sessions in seconds in this version and in the previous one (in the brackets)

Session	#1	#2	#3	#4	#5
Time	221.5	169.86	138.71	121.73	107
	(191)	(123.63)	(132.5)	(94.5)	(86.8)
Std. Dev.	68.44	57.2	40.82	29.04	30.58
	(44.79)	(38.43)	(32.92)	(21.79)	(51.69)
No.	14	14	14	11	9
	(8)	(8)	(8)	(8)	(5)

Execution times for each session are associated by simple correlation ($r = 0.72$, $p < 0.005$ between the first and the third session; $r = 0.63$, $p < 0.05$ between the second and the third session) and these associations remain when running partial correlation ($r = 0.79$, $p < 0.005$ for the first case; $r = 0.71$, $p < 0.01$ for the second).

The number of completed lists correlates inversely with the execution time of the first ($r = -0.69$; $p < 0.01$), the second ($r = -0.64$, $p < 0.05$) and the third session ($r = -0.87$, $p < 0.001$) just running simple correlations.

Repeated measures ANOVA is statistically significant, with $F1.22, 15.91 = 13.163$ and $p = 0.001$. The Bonferroni test confirms a significant difference between the first and the third sessions' execution time only ($p < 0.001$).

The number of completed lists correlates significantly with all the execution times of the supermarket aisle shopping scene of the first three sessions ($r = 0.61$, $p < 0.05$ for the first session, $r = -0.59$, $p < 0.05$ for the second session and $r = -0.86$, $p < 0.001$ for the third one) only with simple correlations. The time spent in the aisles shopping scenes in the first and third session correlates statistically, with $r = 0.63$ and $p < 0.05$ for simple correlation ($r = 0.72$, $p < 0.01$ controlling the effect of the second session); the same thing can be observed for execution times of the cash desk scenes ($r = 0.84$, $p < 0.001$ between the 1st and 3rd session running partial correlation; $r = 0.94$, $p < 0.001$ with simple correlation).

Repeated measures ANOVA for the three execution times in the first three "aisle scenes" is statistically significant ($F1.25, 16,2 = 10.43$, $p < 0.005$). The Bonferroni test confirms a significant difference between the first and the third execution time ($p < 0.001$) and between the second and the third execution time ($p < 0.05$).

Running the same statistical analysis with the time spent in the cash desk scene, ANOVA is statistically significant ($F_{2, 26} = 26.62$, $p < 0.001$) and the difference appears between the first and second session ($p < 0.005$), the first and the third session ($p < 0.001$) and the second and the third session ($p = 0.01$).

Comparing the execution times of the VS's previous version and this one, a substantial difference emerged in the fourth session ($t = 2.228$, $p < 0.05$) and the execution speed was higher in the old VS.

Errors. The average of errors made in the first four sessions are shown in Table 4. Values in the brackets are the errors made in the previous VS version.

In the first session, there is a significant correlation between WE and FE when eliminating the influence of DE ($r = 0.57$, $p < 0.05$). In the third session a significant partial correlation between FE and DE exists ($r = 0.67$, $p < 0.05$).

Table 4. Comparison between errors made in the first four sessions in the new and in the previous VS version (in the brackets)

Session	#1	#2	#3	#4
Wrong Item Error	1.36 ± 1.39 (0.00 ± 0.00)	0.36 ± 0.84 (0.88 ± 1.46)	0.93 ± 1.39 (0.38 ± 0.74)	0.50 ± 0.67 (0.5 ± 0.93)
Fall Error	0.5 ± 0.86 (0.88 ± 1.13)	0.43 ± 0.94 (0.38 ± 0.74)	0.36 ± 0.63 (0.13 ± 0.35)	0.33 ± 0.89 (0.5 ± 0.93)
Drop Error	0.5 ± 0.76 (1.38 ± 1.3)	0.93 ± 1 (0.38 ± 0.744)	0.36 ± 0.63 (1.63 ± 1.77)	0.42 ± 0.67 (3.38 ± 8.37)

Comparing the gathered data of this version of VS with the previous one, WE of the first session were significantly less in the old version of the system ($t = 2.738$, $p < 0.05$), while DE made in the third session are less in the new one ($t = 2.46$, $p < 0.05$).

3.3 Personal Characteristics

None of the answers to the anamnestic questionnaire's queries were related to the usability of the system.

Results concerning the tendency of being involved in the present experience and motion sickness, are shown in Table 5.

Table 5. Immersive tendencies questionnaires and motion sickness susceptibility questionnaires

	Min.	Max.	Mean	Std. Dev.
ITQ focus	32	53	41.79	±6.75
ITQ involvement	18	46	31.14	±8.54
ITQ game	1	13	7.07	±3.47
ITQ total score	58	110	80	±16.31
MMSQ total score	0	126.72	31	±34.33

The total ITQ scores are significantly correlated with three of the subscales of the ITC-SOPI questionnaire in particular Spatial Presence ($r = 0.56$, $p < 0.05$), Engagement ($r = 0.6$, $p < 0.05$) and Realness ($r = 0.64$, $p < 0.05$) with simple correlations only. It is possible to find significant correlations between the subscale of SOPI "Engagement" and the subscales of ITQ "Involvement" ($r = 0.77$, $p < 0.01$) and "Game" ($r = -0.64$, $p < 0.05$).

A significant association is found between SUS scores and the subscale "Involvement", with r = 0.7 and p < 0.05, partializing the effect of the other two subscales of the Immersive Tendencies Questionnaire.

3.4 User Comments and Observations

Spontaneous comments and behaviors manifested during the interaction with the Virtual Environment have been categorized into three groups based on common characteristics: "problems encountered", "emotion" and "realism".

Problems Encountered. Most problems encountered are related to the quality of vision, despite the SSQ subscale "Oculomotor symptoms" does not receive high scores.

An individual performed the whole exercise holding the HMD with one hand, declaring that he could not visualize the environment properly. Another person complained about the environment being blurred, "as if a mesh would stand between me and the scene". Two other subjects complained about the products being blurred.

One person complained about being annoyed by the HTC Vive cable connection; this distracted the user from the VE.

One person suffered from disorientation and vertigo when passing from the aisle shopping environment to the cash desk scene and back.

Emotion. Both positive and negative comments about the subject's emotional state emerged during the test.

For instance, one person stated that the exercise is not motivating enough: the subject would have preferred a preset number of product lists and the opportunity to challenge himself in completing all the shopping sessions as fast as possible. Another user explicitly asked for more salient stimuli, suggesting to add something that would scare people while performing the exercise.

Among the reported positive observations, one person commented that the exercise was fun because he could behave differently from reality. Another subject laughed while positioning all the products in the cart in an opposite way he would have done it in real life. Likewise, another person made intentional errors throwing products on the floor and laughed about the unexpected consequences of such behavior, namely the error sounds and the fact that the objects did not break as expected. One subject decided to stand in the position occupied by the shopping cart in the VE, enjoying the sensation of overlapping reality and virtual objects.

Again, exclamations such as "it's beautiful!", "what a cool supermarket!" and exultant behaviors at the end of each shopping session are observed in three subjects.

Realism. Most of the comments are related to the realism of the virtual environment. Users reported two different categories of comments, those related to graphic realism, and those related to the differences between object behavior in reality and in the VE.

Regarding the first category, there are observations about some of products being unproportioned, in particular fruits.

The lack of details in the some of the 3D models was sometimes criticized by the users. For example, the slice of Parmesan cheese was often mistaken for a slice of cake. One user suggested to include expiration dates and prices on the shelves to increase the realism of the scene.

While the comments about the graphic realism are all negative, those related to the behavior of the objects in the virtual environment arouse amazement and amusement and are related to emotions. One user appreciated the very low final cost of the purchased products and claimed that he would have liked to shop in that supermarket more often. Another person was intrigued by the "strange feeling" of a virtual object on the floor being in the same position as its feet while not producing any physical feedback.

Twelve people commented about objects not breaking into pieces when falling on the floor; one user declared that it was better that way as he would have made a mess in real life. One subject laughed about not being able to kick an item that fell to the floor.

Finally, one person stated that performing the whole task with one hand was very strange, while another user reported that employing the controllers instead of touching the products elicited a strange, almost unpleasant feeling.

4 Discussion

The system obtains an "excellent" usability, with an average score greater than 85 and with a minimum score 70, corresponding to a good usability [24]. The standard deviation obtained in this evaluation is higher than the one found in our previous study, nonetheless this remains much lower than the standard deviation reported in other studies [25]. This fact confirms that the sample evaluates the system as definitely usable.

Some observations can be made about the objective measures.

First of all, the number of completed lists is lower than in the previous study. This is due both to the shorter time given to the participants to complete the task (10 min instead of 15), and to the fact that the products need to be taken in a specific order, while in the previous study there was no restriction about that.

The new system appears to be more usable than the previous one, although not significantly. Moreover, usability does not explain the time spent in completing the task, nor when taking into account a whole session, neither when considering the two scenes separately (aisle scene and cash desk scene). These results differed from those of other studies, including our previous one, evidencing an association between performance and usability of the system [25, 26]. Furthermore, unlike what found in [27], errors are not associated with perceived usability; this result also emerged in our previous study. This could be due to the fact that the sample, decided to spend some time to explore the virtual environment while enjoying the unusual experience. Likewise, the lack of association between usability and errors can be explained by observing some of the users' behaviors such as dropping the products on purpose to experience the systems feedbacks.

The usability of the system is associated with the involvement experienced, although this relationship disappears when removing the influence of spatial presence. This result confirms the fact that usability is associated with a greater sense of presence in a broad sense, as reported in [28, 29], although without correlating to one of its specific aspect. This relationship did not emerge in our previous study, probably because of the very small sample. Furthermore, the usability of the system is correlated with the tendency of being deeply involved in the experience. This can be explained by the fact that, as addressed in several studies, subjects experiencing more emotions during a usability test tend to find less system usability problems [30]. With regard to the sense of presence experienced by the users, our previous research shown scores that were - even if not statistically significant - higher than in this version of VS. This may be due to the fact that the interaction has been simplified causing a reduction of the challenge, and, consequently, a reduction of the level of engagement [31]. Another cause could be the use of a single controller differing from what users are experiencing in real life, when using both arms to perform such a task.

On the other hand, unlike what is described in other studies [32], usability is not associated with cyber sickness, a result emerged in our previous study too. This could be due, again, to the small sample: further studies are necessary to better investigate this discrepancy.

Referring back to performance, it is noticeable that the first session lasted longer than the others, and significantly more time than the third one. This is probably due to the users' tendency, during the first session, to spend some time in exploring the VE without feeling the need of completing the task as quickly as possible. This behavior confirmed the findings of previous studies, including the one adopting a similar version of the virtual supermarket designed for elderlies with MCI [14, 20]. Furthermore, task execution speed was higher in the first VS version.

Results confirmed that the more a person tends to immerse himself in an experience, the more sense of presence he would feel during a task. This is also reported in the study by Witmer and Singer [19].

Unlike in our previous study, none of the users reported problems with the controllers. The already mentioned changes of HTC Vive controller buttons to be employed during the tests are thus positive as they do facilitate the overall interaction with the VE. This will probably facilitate the interaction of future patients with cognitive problems as in the case of post-stroke patients [13].

Since many users complained about blurring problems and difficulties in recognizing the shopping items, these aspects will need to be addressed in a future version of the virtual environment, especially when dealing with patients with cognitive difficulties. Moreover objects out of focus distract people from the experience in the VE and decrease the motivation during the task [33].

The fact that most of the comments addressed realism confirmed how important this aspect is for the general involvement in the exercise [34]; in addition, a lot of comments related to the emotion, as often reported in studies employing virtual reality environments [35]. This fact draws attentions on the importance of creating entertaining scenarios that would stimulate positive emotions in the user, as proposed in studies aiming at the improvement of users functionalities [36].

Despite many suggestions addressing the potential improvements to the application, the overall opinion about the virtual supermarket was positive as users found the scenario and the task realistic, enjoyable and fun: these are considered all essential features in a future enhancement training for post stroke patients [5].

5 Conclusions and Future Works

This article describes the validation of the virtual supermarket application designed for the rehabilitation of post stroke patients, with the aim of assessing its usability and the impact on some users' human factors.

Results show that this virtual environment has almost an "excellent" usability, according to Brooke [24] and that this positive experience increases the sense of presence. This specific results suggest that the application would enhance patients' motivation to exercise.

The modification of the interaction controls (HTC VIVE Controller buttons), planned after the validation of the first version of the Virtual Supermarket, is effective, as no subject in the current study failed to operate the controllers. In our previous work two different interaction procedures confused the sample, generating frustration.

Nonetheless, the presented study has some limitations.

The sample of 15 subjects, although larger than in our previous work, is not sufficient to establish a model that would explain which aspect of the experience influences the performance. It is likely that variables that are not fully evaluated in this research, such as the graphic quality of the environment [37] and the experience of using the device [38], should be included in a future analysis. These variables are taken into account but they are not investigated in depth in this work.

The presence of the HTC Vive pc connection cable is not perceived as problematic by the subjects in this sample. However, it seems essential to remove it in the rehabilitation of patients with physical disorders, especially to avoid problems mentioned in other studies, such as [39]. In this regard, it has to considered that the new versions of Head Mounted Display have wireless connection.

The absence of cyber sickness problems during the tests is a positive finding of this study. However, shopping items have to be in focus and the sensation of unpleasantness caused by the blurred vision must be avoided; this is not possible with the HTC Vive used in this experimental action because of specific of the device, but this problem should be overcome by using the new generation of HMD those have a higher resolution. It is necessary to improve the overall graphic quality of the virtual environment before employing the application on patients. This is especially true in post-stroke patients frequently suffering from vision-related disorders [40]. For the same reason the color and shape of some products must be adapted to best represent the real object.

Another aspect to be taken into account in the future developments of the application is the integration, in the virtual environment, of the user's arm and hand avatar. Several studies report that seeing your own body in the virtual environment increases the sense of presence and, consequently, performance and motivation [41].

References

1. The top 10 causes of death. https://www.who.int/en/news-room/fact-sheets/detail/the-top-10-causes-of-death. Accessed 14 Feb 2019
2. Mayo, N.E., Wood-Dauphinee, S., Ahmed, S., et al.: Disablement following stroke. Disabil. Rehabil. **21**, 258–268 (1999). https://doi.org/10.1080/096382899297684

3. Kwakkel, G., Kollen, B.J., Krebs, H.I.: Effects of robot-assisted therapy on upper limb recovery after stroke: a systematic review. Neurorehabil. Neural Repair 22, 111–121 (2008). https://doi.org/10.1177/1545968307305457

4. Faria, A.L., Andrade, A., Soares, L., i Badia, S.B.: Benefits of virtual reality based cognitive rehabilitation through simulated activities of daily living: a randomized controlled trial with stroke patients. J. Neuroeng. Rehabil. 13, 96 (2016). https://doi.org/10.1186/s12984-016-0204-z

5. Levin, M.F., Weiss, P.L., Keshner, E.A.: Emergence of virtual reality as a tool for upper limb rehabilitation: incorporation of motor control and motor learning principles. Phys. Ther. 95, 415–425 (2015). https://doi.org/10.2522/ptj.20130579

6. Cho, K.H., Lee, K.J., Song, C.H.: Virtual-reality balance training with a video-game system improves dynamic balance in chronic stroke patients. Tohoku J. Exp. Med. 228, 69–74 (2012)

7. Rand, D., Weiss, P.L.T., Katz, N.: Training multitasking in a virtual supermarket: a novel intervention after stroke. Am. J. Occup. Ther. 63, 535–542 (2009)

8. Laver, K., George, S., Ratcliffe, J., Crotty, M.: Virtual reality stroke rehabilitation - hype or hope? Aust. Occup. Ther. J. 58, 215–219 (2011). https://doi.org/10.1111/j.1440-1630.2010.00897.x

9. Zhang, L., Abreu, B.C., Masel, B., et al.: Virtual reality in the assessment of selected cognitive function after brain injury. Am. J. Phys. Med. Rehabil. 80, 597–604 (2001)

10. Rand, D., Katz, N., (Tamar) Weiss, P.L.: Evaluation of virtual shopping in the VMall: comparison of post-stroke participants to healthy control groups. Disabil. Rehabil. 29, 1710–1719 (2007). https://doi.org/10.1080/09638280601107450

11. Edmans, J.A., Gladman, J.R.F., Cobb, S., et al.: Validity of a virtual environment for stroke rehabilitation. Stroke 37, 2770–2775 (2006). https://doi.org/10.1161/01.STR.0000245133.50935.65

12. Jannink, M.J.A., Erren-Wolters, C.V., de Kort, A.C., van der Kooij, H.: An electric scooter simulation program for training the driving skills of stroke patients with mobility problems: a pilot study. Cyberpsychol. Behav. 11, 751–754 (2008). https://doi.org/10.1089/cpb.2007.0271

13. Lee, J.H., Ku, J., Cho, W., et al.: A virtual reality system for the assessment and rehabilitation of the activities of daily living. Cyberpsychol. Behav. 6, 383–388 (2003). https://doi.org/10.1089/109493103322278763

14. Mondellini, M., Arlati, S., Pizzagalli, S., et al.: Assessment of the usability of an immersive virtual supermarket for the cognitive rehabilitation of elderly patients: a pilot study on young adults. In: 2018 IEEE 6th International Conference on Serious Games and Applications for Health (SeGAH), pp. 1–8. IEEE (2018)

15. VIVE™—VIVEVirtual Reality System. https://www.vive.com/us/product/vive-virtual-reality-system/. Accessed 14 Feb 2019

16. Unity 3D. https://unity3d.com/. Accessed 14 Feb 2019

17. Golding, J.F.: Motion sickness susceptibility questionnaire revised and its relationship to other forms of sickness. Brain Res. Bull. 47, 507–516 (1998)

18. Kennedy, R.S., Fowlkes, J.E., Berbaum, K.S., Lilienthal, M.G.: Use of a motion sickness history questionnaire for prediction of simulator sickness. Aviat. Space Environ. Med. 63, 588–593 (1992)

19. Witmer, B.G., Singer, M.J.: Measuring presence in virtual environments: a presence questionnaire. Presence Teleoperators Virtual Environ. 7, 225–240 (1998). https://doi.org/10.1162/105474698565686

20. Brooke, J.: SUS-a quick and dirty usability scale. Usability Eval. Ind. 189(194), 4–7 (1996)

21. Lessiter, J., Freeman, J., Keogh, E., Davidoff, J.: A cross-media presence questionnaire: the ITC-sense of presence inventory. Presence Teleoperators Virtual Environ. **10**, 282–297 (2001). https://doi.org/10.1162/105474601300343612

22. Kennedy, R.S., Lane, N.E., Berbaum, K.S., Lilienthal, M.G.: Simulator sickness questionnaire: an enhanced method for quantifying simulator sickness. Int. J. Aviat. Psychol. **3**, 203–220 (1993). https://doi.org/10.1207/s15327108ijap0303_3

23. Measuring Usability with the System Usability Scale (SUS). https://www.userfocus.co.uk/articles/measuring-usability-with-the-SUS.html. Accessed 14 Feb 2019

24. Brooke, J.: SUS: a retrospective (2013)

25. Bangor, A., Kortum, P.T., Miller, J.T.: An empirical evaluation of the system usability scale. Int. J. Hum. Comput. Interact. **24**, 574–594 (2008). https://doi.org/10.1080/104473108 02205776

26. Sonderegger, A., Sauer, J.: The influence of design aesthetics in usability testing: effects on user performance and perceived usability. Appl. Ergon. **41**, 403–410 (2010). https://doi.org/10.1016/J.APERGO.2009.09.002

27. Nielsen, J., Levy, J.: Measuring usability: preference vs. performance. Commun. ACM **37**, 66–75 (1994). https://doi.org/10.1145/175276.175282

28. Yoon, S.-Y., Laffey, J., Oh, H.: Understanding usability and user experience of web-based 3D graphics technology. Int. J. Hum. Comput. Interact. **24**, 288–306 (2008). https://doi.org/10.1080/10447310801920516

29. Brade, J., Lorenz, M., Busch, M., et al.: Being there again – Presence in real and virtual environments and its relation to usability and user experience using a mobile navigation task. Int. J. Hum Comput Stud. **101**, 76–87 (2017). https://doi.org/10.1016/j.ijhcs.2017.01.004

30. Norman, D.: Emotion & design: attractive things work better. Interactions **9**, 36–42 (2002). https://doi.org/10.1145/543434.543435

31. Ermi, L., Mäyrä, F.: Changing Views: Worlds in Play Fundamental Components of the Gameplay Experience: Analysing Immersion (2005)

32. Mousavi, M., Jen, Y.H., Musa, S.N.B.: A review on cybersickness and usability in virtual environments. Adv. Eng. Forum **10**, 34–39 (2013). https://doi.org/10.4028/www.scientific.net/AEF.10.34

33. Westerink, J., et al.: Motivation in home fitnessing: effects of immersion and movement. In: Jacko, Julie A. (ed.) HCI 2007. LNCS, vol. 4553, pp. 544–548. Springer, Heidelberg (2007). https://doi.org/10.1007/978-3-540-73111-5_62

34. Finkelstein, S., Suma, E.A., Lipps, Z., et al.: Astrojumper: motivating exercise with an immersive virtual reality exergame. Presence Teleoperators Virtual Environ. **20**, 78–92 (2011). https://doi.org/10.1162/pres_a_00036

35. Riva, G., Mantovani, F., Samantha Capideville, C., et al.: Affective interactions using virtual reality: the link between presence and emotions. Cyberpsychol. Behav. **10**, 45–56 (2007). https://doi.org/10.1089/cpb.2006.9993

36. Herrero, R., García-Palacios, A., Castilla, D., et al.: Virtual reality for the induction of positive emotions in the treatment of fibromyalgia: a pilot study over acceptability, satisfaction, and the effect of virtual reality on mood. Cyberpsychol. Behav. Soc. Netw. **17**, 379–384 (2014). https://doi.org/10.1089/cyber.2014.0052

37. Zimmons, P., Panter, A.: The influence of rendering quality on presence and task performance in a virtual environment. In: 2003 Proceedings of the IEEE Virtual Reality, pp. 293–294. IEEE Computer Society (2003)

38. Rosenberg, J., Grantcharov, T.P., Bardram, L., Funch-Jensen, P.: Impact of hand dominance, gender, and experience with computer games on performance in virtual reality laparoscopy. Surg. Endosc. **17**, 1082–1085 (2003). https://doi.org/10.1007/s00464-002-9176-0

39. Coldham, G., Cook, D.M.: VR usability from elderly cohorts: preparatory challenges in overcoming technology rejection. In: 2017 National Information Technology Conference (NITC), pp. 131–135. IEEE (2017)
40. Jones, S.A., Shinton, R.A.: Improving outcome in stroke patients with visual problems. Age Ageing **35**, 560–565 (2006). https://doi.org/10.1093/ageing/afl074
41. Regenbrecht, H., Schubert, T.: Real and illusory interactions enhance presence in virtual environments. Presence: Teleoperators Virtual Environ. **11**(4), 425–434 (2002)

Upper Limb Rehabilitation with Virtual Environments

Gustavo Caiza[1] , Cinthya Calapaqui[2], Fabricio Regalado[2], Lenin F. Saltos[2],
Carlos A. Garcia[2] , and Marcelo V. Garcia[2,3(✉)]

[1] Universidad Politecnica Salesiana, UPS, 170146 Quito, Ecuador
`gcaiza@ups.edu.ec`
[2] Universidad Tecnica de Ambato, UTA, 180103 Ambato, Ecuador
`{ccalapaqui1120,aregalado7193,lf.saltos,ca.garcia,mv.garcia}@uta.edu.ec`
[3] University of Basque Country, UPV/EHU, 48013 Bilbao, Spain
`mgarcia294@ehu.eus`

Abstract. In this article an application is developed based on 3D environments for the upper limbs rehabilitation, with the aim of performing the measurement of rehabilitation movements that the patient makes. A robotic glove is used for virtualized the movements with the hand. The hand movements are sent to a mathematical processing software which runs an algorithm to determine if the rehabilitation movement is right. Through virtual reality environments, the injured patients see the correct way to perform the movement and also shows the movements that the patient makes with the robotic glove prototype. This system allows to evaluate the protocol of upper limbs rehabilitation, with the continuous use of this system the injured patient can see how his condition evolves after performing several times the proposed virtual tasks.

Keywords: Robotic glove · Virtual reality · Virtual rehabilitation ·
Dynamic temporary alignment algorithm (DTW)

1 Introduction

The development of society service technology has been growing at a rapid pace. This statement is possible given the fact that day by day new devices and equipment are created aiming to make tasks execution easier. Whether they are industrial or for assistance. Through the incorporation of different disciplines such as mechanics, electronics, computing, artificial intelligence, control engineering, physics and others. It has been achieved to converge on a common goal by extending the application field of robotics due to the fusion with knowledge areas mentioned [1].

As a result, robotic mechanisms take place and are dedicated to support daily activities of a professional. Whether is on the working space, medical applications, informative tasks or others [1]. Thus, robotics is becoming a main axis for different areas. Being medicine one of the fields that could embrace applications involving any kind of mechanisms. Through the integration of sensors,

L. T. De Paolis and P. Bourdot (Eds.): AVR 2019, LNCS 11613, pp. 330–343, 2019.
https://doi.org/10.1007/978-3-030-25965-5_24

control algorithms and user interfaces it has been developed new systems dedicated to the interaction with patients of diverse pathologies by providing an extensive control to the treating doctor on the execution of surgical techniques or assisting rehabilitation sessions [2].

Robotics in health can support different areas such as telemedicine, visual assistance, basic interaction with patients, rehabilitation, among others. When it comes to telemedicine, it is pretended to use robots, to execute surgical tasks remotely. Hence the treating surgeon can have a robot that commands a slave robot to perform risky operations or actions that need a person expert on a specific area [3].

Additionally, robots can provide visual assistance to guide low qualified persons to perform first aid chores, transport medicine to not much transited spaces, or interact in a basic way with patients in medical centers. On the other hand, rehabilitation tasks are being complementing with prototypes that make interaction with much more sophisticated scenarios easier. One clear example is this is the use of gloves, lenses and helmets that allow the immersion on simulated environments. All this with the purpose of giving recuperation environments different to the traditional ones.

Virtual reality (VR) is a fusion between robotics the use of reality created by computer can broaden application fields and improve rehabilitation techniques. Indeed, several researches have demonstrated the effectiveness in the use of this kind of alternatives. Because the surgeons can modify exercises as they want to and get to know the real advance on the recuperation of an injured member [4].

Specifically, in the recovering of hands diverse techniques had been used. Whether it is for providing feedback or to measure signs generated by them. In that sense, it can be established two kinds of measures. Identified as fine motor and gross motor gestures. Gross motor gestures make reference to the measurement of notorious movements such as closing or opening the hand completely. Unlike them, fine motor gestures make reference to the measurement of precise gestures like open, between open or close a finger. In this perspective it is necessary to determinate the level of rehabilitation that the user can accomplish given the fact that commonly rehabilitation sessions are divided in phases and depends on the level of affectation that the user presents [4].

The present work proposes the use of an electronic glove capable of recognizing gestures executed by the hand of a user. Through wireless communication, the robotic glove is equipped with resistive sensors that emit the information of developed gestures, in order to measure the fulfillment of tasks previously raised. To facilitate rehabilitation tasks, the work proposes the inclusion of environments developed in virtual reality, where the extremity put to the test is visualized, which modifies the simulated environment every time you execute a gesture of gross motor skills: fist or open your hand.

This article is divided into 6 sections including the introduction. In Sect. 2 the related works are detailed. In Sect. 3 the state of technology is analyzed. The proposed system Architecture is presented in Sect. 4. The analysis of the results of this investigation states in Sect. 5. Finally, Sect. 6 develops the conclusions and future work.

2   Related Works

By approaching the section of mechanisms to support the rehabilitation it can be appreciated two large groups such as invasive systems (exoskeletons, prosthesis) or non-invasive (social assistance systems) [5]. Inside the group of invasive assistance, it has been developed investigations like a hand robot with a soft tendon routing system [6]. Here it is presented an Exo-Glove as a robotic portable device with a soft texture in the shape of a glove which is destined to the rehabilitation of hand grip functioning. Through the soft tendon routing system and an adaptive mechanisms of sub-action the robot generates strength so the patient can be able to move the fingers and allow him to grab objects of different sizes.

Following the same line, it can be found the robotic rehabilitation of the hand [7]. Here, a glove named Gloreha is used in the rehabilitation of the hand for patients that have suffered a stroke. The system is made up of an actuator and an elastic load transistor. Gloreha pretends to integrate moves with force and speed to one or more injured fingers with high adjustment capacity and reliability.

On the other hand, non-invasive robotic systems used in therapy sessions are referred to mechanisms that do not have direct contact with the patient. These systems are developed using new and innovative techniques aiming to convert tedious rehabilitation sessions into motivational games. These games allow the patient to keep focus on the session given the high level of interaction as it completes its rehabilitation for upper extremities affected by ictus, neuromuscular illnesses, neurodegenerative, nervous system disorders, among others, allow the patient to stay focused in the session given the high level of interaction.

One of the techniques used on non-invasive prototypes is having interaction systems human- robot. These systems provide virtual reality scenarios or augmented reality [8]. At this point, in [9] implements a human- robot interaction system (HRI) for the treatment of autism problems. These kind of systems are directed by a robotic partner that interacts with the patient in an exercises routine with the aim of grabbing the attention of the patient and teach simple coordinated behaviors. Most of these systems use humanoids to replicate the patient's gestures and moves of the upper and lower extremities.

In addition, a virtual rehabilitation system [10] is used to treat hand injures typical of a post-stroke. This system incorporates virtual scenarios manipulated by a glove from P5 game and an Xbox. However, these kind of proposals interact with the patient at a shallow level by not measuring signs given for the hand reactions but its upper and lower extremities.

By pointing out publications developed for the measure of signals given by the gestures of the hand it can be mentioned several publications made in the last 5 years. For example, in [11] the proposal of a glove based on tendons controlled by a single motor and providing 8 degrees of freedom. Despite the studies of the dynamic effects of the glove, one of the disadvantages is that there is not a movement feedback generated by the patient. Instead of this, the glove forces the patient's fingers to complete daily and common moves.

Additionally, [12] it presents the design, implementation and experimental validation of a haptic portable glove. The robot includes little DC engines for the forces feedback which can be connected to a computer to know the level of gross motor of the patient's hand. The main goal and constraint of this publication is that it is possible to assist a person with the complete grip of objects but it is limited to not being able to execute neutral moves (half open hand). On the same line, it is presented a robotic glove that measures the hand movements through electromyography sensors. Despite the fact that the proposal is very flexible and can be used by any type of person. It is restricted by the kind of signals coming from primary elements which can give gross motor gestures measurements back [13].

The work proposes a novel rehabilitation method that includes the processing of signals generated by resistive force sensors, wireless communication with a computer and the use of detected gestures to modify virtual reality scenarios. In order to interpret the structure of the proposed system more adequately, processing of the glove signals, communication with libraries of information reception and interaction with environments of virtual reality oriented to rehabilitation.

3   State of Technology

3.1   Rehabilitation of Upper Extremities

The different disabilities that occur in people are of different kinds, they may be acquired or you may have been born with a disability. In the upper extremities you can have injuries in what corresponds to arms or wrists, for this type of disparities or injuries a person performs rehabilitation with the purpose of better mobility of their limbs [9]. The rehabilitation of the hands corresponds to exercises that a person must perform in such a way that the exercises are of help to the motor of the person, there are several ways that a person follow instructions to perform an exercise determined by the doctor, nowadays there are applications in which the person can interact to perform a task that is requested by the game or application [8].

3.2   Upper Extremity Conditions

Although various types of conditions have been documented as diseases of the upper extremities, the most common found in rehabilitation are stroke, neuromuscular and neurodegenerative diseases. The term stroke refers to a set of disorders, including cerebral infarction, cerebral hemorrhage and subarachnoid hemorrhage. The worst scenarios resulting from stroke can be an accident or stroke and stroke. On the other hand, hereditary neuromuscular diseases refer to heterogeneous disorders detected at an early age.

Depending on the level, they can be treated with adequate rehabilitation. Finally, neurodegenerative diseases are characterized by the progressive and unstoppable loss of neurons responsible for motor skills [14,15].

These types of diseases can be treated by means of dedicated therapy, where methods such as virtual reality, augmented reality and robotics have been considered to complement traditional rehabilitation methods.

3.3   Virtual Reality

Virtual reality (VR) is the concept that has the reality developed by computer, and may have different advantages and applications of entertainment, space visualization and rehabilitation are the most presented in research. In the medical rehabilitation environment, VR has been playing an important role in the variation of scenarios, where a person can interact with completely immersive and appropriate environments for limb recovery. In addition to visually motivating the patient, virtual reality can stimulate the reaction of limbs in patients, therefore, the natural reaction of a patient can accelerate the recovery process of an affected member.

However, virtual reality is complemented by input devices that allow modifying the simulated environment. For rehabilitation purposes, this type of device is usually electronic gloves or robotic arms that detect the gestures generated by the user's hand.

3.4   Rehabilitation Engineering

The main objective of rehabilitation engineering is to offer assistance aimed at promoting the inclusion of people with special abilities in educational, social and commercial areas. Through rehabilitation engineering, various areas of knowledge have been merged for the development, design, testing and evaluation of technological solutions designed for people who have suffered from diseases related to freedom of motor skills. In this way, people suffering from motor neuron affections, brain traumas, strokes and other diseases can have technology as an ally to complement basic activities such as mobility, listening, speech, vision and recognition [16].

The main benefits of the rehabilitation engineering are: execution of exercises in a precise way, reduction of the physical load to the traffickers, incorporation of new methods based on virtual reality, augmented reality and gathering of relevant information for the optimization of therapy sessions [17].

3.5   Dynamic Temporary Alignment Algorithm (DTW)

The temporal alignment algorithm is one of the most used techniques to find matches between two signals, considering that you have a reference signal and an input signal with which you want to contrast. The main advantages of the DTW include that it can align two signals with a number of different samples, yield a coincidence value between both signals tested and determine a path that visually indicates the similarity between signals. Although the main applications include vector pairing (for speech recognition), the DTW can be used for matching applications between signals, in this case, the recognition of gestures of the upper extremity of a user [18,19].

Originally, DTW was conceived as an alignment algorithm developed for speech recognition. The goal of the algorithm is to align two vector sequences with similar characteristics but displaced in time and with a different number of samples (see Fig. 1). In this way, be $A = (a_1, a_2...a_N)$ of length N and $B = (b_1, b_2...b_M)$ of length M, both $N, M \in \gamma$. The first requirement of the classical DTW is that both signals are discrete, or at least, that they have characteristics with equidistant points in time, that is, they advance in a similar time sampling, a feature space denoted by γ in this work. So, $a_n, b_m \in \gamma$ by $n \in [1 : N]$ and $m \in [1 : M]$. In order to compare the characteristics between the signals tested, it is required to know the local distance measure $x(x, y)$. In this way, if $x(x, y)$ is small, the similarity between the input signals A and the standard signal B is high, while that $x(x, y)$ is big, the compared signals are not similar.

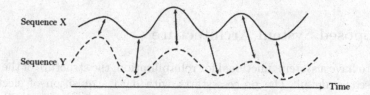

Fig. 1. DTW signal comparison.

To know the path that determines the total weight of the DTW, a cost matrix is generated $C \in c_{NxM}$ defined by $C(n, m) := c(x_n, y_m)$. This matrix locates the input vector on the abscissa axis, while the reference vector is located on the axis of the ordinates.

To obtain the warping path denoted as $WP(N, M) = wp(p_1, p_2, ..., p_L)$, with $p_K = (n_K, m_K)$ and for $K \in [1 : L]$, the boundary condition, monotonicity condition, the step size condition, and the warping window condition must be met. The boundary condition stablish that the begin in the lower left corner and ends in the upper right corner, i.e., $p_1 = (1, 1)$, while $p_L = (N, M)$; the monotonicity condition refers to the fact that path does not go back, so, the abscissa and ordinate indexes should be kept increasing, never decreasing, i.e. $n_1 \leq n_2 \leq ... \leq n_L$, and $m_1 \leq m_2 \leq ... \leq m_L$; the step size condition stablish that the path must go one step at a time, i.e., $p_{K+1} - p_K \in \{(1, 0), (0, 1), (1, 1)\}$ for $K \in [1 : L - 1]$. Finally, the warping window condition ensures that an acceptable path must be very close to the diagonal of the weight matrix $C(n, m)$.

When fulfilling all these requirements, the DTW allows to generate both the path that indicates the alignment, as well as the total weight between the existing relationship between both signals.

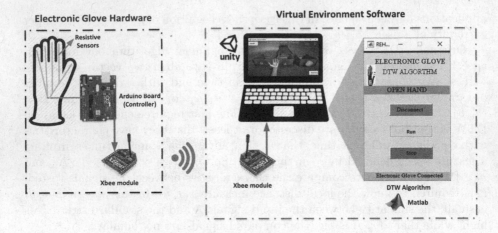

Fig. 2. Proposed architecture

4 Proposed System Architecture

In order to have a system functional for rehabilitation, the structure of the system is proposed in detail. The process starts with data acquisition of electronic's glove using resistive gauges connected an Arduino UNO^{TM} board. This data is transmitted to a virtual environment made in Unity 3D^{TM} software using a xbee^{TM} 802.15.4 protocol module. The virtual environment show to the injured patient different rehabilitation movements, for example: fist, open hand and the extension of the index and thumb. A DTW algorithm determine the correct execution of the tasks given by the system, this work proposes the comparison between the input signals and signals stored as reference standards. In Fig. 2 a diagram of how the system is presented.

4.1 Electronic Glove

The electronic glove is designed with the aim of measuring the mobility of each finger, that is, it has resistive sensors incorporated in each finger, the sensors change their resistance according to the deflection to which they are subjected by the movement of the extremities of the hand.

The variable resistance that the sensors generate are conditioned by means of a voltage divider, it is transform the movement of each finger to voltage values, the voltages are registered into a Arduino Uno^{TM} board through analog ports, to be processed and filtered in order for sending wirelessly using xbee protocol. An interface is implemented in the software to start the data acquisition system of the exercises performed by the patient with the electronic glove, in Fig. 3 the elements that make up the electronic glove are shown.

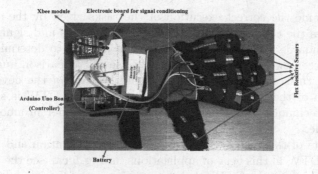

Fig. 3. Details of the electronic glove.

4.2 Data Processing

The reception functions are responsible for capturing the information of the glove to identify the type of movement executed within the MATLAB™ software, these functions belong to a serial communication that is connected through the Xbee™ modules. MATLAB™ is responsible for classifying these movements and recognizing patterns that will serve to modify the behavior of virtual reality scenarios. The libraries have the ability to unlock the information sent by the robotic glove, so that the computer and the developed device can communicate in real time. As mentioned, the signals that the glove sends are interpreted by MATLAB™, in which an algorithm is developed to detect what type of movement the patient is performing with the robotic glove, in Fig. 4 shows the interface developed for rehabilitation system.

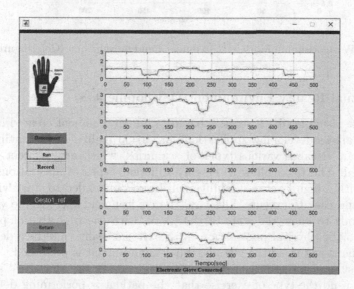

Fig. 4. Matlab™ interface.

To determine the correct execution of the tasks given by the system, this work proposes the comparison between the input signals and signals stored as reference standards. As previously explained, a method to determine the relationship between two signals is through the use of comparison and alignment strategies, in this case, DTW. Knowing that the time of the development of the execution of the tasks can vary between users, the DTW is shown as an appropriate technique to compare two patterns that differ in amount of data and amplitude.

The results of this work corroborate the correct execution and the applicability of the DTW in this type of applications, in Fig. 5 can see the detection of gestures by means of the DTW, the gesture of a closed fist is made. As we can see, there are three signals, the red is reference signal, a signal executed with a correct movement of blue color, while the alignment vectors show magenta color is the reference and the gesture measured is color measured in black, these two is very similar, that is, it has a very large similarity so the weight value of the detection will be within a certain range, the algorithm is used for each gesture that has stored comparing each weight that the algorithm delivers

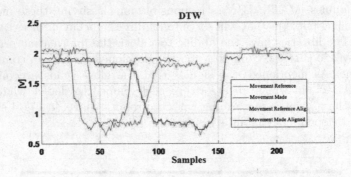

Fig. 5. DTW algorithm in Matlab™ to detect hand movements. (Color figure onlilne)

4.3   Virtual Reality Environment Development

Finally, the visual feedback stage is based on an environment created in a software that allows developing applications in virtual reality. In the environment, the user can visualize various types of scenarios, which are different tasks to be executed. The tasks allow to classify different types of affectations, where patients with a high level of disability will be recommended to start with basic exercises and easy to execute. Likewise, the environment will include visualization methods to know the state of the glove, with the objective of providing feedback to the physical information in the virtual reality interface developed. In Fig. 6 the structure of the virtual reality environment is presented.

The first part is the measurement of gestures using the Matlab™ software, which will send the type of exercise that the patient is performing determined by means of the DTW algorithm. The resulting gesture is sent to Unity™

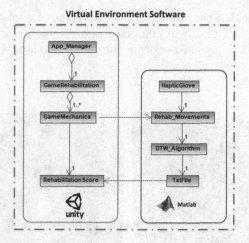

Fig. 6. VR Environment Architecture.

3D software by means of a .txt file which will be continuously read to know the executed movement, the game scenario will execute the animation corresponding to the gesture that the patient performs to complete the tasks programs for the selected game.

System Functionality. The functionality of the developed system is experimented with two applications in virtual reality. The graphical engine used to develop the virtual scenarios is Unity™ 3D, where the instructions to carry out the exercise are displayed in each of the interfaces.

The first scenario shows an environment that motivates the user to execute three types of gestures: fist, open hand and the extension of the index and thumb. As seen in Fig. 7(a), the first game contains a counting method to determine the amount of elements that have been generated by the environment, so that the application will advance only if the expected gesture is generated. Considering the use of the alignment algorithm, the execution time of the gesture can be variant, taking into account that users exposed to rehabilitation may have problems executing the expected movements.

In Fig. 7(b) shows another gesture made by the person, in this case the game asks to make fist to classify the violet object in the correct place.

The second game, unlike the first game that requires the execution of various gestures, the second application is focused on measuring the user's reaction through the generation of two basic gestures. The location of the hand in the form of a pistol and the other making a basic gesture of an "OK" is to move the different characters that show the environment depending on the signal they carry, allowing to measure the reaction time of the user. Figure 7(c) shows the execution of the second application, denoting the applicability of the system in various cases of rehabilitation.

The person executes the movement in the form of a pistol which consists of closing three fingers that will eliminate one zombie at a time, in Fig. 7(d) it is observed how the person performs the movement and Unity interprets it correctly.

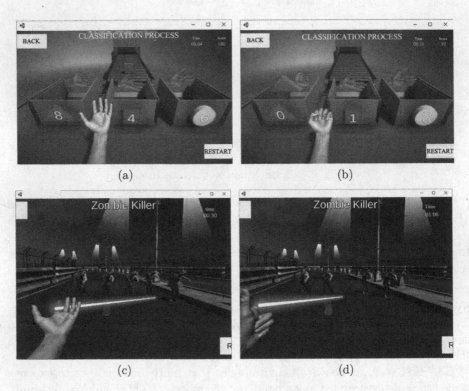

Fig. 7. VR interface. (a) "Palm open" movement to classify the object red. (b) "Fist" movement to classify the object violet. (c) "Palm open" movement in rehabilitation game 2. (d) "thumb up" movement in rehabilitation game 2 (Color figure online)

5   Discussion of Research Findings

The results of performing a correct movement in this case finger 1 is observed in Fig. 8, as the DTW algorithm aligns the two signals, that is, the reference and the new feedback at the time of executing the exercise, as a result it will have a low gain weight, which represents a correct execution of the movement. The red line represents the stored reference signal, the blue signal is the one executed by the person, the magenta signal represents the reference signal, while the black color is the movement signal executed, it can be seen that both signals are almost alienated, showing correct execution as a result.

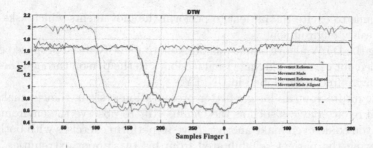

Fig. 8. DTW of signals right movement. (Color figure onlilne)

In Fig. 9 It can be seen how the action is incorrectly executed, the vectors are not aligned, this sample belongs to the thumb, the result of this will give us a very high weight value indicating that the two signals are very Far from being similar (magenta vs black).

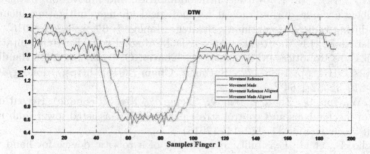

Fig. 9. DTW of different signals. (Color figure onlilne)

6   Conclusion and Ongoing Works

By means of the implementation of a Robotic Glove, it was possible to detect movements produced by a person by hand, these movements were evaluated by the DTW algorithm that allows us to identify if the proposed exercise is performed correctly, the application developed in Unity provides visual feedback to the person, managing to know if it is performing the movement indicated on the screen. With the simultaneous use of this system you can develop motor skills for the hands of people who suffer from affections in the fingers.

With the developed application it is allowed to rehabilitate a person, executing movements continuously, to a future you can see improvement in the capacity of the person when he makes a movement with his hand. The built-in electronic glove captures the movements of each finger of a person this makes the system is not uncomfortable for a patient, resulting in an application that is at the same time a game to excite the patient to perform the movements which asks for the virtual reality environment.

Using the DTW algorithm, it was possible to identify if the proposed exercise was carried out correctly, giving visual feedback to the person through the virtual reality environment developed in UNITY.

In the future you can have a better presentation system, that is, making use of virtual reality devices such as the Oculus Rift to observe the environment in 3D, and you could even make another glove to perform exercises with both hands and allowing the patient rehabilitated from his two upper extremities, adding more complex exercises.

Acknowledgment. This work was financed in part by Universidad Tecnica de Ambato (UTA) and their Research and Development Department under project CONIN-P-0167-2017.

References

1. Roy, R., Sarkar, M.: Knowledge, firm boundaries, and innovation: mitigating the incumbent's curse during radical technological change: mitigating incumbent's curse during radical discontinuity. Strateg. Manag. J. **37**, 835–854 (2016)
2. Van der Loos, H.F.M., Reinkensmeyer, D.J., Guglielmelli, E.: Rehabilitation and Health Care Robotics. In: Siciliano, B., Khatib, O. (eds.) Springer Handbook of Robotics. SHB, pp. 1685–1728. Springer, Cham (2016). https://doi.org/10.1007/978-3-319-32552-1_64
3. Meng, W., Liu, Q., Zhou, Z., Ai, Q., Sheng, B., Xie, S.: (Shane): Recent development of mechanisms and control strategies for robot-assisted lower limb rehabilitation. Mechatronics **31**, 132–145 (2015)
4. Vanoglio, F., et al.: Feasibility and efficacy of a robotic device for hand rehabilitation in hemiplegic stroke patients: a randomized pilot controlled study. Clin. Rehabil. **31**, 351–360 (2017)
5. Chiri, A., Vitiello, N., Giovacchini, F., Roccella, S., Vecchi, F., Carrozza, M.C.: Mechatronic design and characterization of the index finger module of a hand exoskeleton for post-stroke rehabilitation. IEEE/ASME Trans. Mechatron. **17**, 884–894 (2012)
6. In, H., Kang, B.B., Sin, M., Cho, K.-J.: Exo-Glove: a wearable robot for the hand with a soft tendon routing system. IEEE Robot. Autom. Mag. **22**, 97–105 (2015)
7. Borboni, A., Mor, M., Faglia, R.: Gloreha—hand robotic rehabilitation: design, mechanical model, and experiments. J. Dyn. Syst. Meas. Control **138**, 111003 (2016)
8. Mourtzis, D., Zogopoulos, V., Vlachou, E.: Augmented reality application to support remote maintenance as a service in the robotics industry. Procedia CIRP **63**, 46–51 (2017)
9. Zorcec, T., Robins, B., Dautenhahn, K.: Getting engaged: assisted play with a humanoid robot Kaspar for children with severe autism. In: Kalajdziski, S., Ackovska, N. (eds.) ICT 2018. CCIS, vol. 940, pp. 198–207. Springer, Cham (2018). https://doi.org/10.1007/978-3-030-00825-3_17
10. Seo, N.J., Arun Kumar, J., Hur, P., Crocher, V., Motawar, B., Lakshminarayanan, K.: Usability evaluation of low-cost virtual reality hand and arm rehabilitation games. J. Rehabil. Res. Dev. **53**, 321–334 (2016)

11. Xiloyannis, M., Cappello, L., Khanh, D.B., Yen, S.-C., Masia, L.: Modelling and design of a synergy-based actuator for a tendon-driven soft robotic glove. In: 2016 6th IEEE International Conference on Biomedical Robotics and Biomechatronics (BioRob), pp. 1213–1219. IEEE, Singapore (2016)

12. Ben-Tzvi, P., Ma, Z.: Sensing and force-feedback exoskeleton (SAFE) robotic glove. IEEE Trans. Neural Syst. Rehabil. Eng. **23**, 992–1002 (2015)

13. Polygerinos, P., Galloway, K.C., Sanan, S., Herman, M., Walsh, C.J.: EMG controlled soft robotic glove for assistance during activities of daily living. In: 2015 IEEE International Conference on Rehabilitation Robotics (ICORR), pp. 55–60. IEEE, Singapore (2015)

14. Guindo, J., Martínez-Ruiz, M.D., Gusi, G., Punti, J., Bermúdez, P., Martínez-Rubio, A.: Métodos diagnósticos de la enfermedad arterial periférica. Importancia del índice tobillo-brazo como técnica de criba. Revista Española de Cardiología **09**, 11–17 (2009)

15. Chen, D., Liu, H., Ren, Z.: Application of wearable device HTC VIVE in upper limb rehabilitation training. In: 2018 2nd IEEE Advanced Information Management, Communicates, Electronic and Automation Control Conference (IMCEC), pp. 1460–1464. IEEE, Xi'an (2018)

16. Li, Q., Wang, D., Du, Z., Sun, L.: A novel rehabilitation system for upper limbs. In: 2005 27th Annual Conference on IEEE Engineering in Medicine and Biology, pp. 6840–6843. IEEE, Shanghai (2005)

17. Duarte, E., et al.: Rehabilitación del ictus: modelo asistencial. Rehabilitación **44**, 60–68 (2010)

18. Shen, J., Bao, S.-D., Yang, L.-C., Li, Y.: The PLR-DTW method for ECG based biometric identification. In: 2011 Annual International Conference of the IEEE Engineering in Medicine and Biology Society, pp. 5248–5251. IEEE, Boston (2011)

19. Piyush Shanker, A., Rajagopalan, A.N.: Off-line signature verification using DTW. Pattern Recogn. Lett. **28**, 1407–1414 (2007)

Proof of Concept: VR Rehabilitation Game for People with Shoulder Disorders

Rosanna Maria Viglialoro[1(✉)] 📷, Giuseppe Turini[1,2(✉)] 📷,
Sara Condino[1,3] 📷, Vincenzo Ferrari[1,3] 📷, and Marco Gesi[4,5]

[1] EndoCAS Center, Department of Translational Research and of New Surgical
and Medical Technologies, University of Pisa, Pisa, Italy
{rosanna.viglialoro,sara.condino,
vincenzo.ferrari}@endocas.unipi.it
[2] Computer Science Department, Kettering University, Flint, MI, USA
gturini@kettering.edu
[3] Department of Information Engineering, University of Pisa, Pisa, Italy
[4] Department of Translational Research and of New Surgical and Medical
Technologies, University of Pisa, Pisa, Italy
marco.gesi@med.unipi.it
[5] Center for Rehabilitative Medicine "Sport and Anatomy",
University of Pisa, Pisa, Italy

Abstract. Shoulder pain is very common in adult population (16%–26%) and it has several impacts on the Activities of Daily Living (ADLs). In this context, an effective rehabilitation is needed to assist the patients to regain autonomy and improve ADLs.

An alternative approach to monotonous traditional rehabilitation can be offered by Virtual Reality (VR) technology and video game console systems.

In this paper, we present a proof of concept of a computer game for shoulder rehabilitation. The system consists in a VR application developed with the Unity game engine, and it is currently controlled via mouse interactions; however, the software application has been designed for the Nintendo Wii Remote Motion-Plus as the main game controller, but this tool will be integrated in the next development phase.

Our goal is to develop a rehabilitation game: suitable to be used both at home and at a hospital, highly motivating for the patient, and requiring low-cost technology (e.g. a consumer-grade tracking system).

The game has been tested preliminary with three healthy subjects who agreed that the application is engaging, motivating and intuitive.

Keywords: Virtual rehabilitation · Shoulder disorders · Nintendo Wii Remote MotionPlus

1 Introduction

Shoulder pain is very common in adult population with an estimated self-reported prevalence between 16% and 26% [1, 2]. It can be due to local pathologies (e.g. rotator cuff and glenohumeral disorders) but it can be also referred from the neck, from

L. T. De Paolis and P. Bourdot (Eds.): AVR 2019, LNCS 11613, pp. 344–350, 2019.
https://doi.org/10.1007/978-3-030-25965-5_25

abdominal pathologies affecting the diaphragm, liver or other viscera, from polymyalgia rheumatica, alterations in the deep fascia [3, 4] and from malignancy [2]. Regardless of the specific pathology, the shoulder pain has several impacts on Activities of Daily Living (ADLs) (such as eating, holding, wearing etc.) and thus, on the quality of life of patients.

Effective rehabilitation treatments are needed to assist the patients to regain autonomy and improve ADL.

Traditional rehabilitation techniques rely on the involvement of a therapist for each patient and on repeated exercises that are usually carried out with external devices (the nine-hole pegboards, Rolyan Double Curved Shoulder Arc etc.). However, these devices lack digital monitoring which imposes to the therapist to control and monitor the patient during each rehabilitation session. As the consequence, there are enormous costs of manpower and of material resources. Another limitation of traditional methods is the boredom, due to the repetitiveness of exercises, that reduces patient's motivation and greatly influences the effects of rehabilitation.

Virtual Reality (VR) technology is an alternative approach to traditional rehabilitation. VR provides an immersive experience for the user who is within the virtual environment having the ability to interact thanks to the combination of PC and sensor technology [5]. The benefits of rehabilitation based on VR are multiple.

VR offers the possibility to customize the treatments necessary for the patient to maximize success and outcomes. VR can create different game-like scenarios that increase the patient's motivation and thus the chances of fast recovery. In addition, another benefit which contributes to motivate and to encourage the patient during the long-term training session, is the integration of visual and audio rewards, such as gratifying messages which are displayed in real time.

An additional advantage is the possibility to repeat the learning trials and to gradually increase the complexity of the exercises.

Furthermore, VR systems provide objective measures of user's performance which are accessible by the therapist and viewable by the patient.

Other key benefits are the telerehabilitation and the remote data access. The VR system can be used also outside rehabilitation centers, such as patient's home. Thus, the portability is a very important requirement, especially for patients who live in rural and remote areas [6].

The user can interact with the VR environment using input devices, e.g. mouse, computer keyboard and touchscreen or using joystick, haptic interfaces and motion tracking systems.

Recent works have focused on the use of videogames and motion tracking devices to track patient's movements and to correct the motion during the rehabilitation exercises [7]. In particular, several research groups have been exploring the use of video game consoles system like Nintendo Wii® (Nintendo of America Inc., Redmond, WA) which is affordable and readily available in stores and inexpensive., as tool for rehabilitation in different clinical settings [8–10].

In this paper, we present a proof of concept of a computer game system for shoulder rehabilitation. The system is based on a VR-ready application developed with the Unity game engine and it is currently controlled via mouse interactions; however, the software application has been designed for the Nintendo Wii Remote MotionPlus as the main game controller, but this tool will be integrated in the next development phase (Fig. 1).

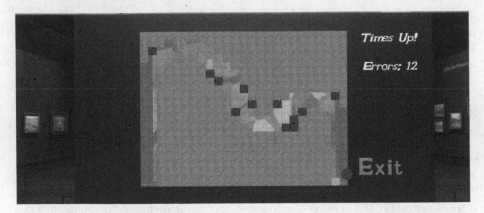

Fig. 1. Overview of game. The user, using the game controller (mouse or Wii Remote Plus) moves the blue circle inside the trajectory. As the circle moves along the trajectory, the "touched" tiles will change their transparency uncovering the painting they are hiding. Once the trajectory is completed, the grid of tiles is destroyed, and the painting is completely shown. Overall, this training exercise improves the shoulder ROM. (Color figure online)

2   Materials and Methods

The purpose of the proposed system is to improve the shoulder Range of Motion (ROM). The Unity game engine (version 5.6.1f) was used to implement the software application. The following paragraphs describe the gamified task and the system architecture.

2.1   Rehabilitation Game: The Game of Painting Discovery

The game of "painting discovery" is played by using either left or right arm, the one affected by shoulder disorders. The aim of the game is to move a blue circle from one end to another following the implemented trajectory. The trajectory consists of number of marked tiles which are located on a grid of tiles covering a painting (Fig. 3). As the circle cursor moves forward within the trajectory, the "touched" tiles will change their transparency to show what is under them. If the circle cursor moves outside the trajectory, the tiles hit will be marked as errors (with a cross symbol), and a sound will trigger to highlight the mistake. A visual counter shows the number of errors on the upper right corner of the game environment together with a countdown timer.

The physiotherapist can set the maximum game time, thus adjusting the difficulty of the game. And once the trajectory is completed, a gratifying message "Congratulations!" will be displayed on the screen, the grid of tiles will be destroyed, and the painting will be shown (Fig. 4).

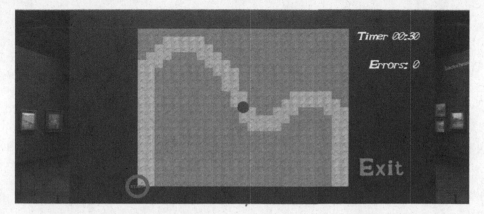

Fig. 2. Screenshot of the video game environment (Color figure online)

On the other hand, if the trajectory is not completed during the game time, the game will stop, and the message "Time's Up!" will be displayed. The implemented trajectory is known as Rolyan Double Curved Shoulder Arc that is used in traditional rehabilitation in order to reach the complete ROM of the shoulder.

Fig. 3. Screenshot of painting discovered.

2.2 System Architecture

The system architecture consists of software and hardware modules (Fig. 2). The hardware modules include: a computer with the game software preloaded, the computer audio speakers to provide the patient audio feedback whenever a mistake is made, a monitor (or a wall projector) to display the game and a mouse (or Wii Remote MotionPlus) as game controller. The software is composed of several modules (C# scripts). The Game Manager script is the heart of the system. It runs the game logic, and together the Grid Manager script, it manages the following actions:

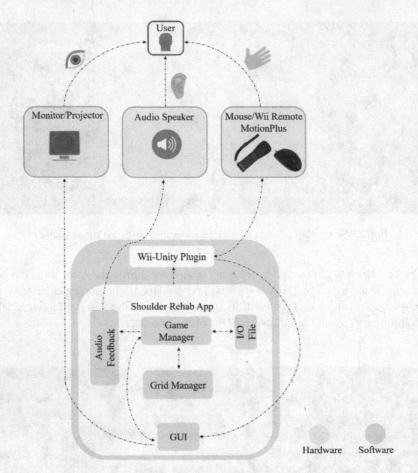

Fig. 4. Scheme of the shoulder rehabilitation application for the Mouse/Wii Remote MotionPlus, illustrating all the interactions between the hardware and software modules (dotted lines).

- loading from file of the game trajectory data (I/O File module),
- providing audio and visual feedback to highlight the mistakes, and
- handling the interactions between virtual environment and game interface.

The mainly elements of the virtual environment are a grid of tiles with a trajectory of marked tiles and a cursor (blue circle) which is moved by using the main game controller (mouse or Wii). The Game Manager is also responsible for game-logic tasks:

- the updating of the countdown timer,
- the handling of the winning/losing conditions,
- the management of the score (number of errors).

Additionally, the Grid Manager script is the module responsible for the creation of the grid of tiles and the handling of all tile-cursor interactions. Finally, the Graphical User Interface (GUI) module provides controls to allow the user to play and customize the game.

Specifically, the interactions tile-cursor (blue circle) are implemented exploiting Unity collision-detection (i.e. trigger colliders) and managed by the Grid Manager script. Trigger events raised by tile-cursor interactions are handled: to mark tiles, to account for mistakes, to play sounds, and to check the completion of the trajectory (winning condition). All others GUI-cursor interactions are implemented through ray-casting and handled using Unity UI events.

3 Preliminary Results

The computer game system was tested on a laptop running Microsoft Windows 10 Pro (UX303UB by ASUS, provided with an Intel Core i5-6200U CPU @ 2.30 GHz core processor, an 8 GB RAM and with a graphics card NVIDIA GeForce 940M) by three healthy subjects. The update frequency was ranging from 70 to 76 fps, whit the physics engine running at 50 fps (default).

A qualitative evaluation was carried out by three healthy subjects to evaluate both the game intuitiveness and game ability to entertain and motivate the user. All subjects have been described the game as fun, motivating and intuitive.

4 Conclusion and Feature Works

In this paper, we present a proof of concept of a computer game system for shoulder rehabilitation. The aim is to develop a rehabilitation system, which is suitable to be used both at home and at hospital, with highly motivating game and low-cost tracking system.

Currently, we defined the overall architecture of framework, implemented the game to be controlled via mouse interactions. Preliminary results are very positive. All subjects have been described the game as fun, motivating and intuitive.

The next development phase will integrate the Wii Remote MotionPlus in the system. Future studies will also focus on the integrating VR headsets, projectors, and even touch screens.

In addition, studies of effectiveness and usefulness will be performed to determine if the use of the game as a training tool improves the ROM shoulder.

Acknowledgment. The research leading to these results has been partially supported by HORIZON 2020 Project VOSTARS (H2020 Call ICT-29-2016 G.A. 731974).

References

1. Mitchell, C., Adebajo, A., Hay, E., Carr, A.: Shoulder pain: diagnosis and management in primary care. BMJ (Clin. Res. Ed.) **331**, 1124–1128 (2005)
2. Linaker, C.H., Walker-Bone, K.: Shoulder disorders and occupation. Best Pract. Res. Clin. Rheumatol. **29**, 405–423 (2015)
3. Natale, G., Condino, S., Soldani, P., Fornai, F., Belmonte, M.M., Gesi, M.: Natale et. al.'s response to Stecco's fascial nomenclature editorial. J. Bodyw. Mov. Ther. **18**, 588–590 (2014)

4. Natale, G., Condino, S., Stecco, A., Soldani, P., Belmonte, M.M., Gesi, M.: Is the cervical fascia an anatomical proteus? Surg. Radiol. Anat. **37**, 1119–1127 (2015)
5. Weiss, P.L., Katz, N.: The potential of virtual reality for rehabilitation. J. Rehabil. Res. Dev. **41**, vii–x (2004)
6. Ines, D.L., Abdelkader, G., Hocine, N.: Mixed reality serious games for post-stroke rehabilitation. In: 2011 5th International Conference on Pervasive Computing Technologies for Healthcare (PervasiveHealth) and Workshops, pp. 530–537 (2011)
7. De Mauro, A.: Virtual reality based rehabilitation and game technology, vol. 727, pp. 341–342 (2011)
8. Lange, B., et al.: Development of an interactive rehabilitation game using the Nintendo Wii Fit Balance Board for people with neurological injuries (2010)
9. Marston, H., Greenlay, S., van Hoof, J.: Understanding the Nintendo Wii console in Long-term Care Facilities (2013)
10. Decker, J., Li, H., Losowyj, D., Prakash, V.: Wiihabilitation: rehabilitation of wrist flexion and extension using a wiimote-based game system (2009)

GY MEDIC v2: Quantification of Facial Asymmetry in Patients with Automated Bell's Palsy by AI

Gissela M. Guanoluisa$^{(\boxtimes)}$, Jimmy A. Pilatasig$^{(\boxtimes)}$,
Leonardo A. Flores$^{(\boxtimes)}$, and Víctor H. Andaluz$^{(\boxtimes)}$

Universidad de las Fuerzas Armadas ESPE, Sangolquí, Ecuador
{gmguanoluisa1, japilatasig2, laflores5,
vhandaluz1}@espe.edu.ec

Abstract. The quantification of Bell's palsy using virtual reality, haptic devices and other technologies has been the subject of numerous investigations in recent years, however those affected with facial palsy have not benefited from efficient diagnoses that would give the physician the confidence to deliver an efficient clinical picture. GYMEDIC in its initial version developed modules of analysis (House-Brackmann System) and rehabilitation (virtual environments), previous research that comes to be the key to implement new Nottingham Scale graduation systems manipulating interactive virtual environments in order to improve figures of the diagnosis of Bell's palsy. This system develops new features for the estimation of facial paralysis: (a) symmetry at rest, (b) symmetry at adjusted rest, (c) symmetry with voluntary facial movements, which allows the inclusion of new measurement scales to the system. In addition, as an interesting point to research, AI is implemented through a simple neuronal network, which allows to adjust figures in three areas: eyebrows, eyes and mouth and therefore improve the diagnosis of patients.

Keywords: Bell's palsy · Quantification · Kinect v2 · Virtual environment · AI

1 Introduction

Bell's palsy accounts for more than 70% of peripheral paralysis cases worldwide with an incidence in different regions, from 10 to 40 per 10,000 people [1]. Paralysis is caused by direct involvement of the seventh cranial nerve characterized by degeneration of the motor and sensory function of the facial nerve [2]. The facial nerve is the most injured of all the cranial pairs causing the neuromuscular disorder called facial paralysis. The injury to the facial nerve prevents the normal movement of different muscles such as: forehead, eyelids and mouth, as well as the expression of emotions [3]. The disorder with slight incidence in children under ten years old; and high in pregnant women, diabetics and people with a previous history of the disease, 67% of patients exhibit an excess of tears in the eyes, 52% present a postauricular pain, 34% have taste disorders and 14% can develop phonophobia, psychological and emotional problems [4–6].

The first step in diagnosis is to determine if facial weakness is due to a problem in the central nervous system or one in the peripheral nervous system [7]. It should be

© Springer Nature Switzerland AG 2019
L. T. De Paolis and P. Bourdot (Eds.): AVR 2019, LNCS 11613, pp. 351–361, 2019.
https://doi.org/10.1007/978-3-030-25965-5_26

based on the existence of peripheral facial paralysis with or without loss of taste of the previous 2/3 of the tongue or an altered secretion of the salivary glands, loss of movement in the eyelid, dysfunctional tearing and nasal obstruction [8].

Given the need to determine a quantitative, reliable prognosis and at the same time plan an adequate treatment according to the severity of the condition, since 1985 numerous classification systems to evaluate facial movements and asymmetry arose, among the most traditional are [9]: *(i) The House-Brackmann* scale, in 1983 was adopted as the North American Standard for the evaluation of facial paralysis values five facial expressions [10]. The assignment of the paralysis classification is from grade I (normal) to grade IV (no movement) ranges on clinical observation and subjective judgment [11]. While *(ii) The Burres-Fisch system* provides a linear measurement index on a continuous graduated scale. All methods quantify the distances between reference points, both at rest and during voluntary movement [4, 8, 12]; *(iii) The Sunnybrook facial classification system* in 2008 was considered a leading tool in the field of clinical use [13]; *(iv) Nottingham System*, the score is obtained in three different stages: the first consists of measuring several lengths (supraorbital [SO] to infraorbital [IO] and CL to M), the second stage evaluates the existence of synkinesias, spasms or contractures (A for absence and P for presence) and the last stage estimates the existence or not of an intermediate nerve deficit through three questions (crocodile tears, dry eye and dysgeusia; Y for yes and N for no) [14] *(v) The Neely-Cheung Rapid Rating System*, a computer-based system that calculates the degree of paralysis using classification methods of up to ten of the traditional standard classification systems [15].

Currently, the introduction of technology in analysis, quantification and rehabilitation processes has given positive results to the traditional paradigm used by physicians dedicated to treating Bell's palsy [16]. Interactive systems are being developed with the intention of capturing the participation and confidence of the user with respect to the specialist in charge of delivering a clinical picture evaluated in facial asymmetry. Virtual and augmented reality systems are applied to motivate the patient in his treatment and at the same time make the process repetitive and interesting in a controlled way [17]. The use of stereo vision cameras (Kinectv2) has made it possible to increase security in facial recognition systems, using stereo images in clinical applications with great success [4]. Interaction with such haptic devices generates feedback and achieves greater immersion of the patient into the virtual environment [18]. This feedback improves patient participation and engagement during motor exercises by activating the multisensory system [19].

With the use of virtual environments, quantification focuses on a semi-immersible world that attracts two important concepts: *(i) Interaction*, the patient is directly involved with the virtual environment in real time, *(ii) Immersion*, through hardware devices, the patient has the sensation of being physically in the virtual environment and feel the security of receiving an effective prognosis from the specialist [4].

Due to the large number of methods found that facilitate the physician to deliver a prognosis and quantification of effective Bell's palsy, it has been considered to integrate to GYMEDICv2 a new Bell's palsy rating system through an analysis module using the Microsoft Kinect sensor. In addition, it is intended to implement IA through a simple neuronal network that allows adjusting percentage figures in the diagnosis of three special areas: eyebrows, eyes and mouth. The research is ongoing and, as such, the proposed work is an additional contribution that allows the system to include more parameters and thus entrust the final diagnosis given to patients affected by Bell's palsy.

2 System Structure

GY MEDIC reveals a versatile design with advances in the area of usability and internal structure, changes aimed at medical specialists responsible for interpreting the data that the system yields, new methods for calculating symmetry, adjustments in mathematical formulas and error correction, are part of version v2. In addition, the current version of the system integrates IA in its operation, with the inclusion of a neural network is intended to adjust the final figures of facial paralysis making more reliable the prognosis that doctors deliver to their patients.

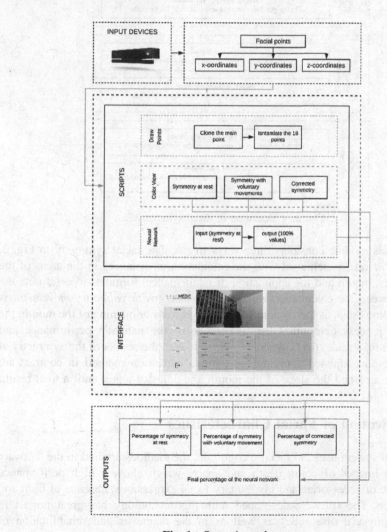

Fig. 1. Operation schema

The Fig. 1, presents the general scheme of operation of the medical system segmented into 4 parts: *(i) The entries*, which manage the acquisition of facial data; It is proposed to use the Kinect device for the detection of the patient's facial coordinates used in the diagnosis, and the recognition of gestures used in the rehabilitation, *(ii) The scripts*, contain several functions to clone and instantiate the points in the 3D plane visible to the users, *(iii) The interface*, contains new elements incorporating all the above-mentioned parts, *(iv) The outputs*, referring to all the values resulting from the execution of point *(iii)*, are displayed on the screen.

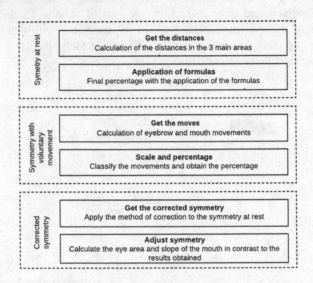

Fig. 2. Calculation of facial symmetry

The analysis module considers three ways to calculate facial symmetry in Fig. 2, *(a) Symmetry at rest*, is achieved with the calculation of distances in the areas of the eyebrows, eyes, mouth and the application of mathematical formulas to determine the difference between the calculated values, *(b) Symmetry with voluntary movement*, is determined actions such as the movement of the eyebrows or opening of the mouth, the mathematical process determines the percentage of the patient's performance and classifies them on a scale, *(c) Corrected symmetry*, with the results of the symmetry at rest the values are adjusted applying the gamma correction method in contrast are calculated eye area and the slope of the mouth and adjusted again until a final result.

3   Quantification of Facial Characteristics

The developed system uses 3D facial recognition, the methods applied to the software algorithm use images of color, depth and space, which allow a high performance against the use of accessories and involuntary facial expressions, absence of light and noise tolerance. The software developed with this technology has great impact in people with physical disabilities as well as in the diagnosis and rehabilitation of pathologies; their cost, performance and robustness make them very effective in the medical support [4, 20, 21].

The software solution incorporates an RGB panel provided by the *Color-FrameReader* attribute for the visualization of the patient's face; a 3D model super-imposed on the real image is generated from the 2D reference points mentioned [22]. Tracking the face in real time, obtaining coordinates and reducing errors when acquiring data implies the use of the following attributes: (a) *CameraSpacePoint* represents a 3D point in the camera space (in meters); (b) *DepthSpacePoint* represents pixel coordinates of a depth image inside the camera [4].

The tracking of the face is a process that allows the visualization and tracking of facial points superimposed on the image of the patient, offering a better interaction with the system in terms of movement, detection of features and diagnosis by a medical professional, the process is detailed in Fig. 3 [4].

Fig. 3. Tracking process

For the initial quantification of the pathology, sixteen reference points located in specific areas of the face detailed in Fig. 4 are captured, points extracted using the *Microsoft.Kinect.Face* library referring to the *HighDetailFacePoints* attribute that provides up to 1300 facial points.

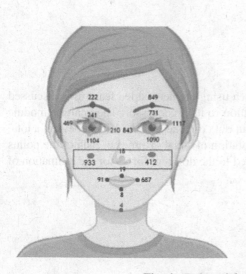

Coordinate	Detail
222	Middle of Eyebrows
849	Middle of Eyebrows
241	Eye Top
469	Eye Outer Corner
210	Eye Inner Corner
1104	Eye Bottom
731	Eye Top
843	Eye Inner Corner
1117	Eye Outer Corner
1090	Eye Bottom
18	Tip of the Nose
19	Mouth Top
91	Mouth Corner
687	Mouth Corner
8	Mouth Bottom
4	Chin Center
933	Rigth Cheeck
412	Left Cheeck

Fig. 4. Referential points of the face

The process of estimating the symmetry index includes calculating the distances between the reference points of the face (see Table 1) and their differences expressed in percentages; these variations decrease or increase according to the level of affection of the patient [4, 9].

Table 1. Distances

Name	Distance	Difference
Right eyebrow - nose	$d_1 = d_{222\_18}$	
Eyebrow left - nose	$d_2 = d_{849\_18}$	
Vertical right eye	$d_1 = d_{241\_1104}$	$\frac{d_1}{d_2}$
Vertical left eye	$d_2 = d_{731\_1090}$	
Left lip - nose	$d_1 = d_{91\_18}$	
Right lip - nose	$d_2 = d_{687\_18}$	

(i) Nottingham Grading System

It is a system for assigning the grade of a tumor for breast cancer. It is also known as the Elston-Ellis modification of the Scarff-Bloom-Richardson system. It is the system that is recommended to be used to determine the degree of Bell's palsy. It consists of assigning a score of 1 to 3 to three characteristics [22]. Once the score is assigned to each characteristic, all of these are added together. Resulting in a final score between 3 and 9. Once this is done, the grade of a tumor is obtained through the following criteria [23]:

- 3 to 5 points: grade 1 or well-differentiated tumor
- 6 to 7 points: grade 2 or moderately differentiated tumor
- 8 to 9 grade 3 points or undifferentiated tumor. Table 1 shows how to assign the score to each characteristic [23].

4 Experimental Tests

4.1 Virtual Environment

The performance of "GY MEDIC V2" when using the new added features is discussed below. Its operability remains even in relation to its predecessor. The analysis module in Fig. 5, starts with the reading of the input data obtained by the Kinect device, a total of eighteen points are assigned to the calculation of facial symmetry, each of the points acquires its 3D coordinates, and are assigned to the different forms for the estimation of facial paralysis.

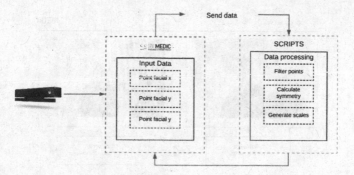

Fig. 5. Data processing

The results of the analysis are displayed on the screen, *(a) The main menu*, which contains some system options such as start the analysis, pause the analysis, finish the analysis, abort the system and navigation between modules described in Fig. 6.

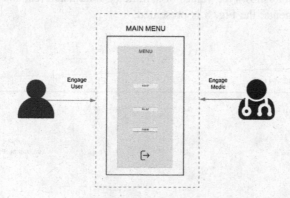

Fig. 6. System main menu

(b) The color view, displays a panel that allows to visualize the patient being analyzed, the superimposition function of the points on the face is maintained, the visual animation of the virtual points is generated in real time, in some situations this animation will be frozen or saturated if the user leaves the shot or the movement of his head is not central to the Kinect as shown in Fig. 7.

Fig. 7. Color view

(c) Results table, The process for calculating facial symmetry gives percentages of 100% for the ease of interpretation of the physician and user, the lower the values in each of the areas analyzed the greater movement dysfunction will exist in the area mentioned, the table shows the values of symmetry at adjusted rest and symmetry with voluntary movements, the Fig. 8 is presented.

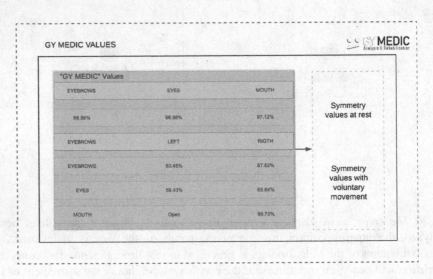

Fig. 8. Results table

(d) Neuronal Network, a simple neuronal network "perceptron" was implemented in the system in assistance to the final medical decision, the network uses as input data the values of symmetry at rest, the outputs are raised as one if the patient is affected and two if the patient does not have sufficient evidence to make a diagnosis, the model is described in the Fig. 9.

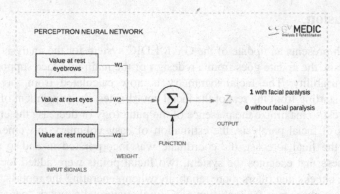

Fig. 9. Neuronal network – perceptron function

The Nottingham rating system was implemented that compares the functional side with the non-dysfunctional side of the face in contrast to the patient's voluntary movements. To estimate a result, samples from several participants were used; it is expected that the percentages will vary if any abnormality is detected both in the symmetry of the face and in its movements (Table 2).

Table 2. Results of the samples according to the Nottingham rating system

Subject 1			Right	Left	Subject 2			Right	Left
1	Rest	Distance 1	97.23	98.53	1	Rest	Distance 1	96.13	97.16
		Distance 2	96.74	97.12			Distance 2	80.91	88.99
2	Raise eyebrows		77.91	71.06	2	Raise eyebrows		60.91	74.76
3	Close eyes		92.82	95.17	3	Close eyes		93.22	94.91
4	Smile		92.3	94.53	4	Smile		89.81	91.92
5	Total		457	456.41	5	Total		420.98	447.74
6	Difference		99.871		6	Difference		94.023	
Subject 3			Right	Left	Subject 4			Right	Left
1	Rest	Distance 1	98.71	99.11	1	Rest	Distance 1	96.39	96.89
		Distance 2	89.91	90.01			Distance 2	95.91	96.15
2	Raise eyebrows		50.12	72.54	2	Raise eyebrows		88.54	89.65
3	Close eyes		67.09	72.46	3	Close eyes		89.19	90.54
4	Smile		79.53	88.31	4	Smile		90.06	93.64
5	Total		385.36	422.43	5	Total		460.09	466.87
6	Difference		91.225		6	Difference		98.548	
Subject 5			Right	Left	Subject 6	Right			Left
1	Rest	Distance 1	87.65	88.94	1	Rest	Distance 1	95.72	94.66
		Distance 2	88.85	84.64			Distance 2	96.45	97.09
2	Raise eyebrows		90.03	89.35	2	Raise eyebrows		80.65	81.06
3	Close eyes		45.67	78.83	3	Close eyes		93.04	94.26
4	Smile		65.79	67.75	4	Smile		77.22	79.43
5	Total		377.99	409.51	5	Total		443.08	446.5
6	Difference		92.303		6	Difference		99.234	

5   Conclusion

This research presents an update of the GY MEDIC system for the analysis module of facial paralysis, the update goes from a redesign of the main interface supported by the concept of usability; The facial symmetry is now calculated from three different approaches, starting from the rest of the facial coordinates, the adjustment of the values obtained at rest to maximize the presence of the pathology or decrease the error if there are no signs of facial paralysis, the estimation of easy voluntary movements in contribution to the final forecast, the coordinate was incorporated in and to the mathematical process that executes the system, two facial points were added located in the center of the cheeks that offers more reliability when generating the reports; One of the most important changes is the inclusion of a simple neuronal network "perceptron" that uses the initial values at rest and decides if the patient suffers or not from facial paralysis; this research continues in development and the addition or subtraction of characteristics in the following advances will depend on the functioning and the results obtained.

Acknowledgements. In addition, the authors would like to thanks to the Corporación Ecuatoriana para el Desarrollo de la Investigación y Academia–CEDIA for the financing given to research, development, and innovation, through the CEPRA projects, especially the project CEPRA-XI-2017-06; Control Coordinado Multi-operador aplicado a un robot Manipulador Aéreo; also to Universidad de las Fuerzas Armadas ESPE, Universidad Técnica de Ambato, Escuela Superior Politécnica de Chimborazo, Universidad Nacional de Chimborazo, and Grupo de Investigación ARSI, for the support to develop this paper.

References

1. Hussain, A., Nduka, C., Moth, P., Malhotra, R.: Bell's facial nerve palsy in pregnancy: a clinical review. J. Obstet. Gynaecol. **37**, 409–415 (2017). 3615
2. Pérez, E., et al.: Guía Clínica para rehabilitación del paciente con parálisis facial perférica. IMSS **5**, 425–436 (2004)
3. Rodríguez-ortiz, M.D., Mangas-martínez, S., Ortiz-reyes, M.G., Rosete-gil, H.S., Valeshidalgo, O., Hinojosa-gonzález, R.: Parálisis facial periférica. Tratamientos y consideraciones **16**, 148–155 (2011)
4. Guanoluisa, G.M., Pilatasig, J.A., Andaluz, V.H.: GY MEDIC: analysis and rehabilitation system for patients with facial paralysis. In: Seki, H., Nguyen, C.H., Huynh, V.-N., Inuiguchi, M. (eds.) IUKM 2019. LNCS, vol. 11471, pp. 63–75. Springer, Cham (2019). https://doi.org/10.1007/978-3-030-14815-7_6
5. Park, J.M., et al.: Effect of age and severity of facial palsy on taste thresholds in Bell's palsy patients. J. Audiol. Otol. **21**, 16–21 (2017)
6. Meléndez, A., Torres, A.: HOSPITAL GENERAL Perfil clínico y epidemiológico de la parálisis facial en el Centro de Rehabilitación y Educación Especial de Durango, México. Hosp. Gen. México **69**, 70–77 (2006)
7. Antonio, J., et al.: Parálisis de Bell: Algoritmo actual y revisión de la literatura. Revista Mexicana de Cirugía Bucal y Maxilofacial **7**, 68–75 (2011)

8. Cid Carro, R., Bonilla Huerta, E., Ramirez Cruz, F., Morales Caporal, R., Perez Corona, C.: Facial expression analysis with kinect for the diagnosis of paralysis using Nottingham grading system. IEEE Lat. Am. Trans. **14**, 3418–3426 (2016)
9. Gaber, A., Member, S., Taher, M.F., Wahed, M.A.: Quantifying facial paralysis using the kinect v2. In: Proceedings of the Annual International Conference of the IEEE Engineering in Medicine and Biology Society, EMBS, pp. 2497–2501 (2015)
10. Quevedo, W.X., et al.: Assistance system for rehabilitation and valuation of motor skills. In: De Paolis, L.T., Bourdot, P., Mongelli, A. (eds.) AVR 2017. LNCS, vol. 10325, pp. 166–174. Springer, Cham (2017). https://doi.org/10.1007/978-3-319-60928-7_14
11. House, J.W., Brackmann, D.E., Angeles, L.: Facial nerve grading system, pp. 146–147
12. Brenner, M.J., Neely, J.G.: Approaches to grading facial nerve function. In: Seminars in plastic surgery, vol. 18, pp. 13–22 (2004)
13. Kanerva, M.: Peripheral Facial Palsy. Grading, Etiology, and Melkersson-Rosenthal Syndrome (2008)
14. Devèze, A., Ambrun, A., Gratacap, M., Céruse, P., Dubreuil, C., Tringali, S.: Parálisis facial periférica. Colloids Surf. A Physicochem. Eng. Asp. **42**, 1–24 (2013)
15. Article, R.: Management of peripheral facial nerve palsy. Eur. Arch. Oto-Rhino-Laryngol. **265**, 743–752 (2008)
16. Da Gama, A., Fallavollita, P.: Motor rehabilitation using kinect: a systematic review. Games Health J. **4**, 123–135 (2015)
17. Valencia, U.P., De Arabia, S.: Virtual reality system for multiple sclerosis rehabilitation using KINECT, pp. 366–369 (2013)
18. Andaluz, V.H., et al.: Virtual reality integration with force feedback in upper limb rehabilitation. In: Bebis, G., et al. (eds.) ISVC 2016. LNCS, vol. 10073, pp. 259–268. Springer, Cham (2016). https://doi.org/10.1007/978-3-319-50832-0_25
19. Ortiz, J.S., et al.: Realism in audiovisual stimuli for phobias treatments through virtual environments. In: De Paolis, L.T., Bourdot, P., Mongelli, A. (eds.) AVR 2017. LNCS, vol. 10325, pp. 188–201. Springer, Cham (2017). https://doi.org/10.1007/978-3-319-60928-7_16
20. Carvajal, C.P., Proaño, L., Pérez, J.A., Pérez, S., Ortiz, J.S., Andaluz, V.H.: Robotic applications in virtual environments for children with autism. In: De Paolis, L.T., Bourdot, P., Mongelli, A. (eds.) AVR 2017. LNCS, vol. 10325, pp. 175–187. Springer, Cham (2017). https://doi.org/10.1007/978-3-319-60928-7_15
21. Arenas, Á.A., Cotacio, B.J., Isaza, E.S., Garcia, J.V., Morales, J.A., Marín, J.I.: Sistema de Reconocimiento de Rostros en 3D usando Kinect. In: Symposium of Image, Signal Processing Artificial Vision XVII (2012)
22. Elston, C.W., Ellis, O.: Pathological prognostic factors in breast cancer. I. The value of histological grade in breast cancer: experience from a large study with long-term follow-up, pp. 403–410 (1991)
23. Felipe, N., Acosta, A.: Proyecto de titulo "Detección automática de células mitóticas en imágenes histológicas usando redes neuronales convolucionales profundas" Índice (2017)

Machine Learning for Acquired Brain Damage Treatment

Yaritza P. Erazo[1](✉), Christian P. Chasi[1](✉), María A. Latta[2](✉), and Víctor H. Andaluz[1](✉)

[1] Universidad de las Fuerzas Armadas ESPE, Sangolquí, Ecuador
{yperazo, cpchasil, vhandaluz1}@espe.edu.ec
[2] Universidad Técnica de Ambato, Ambato, Ecuador
mariaalatta@uta.edu.ec

Abstract. This article presents an alternative rehabilitation system based on a visual feedback system for people suffering from cerebral palsy disorder. The proposed feedback system handles textures and movements in a 3D graphic environment, specially designed to develop skills that improve patient performance. The interface is developed in the Unity3D software, identifying patterns of the body is done through motion Kinect and validation of the correct execution of the exercises sensor is carried out, using the technique of machine learning for training rehabilitation system. The experimental results show the efficiency of the system that generates an improvement in the motor abilities of the upper and lower extremities of the patient.

Keywords: Machine learning · Cerebral palsy · Virtual reality

1 Introduction

The rehabilitation is a treatment that allows to improve and restore certain physical, sensorial and mental capacities [1]. Patients must go to a hospital, or medical center to obtain adequate rehabilitation, the transfer increases the costs of medical care and in some cases the recovery process can take a long time, resulting in a certain number of patients can not finish its treatment and the evolution of its rehabilitation are not as expected, for this reason it has developed rehabilitation applications that can be performed from home [2]. Currently there are studies of the importance and effects of specific diseases that require adequate rehabilitation [1], *e.g.*, Parkinson's disease [3], multiple sclerosis [4, 5], stroke [6], freezing of shoulders [2], among others. Cerebral palsy is a disorder that can occur at the birth or in the course of a person's life, producing a loss in the mobility of the upper and lower body, partial or total difficulty in communication; and even the loss of control of bodily functions. The technological advance of the last years has allowed to develop a series of tools that are very promising and although some are in a research stage, they are all aimed at the recovery of people suffering from this disorder [7].

It has developed several systems focused on cerebrovascular rehabilitation, which can be classified into two areas: rehabilitation through robotic systems and rehabilitation using virtual reality in 3D. In the area of robotic systems there are brain-machine

L. T. De Paolis and P. Bourdot (Eds.): AVR 2019, LNCS 11613, pp. 362–375, 2019.
https://doi.org/10.1007/978-3-030-25965-5_27

interfaces (BCI), which detect brain activity through electrodes implanted in the scalp or brain, read brain activity and make it a signal to perform a certain action, the exoskeletons allow mobility of the upper and lower extremities using signals sent from the brain [8]. The assisted social rehabilitation based on humanoid robots, provide therapeutic tools that allow the realization of physical tasks, and also improves the cognitive function of people, due to the commitment and motivation that is had during the therapy session [9]. A robotic platform is a prototype that moves on the ground for the rehabilitation of the lower extremities using the body weight of the patient as support [10]. The interaction between the brain and robotic devices allows for a movement induced by CIMT stroke therapy, which promotes movement directed to a part of the body, activates the hemisphere of the lesion, through repetitive practice of the affected limb, and as a result gets a better response from the member [11]. The applications developed in the area of virtual reality are based on video games, which were initially developed as entertainment activities, and nowadays are part of people's daily lives, and their applicability can have multiple objectives. A games that allows the user to reach a certain goal whether in education, medicine, communication, military strategy, among others are called serious [12]. For this reason, interactive game interfaces have been developed in Unity 3D that act in conjunction with the Kinect motion sensor for the detection of lower limb movement in order to improve motor function in patients with cerebral palsy [13]. Virtual environments with cartoons have been developed to arouse interest in patients, and using the Kinect movement sensor, inaccurate movements of the participants' upper extremities during rehabilitation activities have been detected, counting only the movements corresponding to the requirements of therapists [14], for these reasons it has been found that interactive games based on movement, are systems that require focus and attention, and can motivate the user to perform coordinated exercises, providing a sense of achievement, even if they can not perform that task in the real world [15].

The aforementioned advances have made it possible to investigate the impact of console games or games developed in virtual reality environments with the help of the interaction of the Kinect motion sensor, to establish patterns of repetition of movement at variable speeds, which result in an improvement in motor impairments of patients who have cerebral palsy [16].

The present work proposes a rehabilitation system based on videogames, which provides the development of motor skills with interactions, immersion and transparency between the patient and the developed interface, generating comfort and an improvement in mobility of the upper and lower extremities. The virtual environment, consists of three video game environments, with the inclusion of an avatar. he movements to be performed by the person to be rehabilitated are based on the Bobath method, which is a technique that search perceptive muscle rehabilitation, and the recovery of patients through motor stimulation of the thick motor limbs [3], is also used the gamification techniques in the interface to motivate the users to carry out the therapy exercises continuously, resulting in an improvement in the ability to move the extremities.

The article is organized in VI Sections including the Introduction. Section 2 presents formulation of the problem. Section 3 presents the pattern identification and validation algorithm to determine if the rehabilitation exercises have been carried out correctly.

Section 4 presents the friendly virtual system-patient environment. The results and their discussion are shown in Sect. 5. Finally, the conclusions are presented in Sect. 6.

2 Problem Formulation

Cerebral palsy CP is a neuromotor disorder that has no cure, is an abnormality that does not present progressive alterations in the brain, but causes motor disabilities that are usually accompanied by cognitive deterioration, irregular behavior and communication problems [7], although CP is a non-progressive and non-curable disorder, it must also be treated in a way that does not get worse over time. The treatment is considered key for this problem, so that at present new technologies emerge for the treatment of CP, one of them is the use of virtual reality.

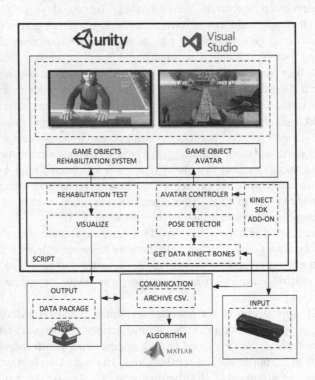

Fig. 1. Diagram of the rehabilitation system

Virtual reality presents an intuitive simulation environment, which through interactive games allows rehabilitation tasks to people who have a light spastic neuromotor disorder, to improve the mobility of the upper and lower extremities, so in this article we present a alternative treatment for CP disorder. The proposed rehabilitation system provides visual feedback for patients with mobility problems to interact with a 3D virtual reality environment through a computer and a visual feedback device given by the Kinect sensor.

The interaction between the virtual interface and the patient is done through the monitoring of the user's skeleton by means of the Kinect sensor, in order to contribute with the entrance to the virtual environment, which must allow the interaction between the patient and an avatar. The avatar executes the movements according to the actions performed by the patient in rehabilitation therapy. Figure 1 shows the block diagram of the proposed system, the interaction between the patient and the device is established by means of a bilateral communication, so that the assisted therapy for the rehabilitation of the upper and lower limbs, generates movements that are captured by the Kinect sensor, where the information is obtained through the three cameras that contains the sensor, which enables the tracking of the body of an avatar that is included in the three interfaces developed in Unity3D, the sensor returns the information of the bones from the body of the skeleton to the avatar created in the virtual reality environment, this allows the interaction of the device with the patient, through a visual feedback the patient can have a technological rehabilitation with better benefits, in friendly environments for the development of rehabilitation activities.

The algorithm of identification and validation of patterns, allows to determine the percentage of learning of the patient with respect to the correct execution of the rehabilitation activities, to perform this function the Maching Learning technique will be used for the training of the network. Figure 1 describes the machine-patient interaction system with visual feedback as the main base for performing upper and lower extremity rehabilitation tasks for patients with cerebral palsy disorder.

3   Identification Algorithm

Classifiers based on neural networks use a learning algorithm to construct an optimal interconnection path, minimizing a loss function established in the processing. The neural network applied in this work is based on three layers Fig. 2, the first layer called input, consists of thirty-six vectors, the second layer is composed of fifty non-linear neurons that are trained to carry out the tasks of classification of therapeutic movements appropriate for the rehabilitation of the upper and lower limbs, the third layer that is formed by three neurons to determine if the movements performed are classified correctly.

This network uses nonlinear neurons and can be trained with any backpropagation algorithm. In this work the algorithm "Gradient Descendant" was used, which allows to find minimum values of convex and differentiable functions in all its domain [17]. The network was trained with input-output patterns given by the positions in the three axes, obtained with the interaction between Kinect and Unity. The activation functions of neurons are:

$$a_1 = \tan sig(w_1 p + b_1) \tag{1}$$

$$a_2 = \tan sig(w_2 a_1 + b_2) \tag{2}$$

$$n = w_3 a_2 + b_3 \tag{3}$$

$$a_3 = \frac{e^n}{\sum\limits_{i=1}^{i} e^n} \tag{4}$$

where a_1 is the response of the first hidden layer; a_2 is the response of the second hidden layer; $n_{[2x1]}$ is the output of the neural network, and a_3 is the probability calculated with $n_{[2x1]}$; in the first layer we have the synaptic weights: $w_{1[60 \times 36]}$, where 60, represents the number of neurons and 36 the input data of the network, $w_{2[30 \times 60]}$ where 30 represents the number of neurons in the second layer and 60 is the neurons used in the previous layer and $w_{3[2x30]}$ where (2) represents the output of the network and 30 the number of neurons used in the previous layer, p represents the system entries in (1); later a_1 and a_2 they become the entries of (2) and (3), $b_{1[60 \times 1]}$, $b_{2[30 \times 1]}$ and $b_{3[30 \times 1]}$ are always active neurons called "bias" that improve the classification of the neural network and are located in the layer before the neurons.

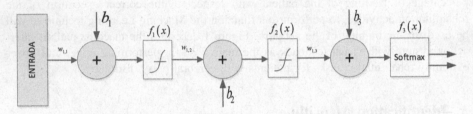

Fig. 2. Neural network of two layers and two non-linear regulators.

The non-linear neural networks are very versatile and in general, their training process is very fast. This type of neural networks has a great advantage over linear ones since they perform input-output assignments unless the network consists of many neurons and many layers. The proposed Neural Network presents thirty-six neurons in the input layer, and for each input we extracted characteristics in the time domain, which are calculated by means of (5) the mean absolute value (MAV) of the X_i signal in a segment i that has N length samples [19].

$$\overline{X_i} = \frac{1}{N} \sum_{k=1}^{N} |x_k|, \qquad i = 1, \ldots, I-1 \tag{5}$$

In the last hidden layer hidden layer there are thirty neurons that perform the classification of characteristics and finally two output neurons that define the classes depending on whether the answer is v_1 [10], the exercises performed by the patient are efficient for recovery therapy, if the recovery exercise performed by the patient is incorrect the output v_2 As a result [01], the outputs of the network are used for each of the environments developed in Unity 3D software. Although a greater number of neurons and layers could be used, with the tests performed it was demonstrated that the algorithm is efficient for the identification and validation of the therapy.

Figure 3 shows the random signals applied to the inputs and outputs determined with MAV. The vectors of inputs and outputs are used to train the network of Fig. 2. The training process was completed in approximately 1000 times and the final average error of the network was 3.3813×10^{-6}. Each epoch corresponds to an iteration with respect to the training of the data of the thirty-six input vectors and the output vector.

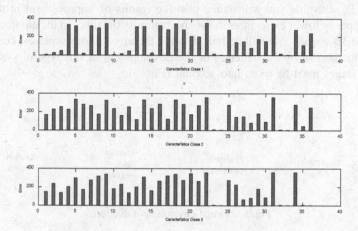

Fig. 3. Characteristic inputs of the neural network.

When training the network with the data obtained by the Kinect sensor in conjunction with the Unity software, it was possible to verify that the algorithm chosen to execute the Neural Network was optimal because it made a correct classification for each class as shown by the Fig. 4 because the error quickly converged to zero and was also very small, means that the neural network is accurate and reliable. By increasing more samples of the movements of the upper and lower extremities, to training. the network will adjust better by reducing the error in a smaller number of times.

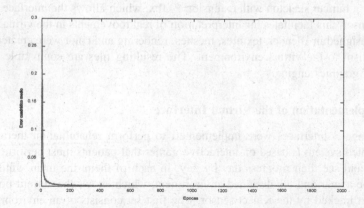

Fig. 4. Average quadratic error of the trained network.

4   Virtual Environment

This section describes the development of the virtual reality environment, here we will detail the construction of a friendly interface for the patient to enter an entertaining environment, since the implemented system is based on an architecture of interaction between the patient and the reality environment virtual through a physical device (Kinect V2), with this you will obtain positive results of improvement in the motor skills of upper or lower extremities in the shortest possible time. In the environment are each of the 3D objects that will be linked to the therapy as: tourist places, country and surroundings known by the patient. For the development of this type of systems, the following stages must be taken into account (Fig. 5).

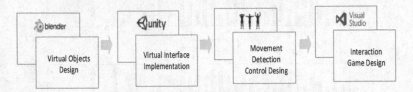

Fig. 5. Virtual environment diagram

4.1   Design of Virtual Objects

In this stage, virtual environments were designed with 3D objects that were linked to rehabilitation therapy, e.g., baskets, fruits, market products, houses, bridges, among others. The aforementioned objects are developed in the Blender design software, since it provides files with extension *. fbx, such format can be imported into the Unity 3D software and used in whatever the programmer needs. The graphic representation that is associated to a user for its identification (avatar) was designed in the software Autodesk-Character Generator, this creates very friendly environments and that adjust to the needs of the users, inside the avatar there are virtual humans that can be move as they have a human skeleton with extension * .fbx, which allows the interface with the Kinect sensor, this facilitates the interpretation of real movements in the virtual. For the objects designed in Blender, textures, meshes, rendering and color were created to give more realism to the virtual environment. The resulting files are compatible with the Unity 3D graphics engine.

4.2   Implementation of the Virtual Interface

In this stage 3 interfaces were implemented to perform rehabilitation therapies, the implemented system is based on interactive games that patients must perform for their recovery and see their progress day by day, in each of them, the main window containing the avatar and the designed environment, which shows the current position of the patient tracked by the Kinect sensor. The first set consists of an environment that allows to perform household chores, where the correct elongation of the arms is evaluated to classify a set of fruits. The second game evaluates the exercises performed with the lower extremities, simulating that the patient walks to cross a bridge in order to reach a tourist area. The third evaluates the exercises performed with the upper extremities to order a set of products from a supermarket. To access each of the established games, a menu is presented at the beginning of the application, which will be an intuitive tool for patients undergoing CP therapy.

4.3 Design of the Motion Detection Algorithm

In the present stage, modules of the Unity 3D Software were used, such as "Kinect Pose Detection" and "Transform. eulerAngles" that were used in programming the script in the Visual Studio 7 software, in order to store each of the data in the form of vectors with positions in x, y, z. These vectors are formed according to the points evaluated in the exercise, considering for the upper extremities 6 points of the body (wrist, elbow, shoulder) and for the lower extremities 6 points (hip, knee, foot). This data is saved in * .txt files for later sending to the mathematical analysis software in this case Matlab. Finally, a database is generated based on the exercises performed of the upper and lower extremities that will be used for training the neural network where they were successfully validated in the virtual reality environment.

4.4 Game Interaction Design

After the construction of the three interfaces, choice of movements of the upper and lower extremities and development of the Machine Learning algorithm, we proceed to elaborate the dynamics of the games. This interaction is done through the Script component, which contains codes developed in C # developed in Visual Studio. Through this component, the virtual rehabilitation system is consolidated, together with the control that verifies and approves if the exercise was carried out correctly.

5 Results and Discussion

This section presents the results of cerebral palsy treatment with the help of Machine Learning. The system of learning through neuronal networks allowed the virtual environment to adapt the rehabilitation games to the current conditions of the patient. Each of the games can be accessed through a menu presented at the beginning of the application Fig. 6. The menu allows access to the configuration of the patient's personal data, the games Fig. 7 and the credits of the application.

The first game corresponds to a routine where the patient takes fruits and places them in their respective basket, in which the correct movement of the arms is evaluated in order to take an object to the objective place programmed in the virtual interface Fig. 8. The second interface evaluates the lower extremities, here the patient crosses a bridge moving from one place to another Fig. 9, with this the progress in recovery can be analyzed. The third exercise evaluates the exercises performed with the upper extremities simulating a shopping routine in the supermarket, where the patient performs the movement to pick up a product and place it in a basket Fig. 10.

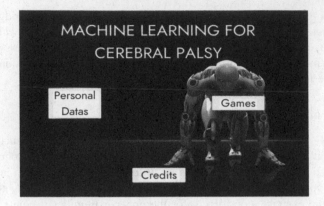

Fig. 6. Initial menu

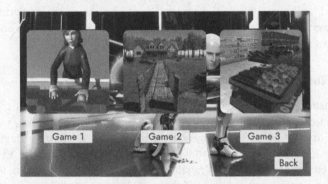

Fig. 7. Application games

Fig. 8. Rehabilitation of upper extremities - environment 1

Fig. 9. Rehabilitation of lower extremities - environment 2

Fig. 10. Rehabilitation of upper extremities - environment 3

5.1   Experimental Tests

To collect data from the experimental tests, three exercises are taken as a basis. In the first recovery exercise, raise your arms to place the fruit in an assigned place. The exercise is performed satisfactorily when the patient begins with a position of the arms forming a right angle with respect to the torso and opens them sequentially, first the right arm and then the left arm, fully stretching the arms, as shown in Fig. 11.

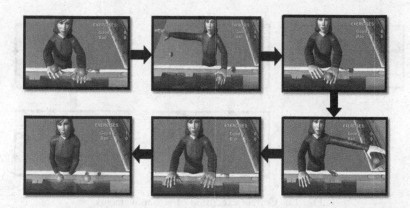

Fig. 11. Correct execution of arm movements.

In the case of Fig. 11, the execution of the task is correctly developed. However, to verify the operation of the exercise in the graphical environment there is an indicator where it is shown if the exercise is completed in an unsatisfactory or satisfactory manner.

In the case of Fig. 12, the execution of the second task can be observed correctly. The development of the task begins in the initial part of the bridge, where the patient must perform the action of moving from the initial part of the bridge through steps carried out properly to reach the other side of the bridge that is his destination. In the same way there is an indicator that allows you to visualize how many steps were carried out correctly and incorrectly.

Fig. 12. Correct execution of leg movements.

In the case of Fig. 13, the execution of the third task can be observed, which is executed correctly when the patient starts with the arms extended parallel to the axis of his body, the patient then raises an arm gradually until it forms an angle of 45 with respect to its line of sight, and finally it descends to its initial position and continues with its next arm until completing the sequence. In the same way you can observe the actions that the patient performed correctly or incorrectly.

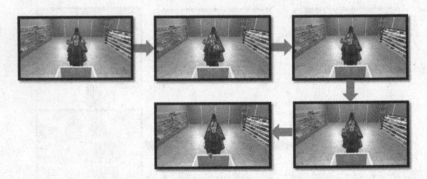

Fig. 13. Correct execution of arm movements.

6 Conclusions

The rehabilitation system based on visual feedback for patients with cerebral palsy disorder was tested with a group of patients with mild spastic neuromotor disorders. To determine the efficiency of the usability of the rehabilitation system, the summary evaluation method SUS was used, using a group of ten questions Table 1, which was answered by six physicians and four affected by the disease. The weighting ranges from one to five, with total disagreement and complete agreement, respectively [20].

Table 1. Questions to evaluate the usability of the system

Questions	
1	How complex was developing the exercises a virtual environment?
2	Do I consider that this system is complex to use?
3	Do I think an induction would be necessary before using the system?
4	Do I think I would need technical support to use the system?
5	Do I think that doing rehabilitation through interactive games helps the patient to complete his therapy?
6	Did I find too many inconsistencies in this system?
7	Do I believe that most rehabilitation routines can be brought to virtual reality?
8	Is it uncomfortable to use the system for rehabilitation of spastic cerebral palsy?
9	Have I found the use of the system very safe?
10	Would the specialist who uses the system need training?

Consequently, the average result of the SUS is 87.5% (Table 2), which indicates that the system is very easy to use, by the patient and the attending physician, the experimental results demonstrate the feasibility and ease of use of the rehabilitation system, with the use of the "Gradient Gradient" algorithm, the sequence of movements established allows to improve the motor skills of the upper and lower extremities, as well as reducing the time of treatment of people who have suffered a CP disorder.

Table 2. SUS tabulation

Preguntas	Doctor 1	Doctor 2	Doctor 3	Doctor 4	Doctor 5	Doctor 6	Paciente 7	Paciente 8	Paciente 9	Paciente 10
1	5	5	4	5	5	4	5	5	4	5
2	1	1	1	1	5	4	4	1	1	1
3	5	5	2	5	5	5	5	5	5	5
4	1	1	2	1	1	1	2	1	2	2
5	5	5	5	5	5	5	5	5	5	5
6	1	1	2	1	1	2	1	1	1	1
7	5	5	4	5	5	2	5	5	5	5
8	1	5	1	1	1	4	1	1	1	1
9	4	5	2	5	5	4	4	5	5	5
10	1	1	4	1	1	5	1	1	5	1
Valor SUS	97,5	90	67,5	100	90	60	87,5	100	85	97,5
Promedio SUS del Sistema	87,5									

Acknowledgements. The authors would like to thanks to the Corporación Ecuatoriana para el Desarrollo de la Investigación y Academia–CEDIA for the financing given to research, development, and innovation, through the CEPRA projects, especially the project CEPRA-XI-2017-06; Control Coordinado Multi-operador aplicado a un robot Manipulador Aéreo; also to Universidad de las Fuerzas Armadas ESPE, Universidad Técnica de Ambato, Escuela Superior Politécnica de Chimborazo, Universidad Nacional de Chimborazo, and Grupo de Investigación ARSI, for the support to develop this work.

References

1. Han, SH., Kim, HG., Choi, H.J.: Rehabilitation posture correction using deep neural network. In: 2017 IEEE International Conference on Big Data and Smart Computing (BigComp) (2017)
2. Ongvisatepaiboon, K., Chan, J., Vanijja, V.: Smartphone-based tele-rehabilitation system for frozen shoulder using a machine learning approach. In: 2015 IEEE Symposium Series on Computational Intelligence (2015)
3. Frazzitta, G., Morelli, M., Bertotti, G., Felicetti, G., Pezzoli, G., Maestri, R.: Intensive rehabilitation treatment in parkinsonian patients with dyskinesias: a preliminary study with 6-month followup. Parkinsons Dis. **2012** (2012)
4. Patti, F., et al.: Effects of a short outpatient rehabilitation treatment on disability of multiple sclerosis patients. J. Neurol. **250**(7), 861–866 (2003)
5. Freeman, J., Langdon, D., Hobart, J., Thompson, A.: The impact of inpatient rehabilitation on progressive multiple sclerosis. Ann. Neurol. **42**(2), 236–244 (1997)
6. Henderson, A., Korner-Bitensky, N., Levin, M.: Virtual reality in stroke rehabilitation: a systematic review of its effectiveness for upper limb motor recovery. Top. Stroke Rehabil. **14**(2), 52–61 (2014)
7. Deep, A., Jaswal, R.: Role of management & virtual space for the rehabilitation of children affected with cerebral palsy: a review. In: 2017 4th International Conference on Signal Processing, Computing and Control (ISPCC) (2017)
8. Bates, M.: From brain to body: new technologies improve paralyzed patients? Quality of life. IEEE Pulse **8**(5), 22–26 (2017)
9. Malik, N.A., Yussof, H., Hanapiah, F.: Potential use of social assistive robot based rehabilitation for children with cerebral palsy. In: 2016 2nd IEEE International Symposium on Robotics and Manufacturing Automation (ROMA) (2016)
10. Bayon, C., et al.: Pilot study of a novel robotic platform for gait rehabilitation in children with cerebral palsy. In: 2016 6th IEEE International Conference on Biomedical Robotics and Biomechatronics (BioRob) (2016)
11. Lu, R., Wu, Y.: Application of brain-computer interface system in stroke patients. Rehabil. Med. **26**(5), 59 (2016)
12. Camara Machado, F., Antunes, P., Souza, J., Santos, A., Levandowski, D., Oliveira, A.: Motor improvement using motion sensing game devices for cerebral palsy rehabilitation. J. Motor Behav. **49**(3), 273–280 (2016)
13. Chang, Y., Han, W., Tsai, Y.: A kinect-based upper limb rehabilitation system to assist people with cerebral palsy. Res. Dev. Disabil. **34**(11), 3654–3659 (2013)
14. Pourazar, M., Mirakhori, F., Hemayattalab, R., Bagherzadeh, F.: Use of virtual reality intervention to improve reaction time in children with cerebral palsy: a randomized controlled trial. Dev. Neurorehabilitation **21**(8), 515–520 (2018)

15. Jaume-i-Capo, A., Martinez-Bueso, P., Moya-Alcover, B., Varona, J.: Interactive rehabilitation system for improvement of balance therapies in people with cerebral palsy. IEEE Trans. Neural Syst. Rehabil. Eng. **22**(2), 419–427 (2014)
16. Nithya, V., Arun, C.: Brain controlled wearable robotic glove for cerebral palsy patients. In: 2017 International Conference on Circuit, Power and Computing Technologies (ICCPCT) (2017)
17. Heravi, A., Hodtani, G.A.: A new correntropy-based conjugate gradient backpropagation algorithm for improving training in neural networks. IEEE Trans. Neural Netw. Learn. Syst. 1–12 (2018). https://doi.org/10.1109/tnnls.2018.2827778
18. Pforte, L.: Extensions of simple modules for $SL_3(2^f)$ and $SU_3(2^f)$. Commun. Algebra **45**(10), 4210–4221 (2016)
19. Yang, J., Zhao, H., Chen, X.: Genetic algorithm optimized training for neural network spectrum 348 prediction. In: Proceedings of the IEEE 2nd International Conference on Computer and Communications (ICCC), vol. 2, no 349 1, pp. 2949–2954 (2016)
20. Sauro, J., Lewis, J.R.: When designing usability questionnaires, does it hurt to be positive? In: Proceedings of the SIGCHI Conference on Human Factors in Computing Systems, pp. 2215–2224. ACM, May 2011

Software Framework for VR-Enabled Transcatheter Valve Implantation in Unity

Giuseppe Turini[1,2] (iD), Sara Condino[2,3(✉)] (iD), Umberto Fontana[2],
Roberta Piazza[2,3] (iD), John E. Howard[4], Simona Celi[5],
Vincenzo Positano[6] (iD), Mauro Ferrari[2], and Vincenzo Ferrari[2,3] (iD)

[1] Computer Science Department, Kettering University, Flint, MI, USA
gturini@kettering.edu
[2] EndoCAS Center, Department of Translational Research and of New Surgical
and Medical Technologies, University of Pisa, Pisa, Italy
{sara.condino,roberta.piazza,
vincenzo.ferrari}@endocas.unipi.it,
mauro.ferrari@med.unipi.it
[3] Department of Information Engineering, University of Pisa, Pisa, Italy
[4] General Motors, Detroit, MI, USA
[5] Fondazione Toscana Gabriele Monasterio, Pisa, Italy
simona.celi@ftgm.it
[6] CMR Unit, Fondazione G. Monasterio CNR-Regione Toscana, Pisa, Italy
vincenzo.positano@ftgm.it

Abstract. VR navigation systems have emerged as a particularly useful tool
for the guidance of minimally invasive surgical procedures to restore the 3D
perception of the surgical field and augment the visual information available to
the surgical team. Over the past decade, X-ray free VR navigation systems based
on Electromagnetic (EM) tracking have been proposed to guide catheter-based
minimally invasive surgical procedures, including transcatheter valve implantation, with a reduced intra-procedural radiation exposure and contrast medium
injections. One of the major limits of current navigation platforms is the lack or
the poor reliability/efficiency of the real-time modeling unit to reproduce the
deformations of the cardiovascular anatomy and surgical instruments. In this
work we propose an EM navigation platform, with software functionalities,
developed using the Unity game engine, to mimic physical behaviors of surgical
instruments inside the patient vasculature, handle the interaction between the
tools and the anatomical model via collision detection, and (in the next version
of the software) deform in real-time the anatomy. Additionally, the technology
used (Unity) can allow a seamless integration of AR headset to further improve
the ergonomics of the 3D scene visualization.

Keywords: Electromagnetic navigation · Virtual reality · Unity game engine ·
Computer-assisted surgery · Minimally-invasive surgery · Surgical simulation

© Springer Nature Switzerland AG 2019
L. T. De Paolis and P. Bourdot (Eds.): AVR 2019, LNCS 11613, pp. 376–384, 2019.
https://doi.org/10.1007/978-3-030-25965-5_28

1   Introduction

Over the past years, minimally invasive interventions based on transcatheter techniques have dramatically changed the treatment of cardiac valve dysfunction avoiding open heart surgery. Transcatheter Valve Implantation (TVI) is now a cornerstone both for congenital and acquired heart valve disease treatment. Transcatheter Pulmonary Valve Replacement (TPVR) is one of the most exciting recent developments in the treatment of congenital heart disease and represents an attractive alternative to surgery in patients with right ventricular outflow tract dysfunction [1].

The acquisition of clinical experience together with technical improvements in valve design and delivery systems can lead to a simplification of the surgical procedure, and thus to a reduction in complication rates and a widespread use of this technique. The development of new Virtual Reality (VR) bioengineering tools for planning and guiding the intervention on a patient-specific basis can further extend TVI to a larger patient population.

VR navigation systems have emerged as a particularly useful tool for the guidance of minimally invasive surgical procedures to restore the 3D perception of the surgical field and augment the visual information available to the surgical team [2]. Over the past decade, X-ray free VR navigation systems based on Electromagnetic (EM) tracking have been proposed to guide catheter-based minimally invasive surgical procedures with a reduced intra-procedural radiation exposure and contrast medium injections [3–10]. One of the major limitations of current systems is the lack or poor reliability/efficiency of the real-time modeling unit to reproduce the deformations of the involved anatomy and surgical instruments. Tools used in TVI are flexible, elongated, cylindrical wires and tubes that cannot be treated as rigid bodies [11]. Moreover, the anatomy deforms due to: the interactions with the catheter-guidewire and the surrounding anatomy, the respiration and heart breathing.

In this work we propose an EM navigation platform for TVI with a virtual catheter able to mimic physical behaviors of surgical instruments inside the patient vasculature, handle the interaction between the tools and the anatomical model via collision detection, and (in the next version of the software) deform in real-time the anatomy.

2   Materials and Methods

The following sections describe in detail our EM navigator (see Fig. 1), emphasizing: the overall design, its main modules, and the extensions/modifications we plan to implement in the near future.

2.1   EM Navigation Platform

The navigation platform is a software-hardware system comprised of several different modules and tools (see Fig. 2).

The main module is a software application developed using the Unity game engine [12]. This application has been designed for a desktop computer and integrates multiple modules/scripts to manage: interactive 3D visualization of the virtual catheter and

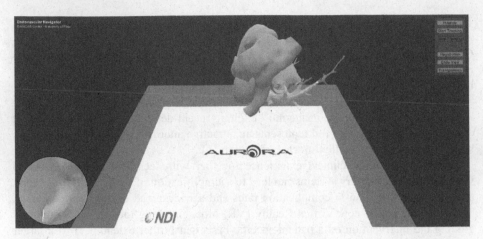

Fig. 1. Overview of EM navigator: virtual anatomy (top center) including heart and pulmonary artery, virtual catheter (blue tube with white tip), NDI Aurora EM tracking system (middle center) in Tabletop Field version, GUI (top right panel), and catheter camera view (bottom left). (Color figure online)

anatomy, the Graphical User Interface (GUI), and the communication with the NDI Aurora real-time EM tracking system (Tabletop Field version) [13].

The principal elements of the virtual 3D environment are the virtual patient anatomy and the virtual catheter. The virtual anatomy includes the heart and pulmonary artery, and it has been generated from medical datasets. The virtual catheter is implemented by a deformable tubular mesh, includes an endoscopic camera on its tip, and integrates Unity components to enable interactions/collisions with the surrounding virtual environment.

The 3D tracking of the real catheter-guidewire prototype is performed by the NDI Aurora system, and relies on 2 sensors integrated inside the catheter body. The communication between the NDI Aurora and the navigator application are handled by a custom Unity plugin.

The last 2 elements of the platform are: the synthetic patient anatomy used to perform a preliminary evaluation of the system, and the NDI Aurora Digitizing Probe necessary to carry out the intraoperative registration of the patient anatomy.

The following diagram in Fig. 2 illustrates all the software/hardware modules and their relative interactions/communications.

2.2 Software Architecture

This section describes in detail the most important modules of the Unity software application of our EM navigator.

Catheter Tracking Integration. The 3D tracking of the real catheter is carried out by the NDI Aurora system, that detects the position and rotation of its sensors at a frequency of 40 Hz. To enable the tracking of the real catheter, two 5 DOF sensors

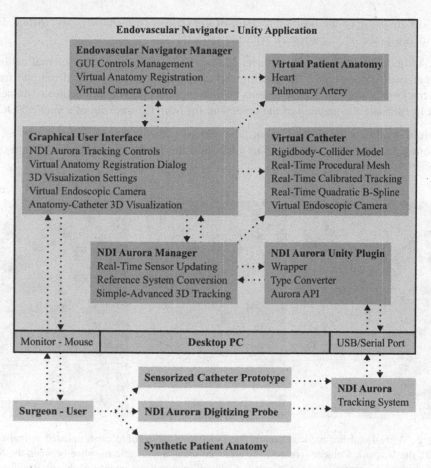

Fig. 2. Diagram of the navigator application, showing: all the software modules (in blue) of the Unity application, all the hardware tools and the user (in red), and all the interactions and data transfers (as dotted directional lines). (Color figure online)

have been embedded into the catheter body: in its distal section, with a distance of 8 cm between each other.

The NDI Aurora system transfers data to the EM navigator via a custom Unity plugin, including 3 modules (see Fig. 2):

- A module (C++ DLL) for the NDI Aurora API.
- A module to perform type conversion from C++ (NDI Aurora API) to C# (Unity).
- A module wrapping the NDI Aurora API functions used in Unity (C# scripts).

The plugin is used by the "NDI Aurora Manager" script (see Fig. 2), handling:

- The acquisition of the 3D tracking information at a constant update rate (40 Hz).
- The conversion of tracking data between reference systems (from Aurora to Unity).

- The updating of the position/rotation of the objects (catheter sections, virtual digitizing probe, etc.) associated with the sensors.

Additionally, the "Virtual Catheter Manager" script integrates an internal calibration algorithm: to estimate the position and orientation of the catheter distal part from the tracking of the two 5 DOF sensors (which are embedded into a dedicated lumen), and to calculate the position of the guidewire tip from the tracking of a single 5 DOF.

Virtual Catheter Implementation. The virtual catheter has been developed to be deformable and to enable interactions against all the surrounding 3D environment (see Fig. 3). Its main components are:

Fig. 3. Virtual catheter implementation: the procedural deformable mesh updated in real-time (left), the "Capsule Collider" components arranged using a quadratic B–spline between the NDI Aurora sensors (center), and the "Capsule Collider" components connected using "Rigidbody" and "Configurable Joint" components exploiting the Unity physics engine (right).

- A child structure consisting of a chain of "Capsule Collider" objects representing catheter sections for Collision Detection (CD) purposes (see Fig. 3 center).
- Each section also includes a "Rigidbody" and a "Configurable Joint" component to link together catheter sections using the Unity physics engine (see Fig. 3 right).
- A virtual endoscopic camera positioned on the virtual catheter tip and providing an endoscopic view to the surgeon-user (see Fig. 1).

The entire virtual catheter structure is handled by the "Virtual Catheter Manager" script, including:

- A procedural triangular mesh: generated at startup, and updated continuously at runtime to match the new position/rotation of the catheter sensors (see Fig. 3 left).
- A quadratic B-spline that, if enabled, continuously rearranges the sections between the catheter sensors accordingly to preconfigured control points (see Fig. 3 center).

- A calibration algorithm to infer the position of the catheter and guidewire from the 3D tracking of the embedded 5 DOF sensors (as mentioned in the previous section and described in detail in [4]).

Virtual Patient Anatomy Setup. The virtual patient anatomy includes the heart and the pulmonary artery, both generated from medical datasets. The 3D models were optimized via decimation and quality enhancement (Siemens NX 11, and Altair HyperMesh 14.0) [14, 15]. Then, after materials and textures were configured for each mesh (Blender 2.79), all virtual organs were imported in Unity [16].

Before beginning with the EM navigation, the virtual anatomy has to be registered with the synthetic anatomy using a point-based rigid-body registration algorithm (see Fig. 4). Once registered, the virtual anatomy is ready to be navigated by the virtual catheter.

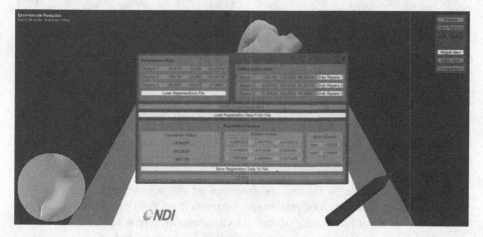

Fig. 4. Virtual anatomy registration, illustrating: the registration dialog (center), the NDI Aurora Digitizing Probe (bottom right), and the registration error during the preliminary evaluation.

It is worth mentioning that all 3D models have been generated/optimized with the final goal of developing a deformable virtual anatomy. For this reason, each surface (triangular) mesh has an associated volumetric (tetrahedral) mesh that will be used in the next phase of the research project to implement deformable virtual organs.

2.3 Preliminary Evaluation

Our navigation platform was tested on a desktop computer running Microsoft Windows 10 (Intel Core i7-7700 – 3.60 GHz, 8 GB RAM, GPU nVidia GeForce GTX 1050), using a virtual 3D environment including: the virtual anatomy composed of approximately 37k vertices and 74k triangles, and the virtual catheter made with 560 vertices and 1080 triangles (for a section of 8 cm in length). The memory required to run the navigator was approximately 96 MB, whereas the frame rate was ranging from 75 to 105 fps (see Fig. 5).

Fig. 5. Preliminary evaluation of the EM navigator: the real catheter-guidewire integrating 2 NDI Aurora sensors during the traversal of the pulmonary artery of the synthetic patient anatomy (left), the virtual catheter emulating the same task on the virtual anatomy relying on the real-time tracking by the NDI Aurora (right).

The registration of the virtual anatomy performed in the preliminary evaluation resulted in a maximum error of ~ 1.2 mm and a Mean Squared Error (MSE) of ~ 0.5 mm (see Fig. 4).

During this test we performed the insertion of the sensorized catheter into the pulmonary artery, focusing our preliminary evaluation on both the accuracy and performance, in terms of frame rates of: the 3D visualization, physics simulation, and 3D tracking (see Fig. 5). The results were satisfactory for both these factors.

3   Conclusions and Future Work

The EM navigation platform for TVI described in this paper allows the surgeon-user to benefit from the 3D visualization of the virtual catheter and virtual anatomy by controlling the sensorized catheter prototype, as well as the endoscopic view that could assist the surgeon to guide the catheter among the several branches of vascular structures.

The preliminary evaluation allowed us to approximately assess the accuracy of the navigation, and the overall performance of the software application, in terms of frame rates of: the 3D visualization, physics simulation, and 3D tracking.

Our system, while assisting in reducing the intraoperative radiation exposure such as standard navigators, is also able to simulate a deformable catheter and in the next phase will also integrate deformable organs to increase the realism of the virtual navigation. In particular, the 3D virtual model of the anatomy showed by the navigator will be up-dated according to the detected collisions with the virtual catheters, for reproducing the actual interaction between the patient anatomy and the surgical instruments. Furthermore, predictive respiratory/cardiac motion models [17] could be used to reproduce the physiology of the cardiovascular district, thus improving the navigation reliability and accuracy.

Additionally, the technology used (Unity) will allow us a seamless integration of AR headset to further improve the ergonomics of the 3D scene visualization.

Acknowledgments. The research leading to these results has been partially supported by the scientific project "VIVIR: Virtual and Augmented Reality Support for Transcatheter Valve Implantation by using Cardiovascular MRI" (PE–2013–02357974) funded by the Italian Ministry of Health through the call "Ricerca Finalizzata 2013".

References

1. Alkashkari, W., Alsubei, A., Hijazi, Z.M.: Transcatheter pulmonary valve replacement: current state of art. Curr. Cardiol. Rep. **20**, 27 (2018)
2. Piazza, R., et al.: Using of 3D virtual reality electromagnetic navigation for challenging cannulation in FEVAR procedure. In: De Paolis, L.T., Bourdot, P., Mongelli, A. (eds.) AVR 2017. LNCS, vol. 10325, pp. 221–229. Springer, Cham (2017). https://doi.org/10.1007/978-3-319-60928-7_19
3. Condino, S., et al.: Simultaneous tracking of catheters and guidewires: comparison to standard fluoroscopic guidance for arterial cannulation. Eur. J. Vasc. Endovasc. Surg. **47**, 53–60 (2014)
4. Condino, S., et al.: Electromagnetic navigation platform for endovascular surgery: how to develop sensorized catheters and guidewires. Int. J. Med. Robot. + Comput. Assist. Surg. MRCAS **8**, 300–310 (2012)
5. Piazza, R., et al.: Design of a sensorized guiding catheter for in situ laser fenestration of endovascular stent. Comput. Assist. Surg. **22**, 27–38 (2017)
6. Hautmann, H., Schneider, A., Pinkau, T., Peltz, F., Feussner, H.: Electromagnetic catheter navigation during bronchoscopy - validation of a novel method by conventional fluoroscopy. Chest **128**, 382–387 (2005)
7. Aufdenblatten, C.A., Altermatt, S.: Intraventricular catheter placement by electromagnetic navigation safely applied in a paediatric major head injury patient. Child Nerv. Syst. **24**, 1047–1050 (2008)
8. Cochennec, F., Riga, C., Hamady, M., Cheshire, N., Bicknell, C.: Improved catheter navigation With 3D electromagnetic guidance. J. Endovasc. Ther. **20**, 39–47 (2013)
9. Jaeger, H.A., et al.: Automated catheter navigation with electromagnetic image guidance. IEEE Trans. Biomed. Eng. **64**, 1972–1979 (2017)
10. Mukherjee, S., Chacey, M.: Diagnostic yield of electromagnetic navigation bronchoscopy using a curved-tip catheter to aid in the diagnosis of pulmonary lesions. J. Bronchol. Internv. Pulmonol. **24**, 35–39 (2017)
11. Wang, Y.Z., Serracino-Inglott, F., Yi, X.D., Yang, X.J., Yuan, X.F.: An interactive computer-based simulation system for endovascular aneurysm repair surgeries. Comput. Animat. Virtual Worlds **27**, 290–300 (2016)
12. Unity Game Engine (homepage). https://unity3d.com. Accessed 07 Mar 2019
13. NDI Aurora (Tabletop Field): real-time electromagnetic tracking system (homepage). https://www.ndigital.com/medical/products/aurora. Accessed 07 Mar 2019

14. Siemens NX (homepage). https://www.plm.automation.siemens.com/global/it/products/nx. Accessed 08 Mar 2019
15. Altair HyperMesh (homepage). https://hyperworks.fr/product/hypermesh. Accessed 08 Mar 2019
16. Blender: the free and open source 3D creation suite (homepage). https://www.blender.org. Accessed 08 Mar 2019
17. Savill, F., Schaeffter, T., King, A.P.: Assessment of input signal positioning for cardiac respiratory motion models during different breathing patterns. In: International Symposium on Biomedical Imaging, pp. 1698–1701 (2011)

Virtual Reality Travel Training Simulator for People with Intellectual Disabilities

David Checa[1](✉), Lydia Ramon[2], and Andres Bustillo[1]

[1] Department of Civil Engineering, University of Burgos, Burgos, Spain
{dcheca, abustillo}@ubu.es
[2] Department of History and Geography, University of Burgos, Burgos, Spain
lrp0031@alu.ubu.es

Abstract. Being able to travel in public transportation autonomously is one daily task that doesn't imply any special effort or learning process for most of the human population. But, for people with intellectual disabilities, it requires an extensive learning process where many possible situations should be considered before the person can perform successfully this task. Currently these people are taught by specialists following supervised learning methodologies in real environments. But, through the use of immersive virtual reality and interaction within a virtual world, they could be trained in a safe environment, increasing their self-confidence in a saver and quicker way. This paper describes the creation of a customized simulator that prepares subjects to use the bus as a means of transportation in a medium-size city. The design of the virtual reality environment and the interaction mechanics is thought to be highly customizable for each subject with a very realistic environment, thanks to the use of photogrammetric digitalization technologies. The environment has been developed in Unreal engine and uses the head mounted display Oculus rift, as well as Oculus touch for interaction with the different assets.

Keywords: Virtual Reality · Training simulator · Intellectual disabilities · Immersive environments · Oculus Rift · Unreal engine · Public transportation

1 Introduction

About 1% of the world's population has some form of intellectual disability [1]. For a person without a disability it is not an especially difficult task to use a public transport system. It involves tasks ranging from identifying the bus line, locating the correct stop, knowing where and how to pay the ticket, as well as recognizing the destination and asking the bus to stop. But such list of tasks can present an extensive variety of cases where different decisions might be taken, that make them especially complex for a person with an intellectual disability. At the same time, the ability to use public transport is one of the primary skills that increases the autonomy and self-confidence of any human being, and especially of people with intellectual disabilities. The training process of such persons is usually long to acquire these skills, including supervised training in real environment, which is extremely time-consuming. This training can be partially replaced by the interaction with a virtual environment that replicates the

L. T. De Paolis and P. Bourdot (Eds.): AVR 2019, LNCS 11613, pp. 385–393, 2019.
https://doi.org/10.1007/978-3-030-25965-5_29

critical aspects of the real-world situations, offering the possibility of repetition and assuring in any time the safety of the subject.

Virtual Reality is ideal for the development of training applications, more specifically in the field of teaching disabled people, because its high interactivity and immersion features are especially suitable for boosting learning rates in daily activities. In a virtual world skills can be practiced and they can be evaluated, deployed, repeated and perfected until they are enough established for real life. Repetition of activities in virtual environments, as some studies suggest [2, 3], increases the probability of generalization in real world situations. In addition, Virtual Reality simulators can help professionals in care centers for the disabled people to evaluate their skills and customized their learning process. Virtual Reality has shown since its origins as a technology that could largely replace traditional training methods.

The effectiveness of virtual reality in learning process of individuals with intellectual disabilities has already been reported in other studies [4], also in some cases for travel training [5] Besides, it has been demonstrated that this training can be generalized to the real world, highlighting its low cost and safe space for the user. Also, other studies point out a series of improvements in relation to memory and attention after the intervention of serious games in Virtual Reality [6]. Some studies have shown the capabilities of Virtual Reality simulators for training to people with intellectual disabilities in daily tasks like shopping autonomously in a virtual supermarket. In this example of virtual training, users have to choose the items, make the purchase and pay [7].

Also, some examples are found in the literature of virtual simulators for bus travel training like the one proposed in this study, although they are not directly related to high-immersive Virtual Reality. In 1998 the creation of a virtual environment for practicing the correct way to pay on a bus was already reported [8], which was expanded in 1999 with the development of "Virtual City" a virtual environment in which teach and practice independent living principles for people with learning disabilities [9]. They found that these skills were correctly transferred to the real world. It seen to be representative of real-world tasks and that users can learn some basic skills. However, authors say that it would be unrealistic to expect transfer of skill if the experimental time is short, demanding therefore long expositions to the virtual environments.

Other studies investigate the effectiveness of virtual reality environments for skills training and how to transfer these competencies to the real world. They conclude that the experience gained in virtual reality can be ported to real-world competencies, adding that people with learning disabilities are more motivated and gain self-confidence [10]. This motivating power of technology must be harnessed to maintain interest in the task at hand and therefore improve and make the proposed objectives more enjoyable.

Virtual reality applications have been shown to be particularly beneficial for people with intellectual disabilities because, first and foremost, they offer users an environment in which to develop skills safely. It gives the opportunity to learn from mistakes without exposing themselves to the consequences they would have in real life. This fact is especially valuable when it helps to motivate these subjects to carry out these activities independently and without fear of suffering accidents, humiliation or exposing to others for these mistakes [7, 11]. Virtual reality also allows people with intellectual disabilities to experience a sense of control over the learning process. This can be highly relevant, taking into account the natural passive tendency of this type of subjects [12].

But the use of virtual reality for training people with intellectual disabilities also has negative aspects to consider that could diminish its effectiveness. Thus, the creation of an experience of this type requires a special analysis of the characteristics of the simulation. It must avoid disorientation of users, always contribute to learning and facilitate its use. Related to this issue, Standen and Brown [13] indicate a series of requirements that these virtual environments must meet. Firstly, for a correct communication of the instructions it should be brief and, preferably, avoided the use of texts messages. Linked to this idea, the use of icons and graphics can be used as reinforcement, but they must be consistent throughout the experience. Secondly, environments must be sufficiently realistic and enhance key aspects necessary for understanding. Thirdly, concerning the interaction mechanics, the use of multiple buttons on the input device should be avoided. It has been reported [14] that users have difficulties with the use of mouse, keyboards or joysticks and these can induce frustration. This fact crashes against the commercial and available means, and therefore most of the reported travel training applications in virtual reality correspond to desktop systems, generally interacting with the virtual world through a monitor and using mouse, keyboards or joysticks for the interaction [15–18]. Recently, low-cost virtual reality interfaces such as HTC VIVE or Oculus Touch have begun to be available in the market at reasonable prices and therefore most recent immersive reality experiences to train these skills [19–21] use this kind of devices.

The purpose of this research is to prepare people with intellectual disabilities to use buses as a means of transportation in their daily lives. Immersive virtual reality is presented as the perfect means to facilitate a safe training space where these people can practice the whole process of taking a bus, from waiting at the stop to choosing the correct bus line, validating the ticket, pressing the stop button and getting off at the correct stop. The experience is designed using immersive virtual reality devices and intuitive interfaces. This will enable them to develop the skills necessary to improve their personal autonomy.

The remaining of this paper is structured as follows: Sect. 2 collects the main objectives of the Virtual Reality simulator; Sect. 3 describes the process to recreate the virtual environments, including both hardware and software, and the process followed to generate and to import the hyper realistic 3D models into the game engine. Finally, Sect. 4 presents the conclusions and future lines of work.

2 Objectives

The simulator aims to improve the skill of using the public transport system by people with intellectual disabilities. Currently, specialized teachers train individually each user in a personalized way for this learning process. This makes the learning process a long and very time-consuming process. Virtual reality offers a new opportunity to develop a cost-effective solution to help these people to learn to travel independently. The objective of the simulator is to provide users a hyper realistic simulator of how to move around the city by means of the public transport system. Its main purpose is to assist people with intellectual disabilities to learn travel skills in a faster and safer way.

The structure of the simulator is divided into three stages: (1) wait at the stop and choose the right bus, (2) pay the ticket and (3) get off at the right stop. First, by waiting at the bus stop for the bus to arrive, the user must be able to recognize the line he wants to get on. To do this, in the bus shelter the user has, as in reality, a map with the lines and destinations. Also, a digital panel indicates the buses that are going to stop at that stop and the time when they are expected to arrive. Each time a bus arrives at the stop the user has to decide whether to get on or let it pass. When the user gets on the bus he will have to pay with his transport card and finally the user will have to detect when the bus is approaching to its destination stop and press the button to request the bus to stop.

3   Development

As we have commented in the introduction, Standen and Brown [13] established some good practices for the creation of this kind of learning experiences, so we have tried to follow them, to a greatest extent, adapting them to the new requirements of immersive virtual reality. For that reason, we have firstly designed a very simple simulator's interface. This interface should facilitate the user's interaction with the virtual environment, avoiding the requirement of pressing buttons or any interactions with keyboard. We use the oculus touch and the interaction will be done through gestures. The first interaction is passive and takes place when the bus is approaching, if the user is outside the bus shelter the bus will stop at the stop. Besides, a second way for the bus to stop has also been programmed: if the user shakes the hand at the top of his head -to make a signal to the driver- the bus will also stop. Once the bus has stopped in front of the user, two transparent spheres are displayed: green (overlap to go up to the bus) and red (let pass and wait for another). Once on the bus, it is time to pay our fare. To increase the user's accessibility, the transport card will appear "stuck" in the hand and he only must approach it to the machine where they can check if he has paid and how much credit they have. Finally, to stop the bus the user only has to press the virtual stop button and the requested stop signal will illuminate, confirming his action.

Also, in order to increase the accessibility, we have not incorporated any type of possibility of movement within the virtual world beyond the natural one delimited by the room scale, that the virtual reality device allows. In this way we avoid navigation problems (Fig. 1).

Fig. 1. Example of interactions proposed in the video game (Color figure online)

To Standen and Brown, the virtual environment must contain enough real-life key elements to allow its recognition by the user, but not too many to distract the user from the learning objective. In our case, one of the core elements of the simulator is the realistic recreation of the situations that the user will find in the real world. These core elements, of course, include the environment; we recreate it in a hyper realistic way thanks to photogrammetry. This technique has been widely used in recent years thanks to the technical advances it has undergone. Basically, very realistic 3D models are created from two-dimensional images using automatic software processes. This technique has been recently used for architectural visualization, urban modeling, heritage conservation or even to recreate crime scenes in police investigations [22]. The capture of entire spaces makes it possible to create virtual reality experiences with six degrees of freedom (6DoF) which allows the user to explore the scenario. The photogrammetric process to create ultra-realistic 3D models to be visualized in virtual reality is divided into four steps: Capture of images, creation of the model in high resolution, reprocessing of the model to create a low resolution model and, finally, its integration in the game engine. In this study case, the data capture is simple and quickly done, because the user will have limited movement (just a few meters), so there is no problem that gaps areas are not captured because they will never be visible to the user (e.g. buildings' roofs). For the acquisition of these data, a Canon 50D reflex camera was used equipped with a 30 mm lens. Ideally and for further process optimization the use of laser station might be required, to have a more detailed points mesh and then mix it with photogrammetry. But such station was not available and this further environment optimization was not performed.

The software processing of these images was done in Reality Capture using the recommended method of image matching, creation of dense point cloud and mesh generation [23]. The 3D model resulting from this calculation is too heavy to be rendered in real time so its polygonal load must be reduced. For this objective, we use the software's own decimate tool. Finally, the model was integrated into the Unreal Engine. We used this engine for its capability of rendering photorealistic images as well as for its easy-to-use visual editor using blueprint for programming (Fig. 2).

Fig. 2. Photogrammetry of a location mixing point data and final image in Unreal engine.

Besides the environment, small details make the difference between an artificial environment and a simulation that can be believed to be the real world. For such elements, they have been manually 3D modelled, replicating real elements. In this category of elements, it might be included the bus and the bus shelter that can be seen in Fig. 3. Of course, in addition to this, many more details were modeled by hand to complement the environment generated by photogrammetry, as well as other resources that were needed for the simulation (e.g. the transport card).

Fig. 3. Example of the personalized elements in the simulator.

Once all 3D models are finished, it is time to create the logic behind the simulation. Unreal Engine provides a visual programming environment that adapts perfectly to this type of project. Figure 4 shows this type of programming that serves, in this case, to create the system that plan the buses planning and deploy the elements of interaction with the user. The use of visual programming systems helps to reduce the costs of the virtual environment interactivity development and to faster-iterate in the game design process. The animations present in the simulation were carried out using timelines controlled by programming, which launch the different animations and contain the logic behind (e.g. the movement of the wheels or the opening and closing of the doors of the bus). In addition, other events were programmed that include information texts using widgets to give more information to the user, such as when user validate the transport card informs of the available credit or, when a stop is requested, the information monitor changes indicating the name of the requested stop.

Finally, the virtual reality device is a Head Mounted Display Oculus Rift with its Oculus Touch controllers. The choice is based on the ease programming and calibration procedure of this hardware. Besides its controls are very ergonomic, which is of special relevance to the audience we're addressing. Finally, this interface can be carried on in an less invasive way that other solution like tracker gloves.

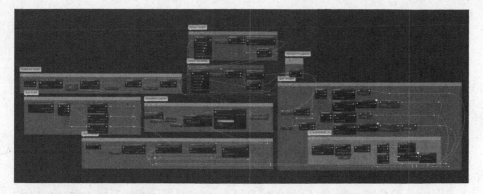

Fig. 4. Example of blueprint used for the creation of interactions.

4   Conclusions

This paper proposed a travel training simulator for people with intellectual disabilities to provide a much more cost-effective solution for helping people to learn how to use public transportation independently. This simulator is based in the following premises: (1) hyper realistic photogrammed reconstruction of real buses stops and travels elements, all of them located in the city of Burgos (Spain) where the simulator is being tested, (2) the use of visual programming to reduce costs of the virtual environment interactivity development and (3) the use of standard but high visual quality HMD Oculus Rift for enhancing the experience.

Future works will be focused in the evaluation of this first prototype and the extension of its interactivity with the user, including an extensive interaction with other users in the virtual reality environment. In this way the teachers can increase the self-confidence of the user and evaluate closely and without pressure if he is ready for real world tests. Besides, once the user is self-confidence with the virtual reality environment, interactivity with avatars can be included to improve his social skills (e.g. old people talking to him in the bus travel, asking him different kind of questions, etc.). Lastly, since this is a high customized simulator, the individual needs of each user must be considered during the evaluation process, in terms of acceptance of virtual reality technology, understanding of the interface usability and the effectiveness of learning transfer to the real world.

Acknowledgments. This work was partially supported through the Program "Impulso de la Industria de Contenidos Digitales desde las Universidades" of the Spanish Ministry of Industry, Tourism and Commerce and funding and we gratefully acknowledge the support of NVIDIA Corporation with the donation of the Titan Xp GPU used for this research.

References

1. McKenzie, K., Milton, M., Smith, G., Ouellette-Kuntz, H.: Systematic review of the prevalence and incidence of intellectual disabilities: current trends and issues. Curr. Dev. Disord. Rep. **3**(2), 104–115 (2016)
2. Tam, S.F., Man, D.W.K., Chan, Y.P., Sze, P.C., Wong, C.M.: Evaluation of a computer-assisted, 2-D virtual reality system for training people with intellectual disabilities on how to shop. Rehabil. Psychol. **50**(3), 285–291 (2005)
3. Zionch, A.: Digital simulations: facilitating transition for students with disabilities. Interv. Sch. Clin. **46**(4), 246–250 (2011)
4. De Oliveira Malaquias, F.F., Malaquias, R.F.: The role of virtual reality in the learning process of individuals with intellectual disabilities. Technol. Disabil. **28**(4), 133–138 (2017)
5. Schwebel, D.C., Combs, T., Rodriguez, D., Severson, J., Sisiopiku, V.: Community-based pedestrian safety training in virtual reality: a pragmatic trial. Accid. Anal. Prev. **86**, 9–15 (2016)
6. Gamito, P., et al.: Cognitive training on stroke patients via virtual reality-based serious games. Disabil. Rehabil. **39**(4), 385–388 (2017)
7. Cromby, J.J., Standen, P.J.P., Newman, J., Tasker, H.: Successful transfer to the real world of skills practised in a virtual environment by students with severe learning difficulties. In: Proceedings of the 1st European Conference on Disability, Virtual Reality and Associated Technologies (1996)
8. Cobb, S.V.G., Neale, H.R., Reynolds, H.: Evaluation of virtual learning environments. In: The 2nd European Conference on Disability, Virtual Reality and Associated Technologies, pp. 17–23 (1998)
9. Brown, D., Neale, H., Cobb, S.: Development and evaluation of the virtual city (1998)
10. Brooks, B.M., Rose, F.D., Attree, E.A., Elliot-Square, A.: An evaluation of the efficacy of training people with learning disabilities in a virtual environment. Disabil. Rehabil. **24**(11–12), 622–626 (2002)
11. Salem-Darrow, M.: Virtual reality's increasing potential for meeting needs of person with disabilities: what about cognitive impairments? In: Proceedings of the 3rd International Conference on Virtual Reality and Persons with Disabilities. California State University, Northridge (1995)
12. Pantelidis, V.S.: Virtual reality in the classroom. Educ. Technol. **33**(4), 23–27 (1993)
13. Standen, P.J., Brown, D.J.: Virtual reality and its role in removing the barriers that turn cognitive impairments into intellectual disability. Virtual Reality **10**(3–4), 241–252 (2006)
14. Standen, P.J., Brown, D.J., Anderton, N., Battersby, S.: Systematic evaluation of current control devices used by people with intellectual disabilities in non-immersive virtual environments. Cyberpsychology Behav. **9**(5), 608–613 (2008)
15. Wu, J.K.Y., Suen, W.W.S., Ho, T.M.F., Yeung, B.K.H., Tam, A.S.F.: Effectiveness of two dimensional virtual reality programme and computer-assisted instructional programme in training Mass Transit Railway (MRT) skills for persons with mental handicap: a pilot study. Hong Kong J. Occup. Ther. **15**(1), 8–15 (2005)
16. Tianwu, Y., Changjiu, Z., Jiayao, S.: Virtual reality based independent travel training system for children with intellectual disability. In: Proceedings of the UKSim-AMSS 2016: 10th European Modelling Symposium on Computer Modelling and Simulation (2017)
17. Fornasari, L., et al.: Navigation and exploration of an urban virtual environment by children with autism spectrum disorder compared to children with typical development. Res. Autism Spectr. Disord. **7**(8), 956–965 (2013)

18. Courbois, Y., Farran, E.K., Lemahieu, A., Blades, M., Mengue-Topio, H., Sockeel, P.: Wayfinding behaviour in down syndrome: a study with virtual environments. Res. Dev. Disabil. **34**(5), 1825–18831 (2013)
19. Simões, M., Bernardes, M., Barros, F., Castelo-Branco, M.: Virtual travel training for autism spectrum disorder: proof-of-concept interventional study. J. Med. Internet Res. **6**(1), e5 (2018)
20. Bernardes, M., Barros, F., Simoes, M., Castelo-Branco, M.: A serious game with virtual reality for travel training with Autism Spectrum Disorder. In: International Conference on Virtual Rehabilitation, ICVR (2015)
21. Sitbon, L., Beaumont, C., Brereton, M., Brown, R., Koplick, S., Fell, L.: Design insights into embedding virtual reality content into life skills training for people with intellectual disability (2017)
22. Buck, U., Naether, S., Räss, B., Jackowski, C., Thali, M.J.: Accident or homicide - Virtual crime scene reconstruction using 3D methods. Forensic Sci. Int. **225**(1–3), 75–84 (2013)
23. Remondino, F., Nocerino, E., Toschi, I., Menna, F.: A critical review of automated photogrammetric processing of large datasets. In: International Archives of the Photogrammetry, Remote Sensing and Spatial Information Sciences - ISPRS Archives (2017)

BRAVO: A Gaming Environment for the Treatment of ADHD

Maria Cristina Barba[1], Attilio Covino[2], Valerio De Luca[1],
Lucio Tommaso De Paolis[1(✉)], Giovanni D'Errico[1], Pierpaolo Di Bitonto[3],
Simona Di Gestore[2], Serena Magliaro[1], Fabrizio Nunnari[1],
Giovanna Ilenia Paladini[1], Ada Potenza[3], and Annamaria Schena[2]

[1] Department of Engineering for Innovation,
University of Salento, Lecce, Italy
lucio.depaolis@unisalento.it
[2] Villa delle Ginestre srl, Volla, NA, Italy
annamariaschena@alice.it
[3] Grifo Multimedia srl, Bari, Italy
{p.dibitonto,a.potenza}@grifomultimedia.it

Abstract. Attention-deficit hyperactivity disorder (ADHD) is a neurodevelopmental disorder that is expressed through different symptoms belonging to three different dimensions: inattention, impulsivity and motor hyperactivity, each of which contributes to the learning and adaptation problems within the different contexts of life. ADHD children have to focus on three main elements: learn to self-control, make and keep friends and feel good about themselves. The BRAVO (Beyond the tReatment of the Attention deficit hyperactiVity disOrder) project aims to realize an immersive therapeutic game context, based on an innovative ICT system, with which improving the relationship between young patients and therapies (administered by means of serious games and gamification). By using wearable equipment and Virtual and Augmented Reality devices, new personalized processes of therapy will be implemented. Such processes will be able to dynamically change in order to follow the patients evolution and support the therapists in the rehabilitation program management.

Keywords: ADHD · Serious game · Virtual reality ·
Augmented reality · Adaptive gamification · Avatar

1 Introduction

Attention Deficit Hyperactivity Disorder (ADHD) is a behavioural and developmental disorder that affects children and teens and includes symptoms such as inattentiveness, hyperactivity and impulsiveness. It has a particularly disruptive effect in daily life activities: affected children have difficulties in planning their homework, estimating time for assignment completion, staying focused on assigned tasks and managing social relationships [1]. Moreover, they have no

© Springer Nature Switzerland AG 2019
L. T. De Paolis and P. Bourdot (Eds.): AVR 2019, LNCS 11613, pp. 394–407, 2019.
https://doi.org/10.1007/978-3-030-25965-5_30

self-esteem and do not have positive thoughts, causing discouragement and emotional explosiveness, poor school achievement, social rejection, and interpersonal relationship problems.

1.1   The Main Challenges

A proper assessment of the child is essential for ADHD treatment planning, which should meet the patient requirements and allow monitoring the symptomatology evolution. From the analysis of the therapeutic needs related to ADHD, it is clear that the therapy should focus on three core elements: learning self-control, making and keeping friends, feeling good about yourself [2]. The acquisition of these macro-skills is the main way to allow the child with ADHD for a quiet life, without the distress and the frustrations due to an underperform in school or a not satisfying social life.

1.2   Serious Games and Gamification for the Treatment of ADHD

Serious games [3] are interactive computer applications that have a challenging goal and an explicit and carefully thought out educational purpose; they are defined as games that are designed to entertain players as they educate, train, or change behaviour. It is clear that the learning function is mixed with the gaming nature of the application; playing a serious game must excite and involve user while ensuring the acquisition of knowledge [4,5].

Recently, new methods have been investigated to improve the daily functional life of a child with ADHD by using a game-based approach to support behavior change [6,7].

Children with ADHD have a motivational deficit and react differently to the gratifications compared to their peers not affected by the same disorder. The playful approach helps to balance motivation and learning; video games are able to motivate children and involve them in the therapeutic process.

Although the literature is full of evidence that these training programs give only near-transfer effects, there is no convincing evidence that even such near-transfer effects are durable [8]. It is advisable, therefore, to work on video games that are able to have a real impact on the life of the child affected by ADHD [9].

1.3   Theoretical Bases

Recent trends are oriented to the use of serious games in supporting the learning of rules and behavioural strategies in everyday life, such as time management, planning/organization, and social skills [10]. These behavioural goals should be translated into appropriate psychological theories in order to have a theoretical reference framework. Indeed, several studies in literature show that serious games based on a psychological framework tend to be more effective than others [11].

The most widespread frameworks for the above mentioned abilities are the self regulation model, the social cognitive theory, and the learning theory [12].

The self-regulation model focuses on how individuals direct and control their activities and emotions in order to achieve their goals. Children with ADHD often have no self-regulation: this prevents them from acquiring all their cognitive and behavioural skills and makes them feel incompetent and insecure in situations where these skills must be used. For this reason, it is important that serious games include components (or game dynamics) that can help children in directing and monitoring their activities, adjusting their actions and emotions, and practicing until the skill has been acquired.

The social cognitive theory argues that children's learning is influenced by interactions with the environment, as well as by personal and behavioural factors. The environment should support the child in behavioural learning by providing reference and gratification models whenever a correct behaviour is shown. Indirect learning, emotional support, and experiential learning are key factors that should be implemented within a successful game design.

According to the learning theory, behaviour is the result of a process of acquiring information from other individuals. Children with ADHD are less sensitive to negative feedback and learn much more if encouraged with repeated positive stimuli. A serious game for children with ADHD should always and immediately reward positive behaviors.

1.4   The BRAVO Project

This paper presents the first results of the BRAVO (Beyond the tReatment of the Attention deficit hyperactiVity disOrder) project, which consists in the development of an immersive serious game aimed at improving the relationship between young patients affected by ADHD and therapies. The rest of the paper is structured as follows. Starting from the goals and the theoretical bases presented so far, in Sect. 2 we will review the technical literature referencing the use of serious game as a therapeutic method for the treatment of ADHD. In Sect. 3, we will discuss the BRAVO project, exploring the functionalities of the system, its hardware and software architecture, and the various choices made. The workflow of the therapeutic process will be described, following the path of the patient in the various interaction steps with the system. Also, the serious games developed for therapeutic purposes will be reviewed. Section 4 describes the project assessment procedure. Finally, Sect. 5 will conclude the paper by describing in detail the already-planned experimentation phase: a final, necessary step for an objective evaluation of the benefits introduced by the system.

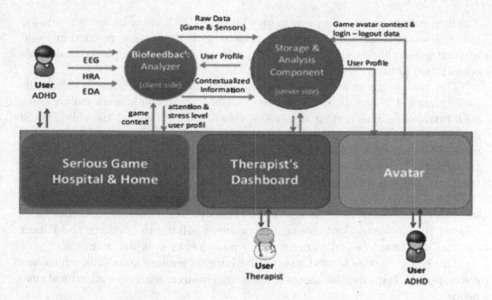

Fig. 1. The BRAVO architecture

2   Related Work

Several works in literature deal with the use of serious games for taking care of children with special educational needs (SEN) and in particular of children affected by ADHD. The fascinating effects of digital games can greatly increase their engagement and concentration, even though the generalization of skills in gamified training to daily life situations is often a challenging issue [13,14]. Serious games have a different impact based on ADHD subcategories, severity of inattention/hyperactivity problems, age and gender of the involved subjects. In a recent experiment [15], girls showed a greater benefit in planning/organizing skills, while among boys those with lower hyperactivity levels and higher levels of Conduct Disorder showed better improvements in planning/organizing skills. Serious games developed for ADHD children can be classified according to the following basic goals:

- general self-regulation skills concerning organizational problems [16], behaviour control [17], and the inhibition of impulsive behaviours [18];
- social and communication skills are stimulated in several games such as the adventures of Pico [19], Alien Health [20], Jumpido Games [21], Little Magic Stories [22], and also in the training platform described in [17];
- academic and cognitive skills such as mathematics, spelling, and pattern matching are stimulated in Games4Learning [23], Jumpido Games [21], and Kaplan Early Learning [24]; moreover, an iPad game [25] was developed to improve the reading quality in terms of attention and comprehension;
- daily life skills are promoted by Alien Health [20] and Pictogram Room [26].

The general structure of serious games for ADHD children includes a mission game, a social community (preferably closed with interaction restrictions) and a game guide that can accommodate, guide or motivate the player during his educational journey [21].

The mission game is a game environment that presents some missions the players should complete and the game actions they should perform to reach such final goals. This system allows also viewing the skills that the child should acquire. Game actions should have increasing difficulty that should be adapted to the child's level of performance. In [22] some examples of games are proposed (Labyrinth, Space Travel Trainer, Explorobot) that have proven effectiveness in teaching behaviour and in their transfer into everyday life.

The ATHYNOS system [27] combines serious games with Augmented Reality to better capture the attention of ADHD children and stimulate cognitive skills related to hand-eye coordination and problem solving. It exploits the Kinect natural user interface [28] to detect body movements as input commands.

The Kinect device is used also by the Kinems gaming suite [29], which was developed to improve the motoric skills of children with special educational needs.

Another type of input used to control or influence games is biofeedback, which consists of data about several aspects of the user physiological state. Moreover, such data usually allow to track the mental state of the children and infer their mood [30].

The ChillFish [31] game exploits the player's breath to control a fish in a 2D underwater environment. The aim is to reduce the stress levels that typically affect ADHD children.

The cognitive behaviour training platform described in [17] exploits Brain Computer Interfaces (BCIs) and the Kinect device to register both the children's brainwaves and their motion during gaming activities. The collected data are stored in a cloud platform to perform both a real-time and an offline analysis. The former allows to control the game flow and train the children's behaviours according to an adaptive intervention scheme. The latter allows evaluating the player's behaviour during the whole session and the treatment effectiveness.

Some research efforts have addressed also the design of tools allowing a classification of ADHD children. In particular, four variables have been proposed to identify children affected by attention deficit [32]: omission error (denoting missing responses to target stimuli), commission error (expressing impulsivity, i.e. the attitude of responding to non-target stimuli), response time, and standard deviation of the response time (as a measure of the irregularity in providing responses).

In more recent research [33–35], electroencephalogram (EEG) signals have been combined with serious games to detect patients affected by ADHD. Neurofeedback data collected during game sessions are analyzed and classified by means of machine learning techniques to detect any ADHD pattern.

Fig. 2. The avatar

3   Materials and Methods

The BRAVO project aims at achieving a therapeutic game environment, based on an innovative ICT system, by:

1. improving the relationship of young patients with therapies;
2. implementing customized therapy processes, able to change dynamically in order to follow the evolution of the patient's attentional levels;
3. supporting therapists in the management of the rehabilitation program.

Behind the project, there is a new generation of serious games able to monitor the patient's behavior by means of wearable sensors, in order to adapt the intervention of the system to the therapy outcomes. The data acquired through a series of devices (Kinect [28], HTC Vive [36] and controller with reference to gaming performance, EEG helmet and bracelet with reference to biofeedback data) allow to draw a user profile useful for evaluating the therapeutic trends. Based on such data, together with child's individual performance and his/her emotional state, the system will suggest to the therapist the most suitable therapeutic game and level difficulty.

By simulating realistic situations, virtual reality-mediated cognitive training is an innovative methodology able to immerse the child affected by behavior disorder, activate one or more sensorial stimulations, and provide a strong sense of presence in the virtual environment. This enables the transfer of acquired or rehabilitated skills during training to the real (non-simulated) environment.

3.1   System Architecture

In the designed platform, therapists and patients have clearly separated roles and tasks. Therefore, there are distinct applications: one to be used by therapists (therapist's dashboard), the other ones reserved for patients/children (serious games, gamification apps, avatar). The former consists of a control interface that allows the monitoring of therapy and the customization of the video games; the other applications consist of a collection of rehabilitative exercises in the form of video games (serious games), which will be performed by the child at the rehabilitation facility during the scheduled therapeutic sessions by using virtual or augmented reality devices or, alternatively, at home (by means of an app) by using mobile device (tablet, pc, smartphone etc).

The avatar is developed for the interaction with the child and will have the task of recognizing and welcoming him on his arrival at the clinic, putting him at ease, and stimulating him to face the therapy exercises.

The interaction among the therapist control interface, the serious games, and the avatar is never direct: in the BRAVO platform all the data exchanged among these components are first transmitted to a server and then analyzed and used.

As shown in Fig. 1, the architecture consists of the modules described in the following subsections.

Storage and Analysis Component. Centralized system for storing patient data. It collects the user interaction data coming from the wearable sensors, the game and the avatar and then it analyzes them.

Biofeedback Analyzer. Module with the specific task of:

- collecting data from sensors (EEG helmet and wristband to detect the stress level) and game performance;
- processing the user emotional state based on the specific context;
- sharing data with other system components.

Serious Game Hospital & Home. Module with the specific task of giving the therapy in two distinct and independent ways: at the clinic, via serious game, connected to the sensors and at home, via the gaming app accessible via tablets and PCs.

Avatar. Module with the task of welcoming and entertaining the child at the entrance and exit of the therapy room and interacting in a personalized way according to the user profile information. The avatar (Fig. 2) represents a virtual friend who will walk the child during the therapeutic process, creating the conditions for a pleasant experience.

Therapist's Dashboard. Module representing the informational dashboard available to the therapist, which can be used to monitor the therapy and receive suggestions regarding the game and level of difficulty for the child. Furthermore, the dashboard allows the therapist to mediate the interaction between avatar and child, just as a puppeteer would do with a virtual puppet (by controlling emotional expressions, non-verbal communication, gestures and dialogues).

3.2   Therapy Workflow

The patient is the main user of the developed system. He/she will be provided with a customized therapy service, which will be able to evolve in a smart way according to the therapy progress.

Once the child/patient is in front of the treatment room, following an automatic face recognition procedure which triggers an authentication in the system, he/she will meet his personal virtual avatar. The avatar will be always consistent with the patient's psychological profile and therapeutic trend; in this way, the child will always feel recognized and welcomed by a familiar figure.

The avatar customization takes place from two independent points of view: an aesthetic one and a behavioural one. The former allows the child to define the avatar look: to this aim, he can choose among several features (physical features, accessories, etc.) available in an avatar creator tool through a touchscreen. The latter allows the therapist to modulate poses, attitudes and voice tones of the avatar by means of a dashboard in accordance with some features highlighted in the child profile. Besides being visualized on a screen at the entrance and at the exit of the therapeutic room, the avatar accompanies the child also during the therapy session in the form of holographic agent. The concept of holographic avatar refers to both the virtual character leading every augmented reality scenario (visualized through Microsoft Hololens [37]) and the hologram displayed through a projection system used as an alternative to the headset.

The main interaction phase consists in a gaming activity through a series of serious games that can be classified into three different categories: topology, attention and respect for rules, action planning.

3.3   Therapeutic Serious Games

BRAVO's game environment consists of some serious games that will be used with a wide range of sensors and actuators, such as EEG headset, VR headset, and wristband, in order to detect and store the patient emotional states, contributing to real-time game adaptation. Serious games were implemented according to well-defined therapeutic goals dealing both with educational/logopedic issues (such as the definition of topological concepts and semantic categories) and with behavioural issues (such as respect for the rules, improvement of the attention span, ability to predict the effect of the actions, improvement of social skills, and so on).

Topological Categories. It is a 7-levels game that aims at supporting the improvement of patient's topological categories skills (e.g. over - under, inside - outside, back - forth, close - far, left - right). The game has been designed to be played in a clinic through an HTC Vive [36] and a controller, allowing the user to move in a virtual environment just like it would move in a real one.

Within the game, the user will explore three different environments (a classroom, a bedroom, and a garden), in which he will be asked to accomplish specific tasks, for example, positioning items or himself in relation to (e.g. near, far, above, under, etc.) other items.

As the levels' difficulty increase, new elements will be added to the level's tasks, such as a countdown or a semaphore. Furthermore, the question complexity will increase too (e.g. multiple topological concepts are expressed in the same question), as well as the complexity of the game scene (Fig. 3).

Fig. 3. Topological categories

Infinite Runner. It is a 8-levels game whose goal is to teach the respect for rules (and how to stand still), the active listening and the awareness of the player's limits. The game will be played by means of a Kinect device [28], which allows to move the game avatar simply by using the human body: in fact, the player's movements are decoded without the need for any additional tool to be worn or linked to the body. Within the game, the user is immersed in a virtual environment depicting a country or a city road, where he is requested to follow a path by running on the spot. As the game goes on, the player has to move in the right direction to avoid several obstacles or collect any requested object. As the levels move forward, the levels' complexity increases, introducing further difficulty elements (such as a flock crossing the road, a traffic light or crosswalks), that will require the player's ability to wait and to improve the attention span. For each level, the therapist can set different parameters, according to the patient's needs: indeed, a specific game mode has been created in order to allow children with motor skills problems to play anyway without running, by simply standing and moving to the right or to the left to accomplish the task (Fig. 4).

Fig. 4. Infinite runner

Space Travel Trainer. It is a 7-levels game that has the educational goal of teaching the patients to plan their actions and manage social relationships.

The game is controlled through the Kinect device [28], which allows detecting the player's hand movements, through which s/he will perform and accomplish the requested tasks. Within the game, the player acts as an astronaut who has the goal to take the spaceship to his friends' planet. During his journey, s/he has to face different tests: in order to pass them, s/he has to learn how to solve complex problem, how to take the right decision at the right moment (improving his decision-making ability), but most of all how to interact with his team members (in order to defeat the enemy) (Fig. 5).

4   Project Assessment

The final phase of BRAVO will include a 27-weeks trial aimed at evaluating its benefits. To such purpose, 60 young patients will be involved, divided into the following age groups:

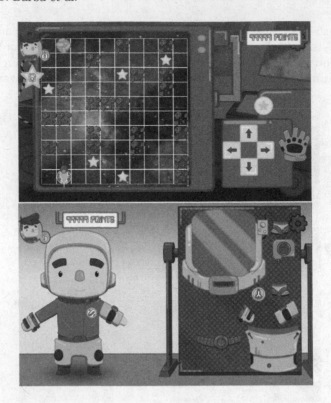

Fig. 5. Space travel trainer

1. Preschoolers: 3–6 years;
2. Primary school age: 6–9 years;
3. Secondary school children: 9–12 years.

Patients will be divided in two groups, equally numerous, randomly. The first group will be used as a control group and will be submitted to the traditional therapy currently used in the clinic. The second group will be instead the active group of the test, receiving therefore the new treatment in the developed game environment. During the therapies administration, both groups will be evaluated in three phases: before the beginning of the test, during, and at the end. In order to assess the real impact of the BRAVO system, usability and effectiveness tests will be implemented. Usability is the degree to which a product can be used by specified consumers to achieve specific objectives with effectiveness, efficiency, and satisfaction in a particular context of use.

The implemented usability tests are based on the expert inspection carried out with Nielsen heuristics. Therefore, these tests will be carried out with target users by means of standardized questionnaires, identified in scientific literature.

To evaluate the system effectiveness on patients, specific tests will be created and implemented before, during and after the administration of therapies: the comparison of the obtained results will allow therapists to determine changes or improvements found in patients. At the beginning of the experimental phase, the users-serious game interaction aspects will be evaluated, in order to consider possible bias (in the clinical results obtained) deriving from system usability defects or from a user experience not fully in line with its expectations.

5 Conclusions and Future Work

In this paper we have described the BRAVO project, which aims at realizing an immersive therapeutic game context to support an alternative modality for the treatment of ADHD. BRAVO is based on an innovative ICT system, whose goal is to improve the relationship between young patients and therapies.

BRAVO aims at achieving its goal through the combined use of ad-hoc educational games and entertaining personalized avatars. Children/patients will be monitored using on-body sensors.

Through an analysis of the gathered sensor data and gaming scores, the impact of the BRAVO system will be evaluated in terms of clinical improvements, speed of improvement, patient involvement and empowerment.

The first results already show a great potential of serious games in attracting young patients even compared to standard therapy.

At the end of the experimental phase it will be evaluated how to improve the performance of the system compared to the defined KPIs.

References

1. Abikoff, H., et al.: Remediating organizational functioning in children with ADHD: immediate and long-term effects from a randomized controlled trial. J. Consult. Clin. Psychol. **81**, 113 (2013)
2. Shapiro, L.: ADHD: il mio libro di esercizi: Attività per sviluppare la fiducia in se stessi, le abilità sociali e l'autocontrollo. No. v. 6–12 in I materiali Erickson. Strumenti per la didattica, l'educazione, la riabilitazione, il recupero e il sostegno, Erickson (2015). https://books.google.it/books?id=5HPmDAAAQBAJ
3. Bergeron, B.: Developing Serious Games (Game Development Series) (2006)
4. De Paolis, L.T., Aloisio, G., Celentano, M.G., Oliva, L., Vecchio, P.: MediaEvo project: a serious game for the edutainment. In: ICCRD2011 - 2011 3rd International Conference on Computer Research and Development (2011)
5. De Paolis, L.T., Aloisio, G., Celentano, M.G., Oliva, L., Vecchio, P.: Experiencing a town of the middle ages: an application for the edutainment in cultural heritage. In: 2011 IEEE 3rd International Conference on Communication Software and Networks, ICCSN 2011 (2011)
6. Luman, M., Oosterlaan, J., Sergeant, J.A.: The impact of reinforcement contingencies on AD/HD: a review and theoretical appraisal. Clin. Psychol. Rev. **25**, 183–213 (2005)
7. Sagvolden, T., Aase, H., Zeiner, P., Berger, D.: Altered reinforcement mechanisms in attention-deficit/hyperactivity disorder. Behav. Brain Res. **94**, 61–71 (1998)
8. Melby-Lervåg, M., Hulme, C.: Is working memory training effective? A meta-analytic review. Dev. Psychol. **49**, 270 (2013)

9. DeSmet, A., et al.: A meta-analysis of serious digital games for healthy lifestyle promotion. Prev. Med. **69**, 95–107 (2014)
10. Ricciardi, F., De Paolis, L.T.: A comprehensive review of serious games in health professions. Int. J. Comput. Technol. **2014**, 11 (2014)
11. Baranowski, T., Buday, R., Thompson, D.I., Baranowski, J.: Playing for real: video games and stories for health-related behavior change. Am. J. Prev. Med. **34**, 74–82 (2008)
12. Bandura, A., National Institute of Mental Health: Social Foundations of Thought and Action: A Social Cognitive Theory. Prentice-Hall Series in Social Learning Theory (1986)
13. Bul, K.C., et al.: Development and user satisfaction of "Plan-It Commander", a serious game for children with ADHD. Games Health J. **4**, 502–512 (2015)
14. Bul, K.C., et al.: Behavioral outcome effects of serious gaming as an adjunct to treatment for children with attention-deficit/hyperactivity disorder: a randomized controlled trial. J. Med. Internet Res. **18**, e26 (2016)
15. Bul, K.C., Doove, L.L., Franken, I.H., Van Der Oord, S., Kato, P.M., Maras, A.: A serious game for children with attention deficit hyperactivity disorder: who benefits the most? PLoS ONE **13**, e0193681 (2018)
16. Frutos-Pascual, M., Zapirain, B., Zorrilla, A.: Adaptive tele-therapies based on serious games for health for people with time-management and organisational problems: preliminary results. Int. J. Environ. Res. Public Health **11**, 749–772 (2014)
17. Park, K., Kihl, T., Park, S., Kim, M.J., Chang, J.: Narratives and sensor driven cognitive behavior training game platform. In: 2016 IEEE 14th International Conference on Software Engineering Research, Management and Applications (SERA) (2016)
18. Colombo, V., Baldassini, D., Mottura, S., Sacco, M., Crepaldi, M., Antonietti, A.: Antonyms: a serious game for enhancing inhibition mechanisms in children with Attention Deficit/Hyperactivity Disorder (ADHD). In: International Conference on Virtual Rehabilitation, ICVR (2017)
19. Malinverni, L., Mora-Guiard, J., Padillo, V., Valero, L., Hervás, A., Pares, N.: An inclusive design approach for developing video games for children with autism spectrum disorder. Comput. Hum. Behav. **71**, 535–549 (2017)
20. Johnson-Glenberg, M.C., Savio-Ramos, C., Henry, H.: "Alien Health": a nutrition instruction exergame using the Kinect sensor. Games Health J. **3**, 241–251 (2014)
21. Jumpido: Educational games for Kinect (2019). http://www.jumpido.com/en
22. Little Magic Stories (2019). http://www.chrisoshea.org/little-magic-stories
23. Games 4 Learning (2019). http://games4learning.com
24. Kaplan Early Learning (2019). https://www.kaplanco.com
25. Wrońska, N., Garcia-Zapirain, B., Mendez-Zorrilla, A.: An iPad-based tool for improving the skills of children with attention deficit disorder. Int. J. Environ. Res. Public Health **12**, 6261–6280 (2015)
26. Pictogram Room (2019). https://autismodiario.org/2012/03/12/ya-esta-disponible-pictogram-room
27. Avila-Pesantez, D., Rivera, L.A., Vaca-Cardenas, L., Aguayo, S., Zuniga, L.: Towards the improvement of ADHD children through augmented reality serious games: preliminary results. In: IEEE Global Engineering Education Conference, EDUCON (2018)
28. Microsoft Kinect (2019). http://www.xbox.com/kinect

29. Kourakli, M., Altanis, I., Retalis, S., Boloudakis, M., Zbainos, D., Antonopoulou, K.: Towards the improvement of the cognitive, motoric and academic skills of students with special educational needs using Kinect learning games. Int. J. Child-Comput. Interact. **11**, 28–39 (2017)
30. Bodolai, D., Gazdi, L., Forstner, B., Szegletes, L.: Supervising biofeedback-based serious games. In: Proceedings of 6th IEEE Conference on Cognitive Infocommunications, CogInfoCom 2015 (2016)
31. Sonne, T., Jensen, M.M.: ChillFish: a respiration game for children with ADHD. In: Proceedings of the TEI 2016: Tenth International Conference on Tangible, Embedded, and Embodied Interaction (2016)
32. Roh, C.H., Lee, W.B.: A study of the attention measurement variables of a serious game as a treatment for ADHD. Wirel. Pers. Commun. **79**, 2485–2498 (2014)
33. Alchalabi, A.E., Shirmohammadi, S., Eddin, A.N., Elsharnouby, M.: FOCUS: detecting ADHD patients by an EEG-based serious game. IEEE Trans. Instrum. Meas. **67**, 1512–1520 (2018)
34. Alchalabi, A.E., Elsharnouby, M., Shirmohammadi, S., Eddin, A.N.: Feasibility of detecting ADHD patients' attention levels by classifying their EEG signals. In: Proceedings of 2017 IEEE International Symposium on Medical Measurements and Applications, MeMeA 2017 (2017)
35. Alchalcabi, A.E., Eddin, A.N., Shirmohammadi, S.: More attention, less deficit: wearable EEG-based serious game for focus improvement. In: 2017 IEEE 5th International Conference on Serious Games and Applications for Health, SeGAH 2017 (2017)
36. HTC VIVE (2019). https://www.vive.com
37. Microsoft Hololens (2019). https://www.microsoft.com/it-it/hololens

Author Index